The Chemistry of Powder and Explosives

COMPLETE IN ONE VOLUME

BY

TENNEY L. DAVIS, Ph.D.

Emeritus Professor of Organic Chemistry
Massachusetts Institute of Technology
Director of Research and Development
National Fireworks, Inc.

ANGRIFF PRESS
P.O. Box 208
Las Vegas, Nevada 89125
ISBN-0913022-00-4

PREFACE

The present volume contains in one binding the whole contents of Volume I, first published in May, 1941, and the whole contents of Volume II which was published in March, 1943.

The book is primarily for chemists. The writing of it was commenced in order that a textbook might be available for the use of students in the course in powder and explosives which the author gave for about twenty years (nearly every year since the first World War) to fourth-year and graduate students of chemistry and of chemical engineering at the Massachusetts Institute of Technology. The writing of it has been completed while the nation is at war and while many chemists, not previously trained in powder and explosives, are at work preparing, studying, and testing these materials. The purpose of the book is to supply chemists with information concerning the modes of behavior of explosive substances and concerning the phenomena, both chemical and physical, which they exhibit. No effort has been made to describe the use of explosives in ammunition and in blasting beyond the minimum of description which is needed to make clear the modes of their behavior, and no account has been included of the chemical engineering aspects of their manufacture.

The book brings together much material which has never before been collected in one volume, and it sets down some of the facts in what are probably new relationships to one another, but it contains nothing which is not already known to those who are skilled either in chemistry or in the manufacture and use of explosives. It is elementary in the sense that it supplies a background of knowledge for the chemist who wishes to become an expert in any one of several technical branches.

The chapter on pyrotechnics has been made as full as possible. It contains much which will not be found in print elsewhere, but it limits its discussions to civil pyrotechnics—for several reasons. Civil pyrotechnics is a much broader subject than military pyrotechnics. Military pyrotechnic devices differ in no important respect from similar devices for civil and recreational purposes. Their varieties are few. Artifices which are not now used for military purposes may some day be applied to the uses of war, and a broad knowledge of civil pyrotechnics plus an acquaintance with the military necessities will determine the applications.

Workers with explosives will perhaps think that I have included in the chapter on pyrotechnics too much material on the construction of fireworks pieces, but chemists, interested in the manner in which their substances behave, will be tempted to try their hands at making the artifices, and the fireworks makers, aware of the importance of these details, will probably think that the account of them is too meager.

The chapter on aromatic nitro compounds deals with the chemistry of a large and important class of explosive substances, among which TNT stands as the most important of the military high explosives, with tetryl second in importance, while the whole class includes substances which are used, or may be used, in shells, bombs, grenades, and other devices of war. The precise manner in which they are loaded, the amounts which are used, the details of the construction of the devices, etc., are known to those who are concerned with such matters. But the practices change. The principles of the use of the materials and the physical and chemical properties of the substances upon which the principles depend do not change—and they are the proper subject matter of the present book.

The aim of the book has been to describe as clearly and interestingly as possible and as fully as seemed profitable the modes of behavior, both physical and chemical, of explosive substances, whether these modes find practical application or not. Historical material has been included where it was thought that it contributed to this end, and it has not been included elsewhere or for any other reason. It is a fact that a knowledge of the history of ideas, of persons, or of things produces something of the same sympathetic understanding of them that living with them and working with them does.

I am indebted to many friends to whom I wish to make grateful acknowledgment of information, of pictures, and of criticism. Dr. C. G. Storm has read the entire manuscript and has made many helpful suggestions. Dr. Walter O. Snelling has read the chapter on aromatic nitro compounds, has kindly prepared the specimen to illustrate the Munroe effect, and has made the photo-

PREFACE

graphs from which the two figures are reproduced. Allen F. Clark
of Bridgewater, Massachusetts, has read the chapter on pyrotechnics and has supplied many formulas and much other information. His brother, George J. J. Clark of Whitman, Massachusetts,
has furnished several pictures of operations at the plant of the
National Fireworks Company, while the *Boston Globe* with his
consent has supplied others, and a third brother, Wallace Clark
of Chicago, Illinois, has given information relative to the manufacture of Chinese firecrackers and has furnished the pictures which
illustrate the process. George W. Weingart of New Orleans, Lou-
isiana, has supplied information in his letters and has given permission to use material from his "Dictionary and Manual of
Fireworks," and A. St. H. Brock of London, England, has given
permission to quote from his "Pyrotechnics: The History and Art
of Firework Making." I am indebted to Dr. E. Berl, to Dr. Émile
Monnin Chamot, to the U. S. Bureau of Mines, to the Atlas
Powder Company, to E. I. du Pont de Nemours and Company,
Inc., to the Ensign-Bickford Company, to the Hercules Powder
Company, to the Trojan Powder Company, and to the Western
Cartridge Company for pictures; to the Williams and Wilkins
Company of Baltimore for permission to cite and to quote from
Symmes' translation of Naoúm's "Nitroglycerine and Nitro-
glycerine Explosives"; to Sir Isaac Pitman & Sons, Limited, of
London, for permission to cite and to quote from MacDonald's
"Historical Papers on Modern Explosives," published originally
by Whittaker & Co., whose business this firm later took over, and
to the *Journal of the American Chemical Society*, to *Industrial and
Engineering Chemistry*, to the *Journal of Chemical Education*, to
the *Journal of the Franklin Institute*, to *Army Ordnance*, to the
U. S. Bureau of Mines, and to the University of Pennsylvania
Press for permission to cite and to quote from their publications.
My thanks are also due to Professor Warren K. Lewis who has
read the proof of Volume I and to Dr. Joseph Ackerman, Jr., who
has read the proof of Volume II.

TENNEY L. DAVIS

NORWELL, MASSACHUSETTS
7 March 1943

v

CONTENTS

CONTENTS

CONTENTS

CONTENTS

CONTENTS

CONTENTS

FIRST PRINTED IN 1943

CHAPTER I

PROPERTIES OF EXPLOSIVES

Definition

An explosive is a material, either a pure single substance or a mixture of substances, which is capable of producing an explosion by its own energy.

It seems unnecessary to define an explosion, for everyone knows what it is—a loud noise and the sudden going away of things from the place where they have been. Sometimes it may only be the air in the neighborhood of the material or the gas from the explosion which goes away. Our simple definition makes mention of the one single attribute which all explosives possess. It will be necessary to add other ideas to it if we wish to describe the explosive properties of any particular substance. For example, it is not proper to define an explosive as a substance, or a mixture of substances, which is capable of undergoing a sudden transformation with the production of heat *and* gas. The production of heat alone by the inherent energy of the substance which produces it will be enough to constitute the substance an explosive. Cuprous acetylide explodes by decomposing into copper and carbon and heat, no gas whatever, but the sudden heat causes a sudden expansion of the air in the neighborhood, and the result is an unequivocal explosion. All explosive substances produce heat; nearly all of them produce gas. The change is invariably accompanied by the liberation of energy. The products of the explosion represent a lower energy level than did the explosive before it had produced the explosion. Explosives commonly require some stimulus, like a blow or a spark, to provoke them to liberate their energy, that is, to undergo the change which produces the explosion, but the stimulus which "sets off" the explosive does not contribute to the energy of the explosion. The various stimuli to which explosives respond and the manners of their responses in producing explosions provide a convenient basis for the classification of these interesting materials.

1

Since we understand an explosive material to be one which is capable of producing an explosion by its own energy, we have opened the way to a consideration of diverse possibilities. An explosive perfectly capable of producing an explosion may liberate its energy without producing one. Black powder, for example, may burn in the open air. An explosion may occur without an explosive, that is, without any material which contains intrinsically the energy needful to produce the explosion. A steam boiler may explode because of the heat energy which has been put into the water which it contains. But the energy is not intrinsic to water, and water is not an explosive. Also, we have explosives which do not themselves explode. The explosions consist in the sudden ruptures of the containers which confine them, as happens in a Chinese firecracker. Fire, traveling along the fuse (note the spelling) reaches the black powder—mixture of potassium nitrate, sulfur, and charcoal—which is wrapped tightly within many layers of paper; the powder burns rapidly and produces gas. It burns very rapidly, for the heat resulting from the burning of the first portion cannot get away, but raises the temperature of the next portion of powder, and a rise of temperature of 10°C. more than doubles the velocity of a chemical reaction. The temperature mounts rapidly; gas is produced suddenly; an explosion ensues. The powder burns; the cracker explodes. And in still other cases we have materials which themselves explode. The molecules undergo such a sudden transformation with the liberation of heat, or of heat and gas, that the effect is an explosion.

Classification of Explosives

I. **Propellants** or *low explosives* are combustible materials, containing within themselves all oxygen needful for their combustion, which burn but do not explode, and function by producing gas which produces an explosion. Examples: black powder, smokeless powder. Explosives of this class differ widely among themselves in the rate at which they deliver their energy. There are slow powders and fast powders for different uses. The kick of a shotgun is quite different from the persistent push against the shoulder of a high-powered military rifle in which a slower-burning and more powerful powder is used.

II. **Primary explosives** or *initiators* explode or detonate when they are heated or subjected to shock. They do not burn; sometimes they do not even contain the elements necessary for combustion. The materials themselves explode, and the explosion results whether they are confined or not. They differ considerably in their sensitivity to heat, in the amount of heat which they give off, and in their *brisance*, that is, in the shock which they produce when they explode. Not all of them are brisant enough to initiate the explosion of a high explosive. Examples: mercury fulminate, lead azide, the lead salts of picric acid and trinitroresorcinol, *m*-nitrophenyldiazonium perchlorate, tetracene, nitrogen sulfide, copper acetylide, fulminating gold, nitrosoguanidine, mixtures of potassium chlorate with red phosphorus or with various other substances, the tartarates and oxalates of mercury and silver.

III. **High explosives** *detonate* under the influence of the shock of the explosion of a suitable primary explosive. They do not function by burning; in fact, not all of them are combustible, but most of them can be ignited by a flame and in small amount generally burn tranquilly and can be extinguished easily. If heated to a high temperature by external heat or by their own combustion, they sometimes explode. They differ from primary explosives in not being exploded readily by heat or by shock, and generally in being more brisant and powerful. They exert a mechanical effect upon whatever is near them when they explode, whether they are confined or not. Examples: dynamite, trinitrotoluene, tetryl, picric acid, nitrocellulose, nitroglycerin, liquid oxygen mixed with wood pulp, fuming nitric acid mixed with nitrobenzene, compressed acetylene and cyanogen, ammonium nitrate and perchlorate, nitroguanidine.

It is evident that we cannot describe a substance by saying that it is "very explosive." We must specify whether it is sensitive to fire and to shock, whether it is really powerful or merely brisant, or both, whether it is fast or slow. Likewise, in the discussions in the present book, we must distinguish carefully between sensitivity, stability, and reactivity. A substance may be extremely reactive chemically but perfectly stable in the absence of anything with which it may react. A substance may be exploded readily by a slight shock, but it may be stable if left to itself. Another may require the shock of a powerful detonator to make it explode but may be subject to spontaneous decomposition.

The three classes of explosive materials overlap somewhat, for the behavior of a number of them is determined by the nature of the stimuli to which they are subjected and by the manner in which they are used. Black powder has probably never been known, even in the hideous explosions which have sometimes occurred at black powder mills, to do anything but burn. Smokeless powder which is made from colloided nitrocellulose, especially if it exists in a state of fine subdivision, is a vigorous high explosive and may be detonated by means of a sufficiently powerful initiator. In the gun it is lighted by a flame and functions as a propellant. Nitroglycerin, trinitrotoluene, nitroguanidine, and other high explosives are used in admixture with nitrocellulose in smokeless powders. Fulminate of mercury if compressed very strongly becomes "dead pressed" and loses its power to detonate from flame, but retains its power to burn, and will detonate from the shock of the explosion of less highly compressed mercury fulminate. Lead azide, however, always explodes from shock, from fire, and from friction.

Some of the properties characteristic of explosives may be demonstrated safely by experiment.

A sample of commercial black powder of moderately fine granulation, say FFF, may be poured out in a narrow train, 6 inches or a foot long, on a sheet of asbestos paper or a wooden board. When one end of the train is ignited, the whole of it appears to burn at one time, for the flame travels along it faster than the eye can follow. Commercial black powder is an extremely intimate mixture; the rate of its burning is evidence of the effect of intimacy of contact upon the rate of a chemical reaction. The same materials, mixed together as intimately as it is possible to mix them in the laboratory, will burn much more slowly.

Six parts by weight of potassium nitrate, one of sulfur (roll brimstone), and one of soft wood (willow) charcoal are powdered separately and passed through a silk bolting-cloth. They are then mixed, ground together in a mortar, and again passed through the cloth; and this process is repeated. The resulting mixture, made into a train, burns fairly rapidly but by no means in a single flash. The experiment is most convincing if a train of commercial black powder leads into a train of this laboratory powder, and the black powder is ignited by means of a piece of *black match* leading from the end of the train and extending beyond the edge of the surface on which the powder is placed. **The** black match may be ignited easily by a flame, whereas black powder on a flat surface is often surprisingly difficult to light.

Black match may be made conveniently by twisting three or four strands of fine soft cotton twine together, impregnating the resulting cord with a paste made by moistening *meal powder*[1] with water, wiping off the excess of the paste, and drying while the cord is stretched over a frame. A slower-burning black match may be made from the laboratory powder described above, and is satisfactory for experiments with explosives. The effect of temperature on the rate of a chemical reaction may be demonstrated strikingly by introducing a 12-inch length of black match into a 10-inch glass or paper tube (which need not fit it tightly); when the match is ignited, it burns in the open air at a moderate rate, but, as soon as the fire reaches the point where the tube prevents the escape of heat, the flame darts through the tube almost instantaneously, and the gases generally shoot the burning match out of the tube.

Cuprous acetylide, of which only a very small quantity may be prepared safely at one time, is procured by bubbling acetylene into an ammoniacal solution of cuprous chloride. It precipitates as a brick-red powder. The powder is collected on a small paper filter and washed with water. About 0.1 gram of the material, still moist, is transferred to a small iron crucible—the rest of the cuprous acetylide ought to be destroyed by dissolving in dilute nitric acid—and the crucible is placed on a triangle over a small flame. As soon as the material has dried out, it explodes, with a loud report, causing a dent in the bottom of the crucible.

A 4-inch filter paper is folded as if for filtration, about a gram of FFF black powder is introduced, a 3-inch piece of black match is inserted, and the paper is twisted in such manner as to hold the powder together in one place in contact with the end of the match. The black match is lighted and the package is dropped, conveniently, into an empty pail. The powder burns with a hissing sound, but there is no explosion for the powder was not really confined. The same experiment with about 1 gram of potassium picrate gives a loud explosion. All metallic picrates are primary explosives, those of the alkali metals being the least violent. Potassium picrate may be prepared by dissolving potassium carbonate in a convenient amount of water, warming almost to boiling, adding picric acid in small portions at a time as long as it dissolves with effervescence, cooling the solution, and collecting the crystals and drying them by exposure to the air. For safety's sake,

---

[1] Corning mill dust, the most finely divided and intimately incorporated black powder which it is possible to procure. Lacking this, black sporting powder may be ground up in small portions at a time in a porcelain mortar.

quantities of more than a few grams ought to be kept under water, in which the substance is only slightly soluble at ordinary temperatures.

About a gram of trinitrotoluene or of picric acid is heated in a porcelain crucible. The substance first melts and gives off combustible vapors which burn when a flame is applied but go out when the flame is removed. A small quantity of trinitrotoluene, say 0.1 gram, may actually be sublimed if heated cautiously in a test tube. If heated quickly and strongly, it decomposes or explodes mildly with a "zishing" sound and with the liberation of soot.

One gram of powdered picric acid and as much by volume of litharge (PbO) are mixed carefully on a piece of paper by turning the powders over upon themselves (not by stirring). The mixture is then poured in a small heap in the center of a clean iron sand-bath dish. This is set upon a tripod, a lighted burner is placed beneath it, and the operator retires to a distance. As soon as the picric acid melts and lead picrate forms, the material explodes with an astonishing report. The dish is badly dented or possibly punctured.

A Complete Round of Ammunition

The manner in which explosives of all three classes are brought into use will be made clearer by a consideration of the things

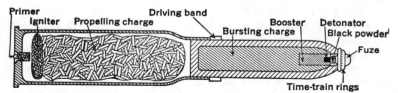

Figure 1. Diagram of an Assembled Round of High-Explosive Ammunition. The picture is diagrammatic, for the purpose of illustrating the functions of the various parts, and does not correspond exactly to any particular piece of ammunition.

which happen when a round of H.E. (high-explosive) ammunition is fired. The brass cartridge case, the steel shell with its copper driving band and the fuze screwed into its nose are represented diagrammatically in the accompanying sketch. Note the spelling of fuze: a *fuze* is a device for initiating the explosion of high-explosive shells or of bombs, shrapnel, mines, grenades, etc.; a *fuse* is a device for communicating fire. In cases where the shell is expected to penetrate armor plate or other obstruction, and not to explode until after it has penetrated its target, the nose of the shell is pointed and of solid steel, and the fuze is screwed into the base of the shell—a base-detonating fuze. The fuze which we wish here to discuss is a point combination fuze, *point* because it is at the nose of the shell, and *combination* because it is designed to explode the shell either after a definite interval of flight or immediately on impact with the target.

The impact of the *firing pin* or trigger upon the *primer cap* in the base of the cartridge case produces fire, a quick small spurt of flame which sets fire to the black powder which is also within the primer. This sets fire to the powder or, in the case of bagged charges, to the *igniter*—and this produces a large hot flame which sweeps out into the chamber of the gun or cartridge, sweeps around the large grains of smokeless powder, and sets fire to them all over their surface. In a typical case the primer cap contains a mixture of mercury fulminate with antimony sulfide and potassium chlorate. The fulminate explodes when the mixture is crushed; it produces fire, and the other ingredients of the composition maintain the fire for a short interval. The igniter bag in our diagram is a silk bag containing black powder which takes fire readily and burns rapidly. The igniter and the bag containing the smokeless powder are made from silk because silk either burns or goes out—and leaves no smoldering residue in the barrel of .the gun after the shot has been fired. For different guns and among different nations the igniters are designed in a variety of ways, many of which are described in the books which deal with guns, gunnery, and ammunition. Sometimes the igniter powder is contained in an integral part of the cartridge case. For small arms no igniter is needed; the primer ignites the propellant. For large guns no cartridge case is used; the projectile and the propelling charge are loaded from the breech, the igniter bag being sewed or tied to the base end of the bag which contains the powder, and the *primer* being fitted in a hole in the breech-block by which the gun is closed.

The smokeless powder in our diagram is a dense, progressive-burning, colloided straight nitrocellulose powder, in cylindrical grains with one or with seven longitudinal perforations. The flame from the igniter lights the grains, both on the outer surfaces which commence to burn inward and in the perforations which commence to enlarge, burning outward. The burning at first is slow. As the pressure increases, the projectile starts to move.

The rifling in the barrel of the gun bites into the soft copper driving band, imparting a rotation to the projectile, and the rate of rotation increases as the projectile approaches the muzzle. As heat accumulates in the chamber of the gun, the powder burns faster and faster; gas and heat and pressure are produced for some time at an accelerated rate, and the projectile acquires acceleration continuously. It has its greatest velocity at the moment when it leaves the muzzle. The greatest pressure, however, occurs at a point far back from the muzzle where the gun is of correspondingly stronger construction than at its open end. The duration of the burning of the powder depends upon its *web thickness,* that is, upon the thickness between the single central perforation and the sides of the cylindrical grain, or, in the multiperforated powders, upon the thickness between the perforations. The powder, if properly designed, is burned completely at the moment when the projectile emerges from the muzzle.

The combination fuze contains two primer caps, and devices, more or less free to move within the fuze, by which these may be fired. When the shell starts to move, everything within it undergoes *setback,* and tends to lag because of its inertia. The fuze contains a piece of metal with a point or firing pin on its rearmost end, held in place by an almost complete ring set into its sides and in the sides of the cylindrical space through which it might otherwise move freely. This, with its primer cap, constitutes the *concussion* element. The setback causes it to pull through the ring; the pin strikes the cap; fire is produced and communicates with a train of slow-burning black powder of special composition (fuze powder) the length of which has been previously adjusted by turning the *time-train rings* in the head of the fuze. The powder train, in a typical case, may burn for any particular interval up to 21 seconds, at the end of which time the fire reaches a chamber or magazine which is filled with ordinary black powder. This burns rapidly and produces a large flame which strikes through to the detonator, containing mercury fulminate or lead azide, which explodes and causes the shell to detonate while it is in flight. The head of the fuze may also be adjusted in such manner that the fire produced by the concussion element will finally burn to a dead end, and the shell in that case will explode only in consequence of the action of the *percussion* element when it hits the target.

When the shell strikes any object and loses velocity, everything within it still tends to move forward. The percussion element consists of a metal cylinder, free to move backward and forward through a short distance, and of a primer cap, opposite the forward end of the cylinder and set into the metal in such fashion that the end of the cylinder cannot quite touch it. If this end of the cylinder should carry a firing pin, then it would fire the cap, and this might happen if the shell were dropped accidentally—with unfortunate results. When the shell starts to move in the gun, the cylinder lags back in the short space which is allotted to it. The shell rotates during flight. Centrifugal force, acting upon a mechanism within the cylinder, causes a firing pin to rise up out of its forward end. The fuze becomes *armed*. When the shell meets an obstacle, the cylinder rushes forward, the pin strikes the cap, fire is produced and communicates directly to the black powder magazine and to the detonator—and the shell is exploded forthwith.

FIGURE 2. Cross Section of a 155-mm. High-Explosive Shell Loaded with TNT.

The high explosive in the shell must be so insensitive that it will tolerate the shock of setback without exploding. Trinitrotoluene (TNT) is generally considered to be satisfactory for all military purposes, except for armor-piercing shells. The explosive must be tightly packed within the shell. There must be no cavities, lest the setback cause the explosive to move violently across the gap and to explode prematurely while the shell is still within the barrel of the gun, or as is more likely, to pull away from the detonator and fail to be exploded by it.

Trinitrotoluene, which melts below the boiling point of water, is generally loaded by pouring the liquid explosive into the shell. Since the liquid contracts when it freezes, and in order to prevent cavities, the shell standing upon its base is supplied at its open end with a paper funnel, like the neck of a bottle, and the liquid TNT is poured until the shell and the paper funnel are both full. After the whole has cooled, the funnel and any TNT which is in it are removed, and the space for the *booster* is bored out with a drill. Cast TNT is not exploded by the explosion of fulminate, which, however, does cause the explosion of granular and compressed TNT. The explosion of granular TNT will initiate the explosion of cast TNT, and the granular material may be used as a booster for that purpose. In practice, tetryl is generally preferred as a booster for military use. It is more easily detonated than TNT, more brisant, and a better initiator. Boosters are used even with high explosives which are detonated by fulminate, for they make it possible to get along with smaller quantities of this dangerous material.

Propagation of Explosion

When black powder burns, the first portion to receive the fire undergoes a chemical reaction which results in the production of hot gas. The gas, tending to expand in all directions from the place where it is produced, warms the next portion of black powder to the *kindling* temperature. This then takes fire and burns with the production of more hot gas which raises the temperature of the next adjacent material. If the black powder is confined, the pressure rises, and the heat, since it cannot escape, is communicated more rapidly through the mass. Further, the gas- and heat-producing chemical reaction, like any other chemical reaction, doubles its rate for every 10° (approximate) rise of temperature. In a confined space the combustion becomes extremely rapid, but it is still believed to be combustion in the sense that it is a phenomenon dependent upon the transmission of heat.

The explosion of a primary explosive or of a high explosive, on the other hand, is believed to be a phenomenon which is dependent upon the transmission of pressure or, perhaps more properly, upon the transmission of shock.[2] Fire, friction, or shock,

---

[2] The effects of static pressure and of the rate of production of the pressure have not yet been studied much, nor is there information concerning the pressures which occur within the mass of the explosive while it is exploding.

acting upon, say, fulminate, in the first instance cause it to undergo a rapid chemical transformation which produces hot gas, and the transformation is so rapid that the advancing front of the mass of hot gas amounts to a wave of pressure capable of initiating by its shock the explosion of the next portion of fulminate. This explodes to furnish additional shock which explodes the next adjacent portion of fulminate, and so on, the explosion advancing through the mass with incredible quickness. In a standard No. 6 blasting cap the explosion proceeds with a velocity of about 3500 meters per second.

If a sufficient quantity of fulminate is exploded in contact with trinitrotoluene, the shock induces the trinitrotoluene to explode, producing a shock adequate to initiate the explosion of a further portion. The explosive wave traverses the trinitrotoluene with a velocity which is actually greater than the velocity of the initiating wave in the fulminate. Because this sort of thing happens, the application of the principle of the booster is possible. If the quantity of fulminate is not sufficient, the trinitrotoluene either does not detonate at all or detonates incompletely and only part way into its mass. For every high explosive there is a minimum quantity of each primary explosive which is needed to secure its certain and complete denotation. The best initiator for one high explosive is not necessarily the best initiator for another. A high explosive is generally not its own best initiator unless it happens to be used under conditions in which it is exploding with its maximum *velocity of detonation.*

Detonating Fuse

Detonating fuse consists of a narrow tube filled with high explosive. When an explosion is initiated at one end by means of a detonator, the explosive wave travels along the fuse with a high velocity and causes the detonation of other high explosives which lie in its path. Detonating fuse is used for procuring the almost simultaneous explosion of a number of charges.

Detonating fuse is called *cordeau détonant* in the French language, and *cordeau* has become the common American designation for it. Cordeau has been made from lead tubes filled with trinitrotoluene, from aluminum or block tin tubes filled with picric acid, and from tubes of woven fabric filled with nitrocellulose or with pentaerythrite tetranitrate (PETN). In this country the Ensign-Bickford Company, at Simsbury, Connecticut, manufactures *Cordeau-Bickford*, a lead tube filled with TNT, and *Prima-cord-Bickford*,[3] a tube of waterproof textile filled with finely powdered PETN. The cordeau is made by filling a large lead pipe (about 1 inch in diameter) with molten TNT, allowing to cool, and drawing down in the manner that wire is drawn. The finished tube is tightly packed with finely divided crystalline TNT. Cordeau-Bickford detonates with a velocity of about 5200 meters per second (17,056 feet or 3.23 miles), Primacord-Bickford with a velocity of about 6200 meters per second (20,350 feet or 3.85 miles). These are not the maximum velocities of detonation of the explosives in question. The velocities would be greater if the tubes were wider.

Detonating fuse is fired by means of a blasting cap held snugly and firmly against its end by a *union* of thin copper tubing crimped into place. Similarly, two ends are spliced by holding them in contact within a *coupling*. The ends ought to touch each other, or at least to be separated by not more than a very small space, for the explosive wave of the detonating fuse cannot be depended upon to throw its initiating power across a gap of much more than ⅛ inch.

When several charges are to be fired, a single main line of detonating fuse is laid and branch lines to the several charges are connected to it. The method by which a branch is connected to a main line of cordeau is shown in Figures 3, 4, 5, 6, and 7. The main line is not cut or bent. The end of the branch is slit in two (with a special instrument designed for this purpose) and is opened to form a V in the point of which the main line is laid— and there it is held in place by the two halves of the slit branch cordeau, still filled with TNT, wound around it in opposite directions. The connection is made in this manner in order that the explosive wave, traveling along the main line, may strike the

---

[3] These are not to be confused with *Bickford fuse* or *safety fuse* manufactured by the same company, which consists of a central thread surrounded by a core of black powder enclosed within a tube of woven threads, surrounded by various layers of textile, waterproof material, sheathing, etc. This is *miner's fuse*, and is everywhere known as Bickford fuse after the Englishman who invented the machine by which such fuse was first woven. The most common variety burns with a velocity of about 1 foot per minute. When the fire reaches its end, a spurt of flame about an inch long shoots out for igniting black powder or for firing a blasting cap.

branch line squarely against the length of the column of TNT, and so provoke its detonation. If the explosive wave were traveling from the branch against the main line (as laid), it would

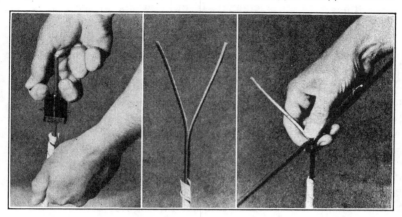

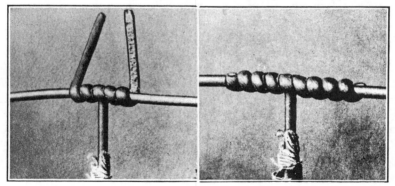

FIGURES 3, 4, 5, 6, and 7. Method of Connecting a Branch to a Main Line of Cordeau. (Courtesy Ensign-Bickford Company.) FIGURE 3. Slitting the Branch Line. FIGURE 4. The Slit End Open. FIGURE 5. The Main Line in Place. FIGURE 6. Winding the Splice. FIGURE 7. The Finished Junction.

strike across the column of TNT and would shatter it, but would be less likely to make it explode. For connecting a branch line of Primacord, it is satisfactory to make a half hitch of the end around the main line.

A circle of detonating fuse around a tree will rapidly strip off a belt of heavy bark, a device which is sometimes useful in the control of insect pests. If the detonating fuse is looped successively around a few blocks of TNT or cartridges of dynamite, and if these are strung around a large tree, the tree may be felled very quickly in an emergency. In military operations it may be desirable to "deny a terrain to the enemy" without occupying it oneself, and the result may be accomplished by scattering mustard gas over the area. For this purpose, perhaps during the night, a long piece of Primacord may be laid through the area, looped here and there in circles upon which tin cans of mustard gas (actually a liquid) are placed. The whole may be fired, when desired, by a single detonator, and the gas adequately dispersed.

Velocity of Detonation

If the quantity of the primary explosive used to initiate the explosion of a high explosive is increased beyond the minimum necessary for that result, the velocity with which the resulting explosion propagates itself through the high explosive is correspondingly increased, until a certain optimum is reached, depending upon the physical state of the explosive, whether cast or powdered, whether compressed much or little, upon the width of the column and the strength of the material which confines it, and of course upon the particular explosive which is used. By proper adjustment of these conditions, by pressing the powdered explosive to the optimum density (which must be determined by experiment) in steel tubes of sufficiently large diameter, and by initiating the explosion with a large enough charge of dynamite or other booster (itself exploded by a blasting cap), it is possible to secure the maximum velocity of detonation. This ultimate maximum is of less interest to workers with explosives than the maximum found while experimenting with paper cartridges, and it is the latter maximum which is generally reported. The physical state and density of the explosive, and the temperature at which the determinations were made, must also be noted if the figures for the velocity of detonation are to be reproducible.

Velocities of detonation were first measured by Berthelot and Vieille,[4] who worked first with gaseous explosives and later with liquids and solids. They used a Boulengé chronograph the precision of which was such that they were obliged to employ long

[4] Berthelot, "Sur la force des matières explosives," 2 vols., third edition, Paris, 1883, Vol. 1, p. 133. Cf. *Mém. poudres*, **4, 7** (1891).

Figure 8. Pierre-Eugène Marcellin Berthelot (1827-1907) (Photo by P. Nadar, Paris). Founder of thermochemistry and the science of explosives. He synthesized acetylene and benzene from their elements, and alcohol from ethylene, studied the polyatomic alcohols and acids, the fixation of nitrogen, the chemistry of agriculture, and the history of Greek, Syriac, Arabic, and medieval chemistry. He was a Senator of France, Minister of Public Instruction, Minister of Foreign Affairs, and Secretary of the Academy of Sciences, and is buried in the Panthéon at Paris.

columns of the explosives. The Mettegang recorder now commonly used for these measurements is an instrument of greater precision and makes it possible to work with much shorter cartridges of the explosive materials. This apparatus consists essentially of a strong, well-turned and balanced, heavy cylinder of steel which is rotated by an electric motor at a high but exactly known velocity. The velocity of its smoked surface relative to a platinum point which almost touches it may be as much as 100 meters per second. The explosive to be tested is loaded in a cylindrical cartridge. At a known distance apart two thin copper wires are passed through the explosive at right angles to the axis of the cartridge. If the explosive has been cast, the wires are bound tightly to its surface. Each of the wires is part of a closed circuit through an inductance, so arranged that, when the circuit is broken, a spark passes between the platinum point and the steel drum of the chronograph. The spark makes a mark upon the smoked surface. When the explosive is now fired by means of a detonator at the end of the cartridge, first one and then the other of the two wires is broken by the explosion, and two marks are made on the rotating drum. The distance between these marks is measured with a micrometer microscope. The duration of time which corresponds to the movement of the surface of the rotating drum through this distance is calculated, and this is the time which was required for the detonation of the column of known length of explosive which lay between the two wires. From this, the velocity of detonation in meters per second is computed easily.

Since a chronograph is expensive and time-consuming to use, the much simpler method of Dautriche,[5] which depends upon a comparison of the unknown with a standard previously measured by the chronograph, finds wide application. Commercial cordeau is remarkably uniform. An accurately measured length, say 2 meters, of cordeau of known velocity of detonation is taken, its midpoint is marked, and its ends are inserted into the cartridge of the explosive which is being tested, at a known distance apart, like the copper wires in the absolute method (Figure 9). The middle portion of the loop of cordeau is made straight and is laid upon a sheet of lead (6-8 mm. thick), the marked midpoint being

[5] *Mém. poudres,* 14, 216 (1907); *Comp. rend.,* 143, 641 (1906).

placed upon a line scratched in the lead plate at right angles to the direction of the cordeau. When the detonator in the end of the cartridge of explosive is fired, the explosive wave first encounters one end of the cordeau and initiates its explosion from this end, then proceeds through the cartridge, encounters the other end of the cordeau, and initiates its explosion from that end. The explosive waves from the two ends of the cordeau meet one another and mark the point of their meeting by an extra-deep, sharp furrow in the lead plate, perhaps by a hole punched

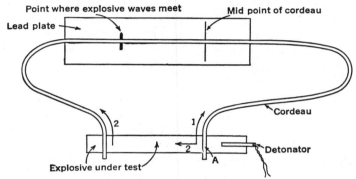

FIGURE 9. Dautriche Method of Measuring Velocity of Detonation. From the point A the explosion proceeds in two directions: (1) along the cordeau (of known velocity of detonation), and (2) through the cartridge of explosive which is being tested and then along the cordeau. When the two waves in the cordeau meet, they make a mark in the lead plate upon which the cordeau is resting.

through it. The distance of this point is measured from the line where the midpoint of the cordeau was placed. Call this distance d. It is evident that, from the moment when the near end of the cordeau started to detonate, one explosive wave traveled in the cordeau for a distance equal to one-half the length of the cordeau plus the distance d, while the other explosive wave, during the same interval of time, traveled in the explosive under examination a distance equal to the distance between the inserted ends of cordeau, then in the cordeau a distance equal to one-half its length minus the distance d. The times required for the passage of the explosive waves in the cordeau are calculated from the known velocity of detonation of the cordeau used; thence the time required for the detonation of the column of explosive which stood between the ends of the cordeau; thence the velocity of detonation in meters per second.

Velocities of detonation have recently been measured by high-speed photography of the explosions through a slit, and by other devices in which the elapsed times are measured by means of a cathode-ray oscillograph.

The Munroe Effect

The mark which explosive waves, traveling toward each other on the same piece of cordeau, make at the point where they meet is evidently due to the fact that they spread out sideways at the point of their encounter. Their combined forces produce an effect greater than either alone could give. The behavior of jets of water, shot against each other under high pressure, supplies a very good qualitative picture of the impact of explosive waves. If the waves meet at an angle, the resultant wave, stronger than either, goes off in a direction which could be predicted from a consideration of the parallelogram of forces. This is the explanation of the Munroe effect.

Charles Edward Munroe,[6] while working at the Naval Torpedo Station at Newport, discovered in 1888 that if a block of guncotton with letters countersunk into its surface is detonated with its lettered surface against a steel plate, the letters are indented into the surface of the steel. Similarly, if the letters are raised above the surface of the guncotton, by the detonation they are reproduced in relief on the steel plate, embossed and raised above the neighboring surface. In short, the greatest effects are produced on the steel plate at the points where the explosive material stands away from it, at the points precisely where explosive waves from different directions meet and reinforce each other. Munroe found that by increasing the depth of the concavity in the explosive he was able to produce greater and greater effects on the plate, until finally, with a charge which was pierced completely through, he was able to puncture a hole through it.[7] By introducing lace, ferns, coins, etc., between the flat surface of a

---

[6] For biographical notice by C. A. Browne, see *J. Am. Chem. Soc.*, **61**, 731 (1939).

[7] Cf. article by Marshall, "The Detonation of Hollow Charges," *J. Soc. Chem. Ind.*, **29**, 35 (1920).

FIGURE 10. Charles Edward Munroe (1849-1938). Leader in the development of explosives in the United States. Invented *indurite*, a variety of smokeless powder, and discovered the Munroe effect. Professor of chemistry at the U. S. Naval Academy, Annapolis, Maryland, 1874-1886; chemist at the Naval Torpedo Station and Naval War College, Newport, Rhode Island, 1886-1892; professor of chemistry at George Washington University, 1892-1917; and chief explosives chemist of the U. S. Bureau of Mines in Washington, 1919-1933. Author and co-author of many very valuable publications of the Bureau of Mines.

block of explosive and pieces of armor plate, Munroe was able to secure embossed reproductions of these delicate materials. Several fine examples of the Munroe effect, prepared by Munroe himself, are preserved in a fire screen at the Cosmos Club in Washington.

The effect of hollowed charges appears to have been rediscovered, probably independently, by Egon Neumann, who claimed it as an original discovery, and its application in explosive technique was patented by the Westfälisch-Anhaltische Sprengstoff-

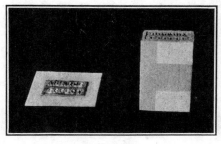 

Fig. 11   Fig. 12

FIGURES 11 and 12. Munroe Effect. (Courtesy Trojan Powder Company). FIGURE 11. Explosive Enclosed in Pasteboard Wrapper. Note that the letters incised into the surface of the explosive are in mirror writing, like words set in type, in order that the printing may be normal. A steel plate after a charge like that pictured was exploded against it, the incised surface being next to the plate. FIGURE 12. A section of steel shafting after a charge like that represented in FIGURE 11 had been exploded upon it, the incised surface of the explosive being next to the steel.

A. G. in 1910.[8] Neumann, working with blocks of TNT having conical indentations but not complete perforations, found that such blocks blew holes through wrought-iron plates, whereas solid blocks of greater actual weight only bent or dented them.

It has been recommended that torpedoes be loaded with charges hollowed in their forward ends. Advantage is taken of the Munroe effect in the routine blasting of oil wells, and, intentionally or not, by every explosives engineer who initiates an explosion by means of two or more electric blasting caps, fired simultaneously, at different positions within the same charge.

[8] Ger. Pat. Anm. W. 36,269 (1910); Brit. Pat. 28,030 (1911). Neumann, *Z. angew. Chem.*, 2238 (1911); *Z. ges. Schiess- u. Sprengstoffw.*, 183 (1914).

Sensitivity Tests

Among the important tests which are made on explosives are the determinations of their sensitivity to impact and to temperature, that is, of the distance through which a falling weight must drop upon them to cause them to explode or to inflame, and of the temperatures at which they inflame, explode, or "puff" spontaneously. At different places different machines and apparatus are used, and the numerical results differ in consequence from laboratory to laboratory.

For the falling weight or *impact* or *drop test* a 2-kilogram weight is generally used. In a typical apparatus the explosive undergoing the test is contained in a hole in a steel block, a steel plunger or piston is pressed down firmly upon it, and it is directly upon this plunger that the weight is dropped. A fresh sample is taken each time, and material which has not exploded from a single impact is discarded. A drop of 2 to 4 cm. will explode mercury fulminate, one of about 70 to 80 cm. will cause the inflammation of black powder, and one of 60 to 180 cm. will cause the explosion of TNT according to the physical state of the sample.

For determining the *temperature of ignition*, a weighed amount of the material is introduced into a copper capsule (a blasting cap shell) and this is thrust into a bath of Wood's metal previously heated to a known temperature. If no explosion occurs within 5 seconds (or other definite interval), the sample is removed, the temperature of the bath is raised 5° (usually), and a fresh sample in a fresh copper capsule is tried. Under these conditions (that is, within a 5-second interval), 4F black powder takes fire at $190° \pm 5°$, and 30-caliber straight nitrocellulose smokeless powder at $315° \pm 5°$. In another method of carrying out the test, the capsule containing the explosive is introduced into the metal bath at 100°, the temperature is raised at a steady and regulated rate, and the temperature at which the explosive decomposition occurs is noted. When the temperature is raised more rapidly, the inflammation occurs at a higher temperature, as indicated by the following table.[9] The fact that explosives are more sensitive to shock and to friction when they are warm is doubtless due to the same ultimate causes.

[9] van Duin, Dissertation, Utrecht, 1918, p. 89. The experiments were carried out with 0.1-gram samples in glass tubes.

TEMPERATURE OF IGNITION
Heated from 100°

	at 20° per minute	at 5° per minute
Trinitrotoluene...............	321°	304°
Picric acid...................	316°	309°
Tetryl......................	196°	187°
Hexanitrodiphenylamine......	258°	250°
Hexanitrodiphenyl sulfide......	319°	302°
Hexanitrodiphenyl sulfone.....	308°	297°

Substances like trinitrotoluene, picric acid, and tetryl, which are intrinsically stable at ordinary temperatures, decompose slowly if they are heated for considerable periods of time at temperatures below those at which they inflame. This, of course, is a matter of interest, but it is a property of all samples of the substance, does not vary greatly between them, and is not made the object of routine testing. Nitrocellulose and many nitric esters, however, appear to be intrinsically unstable, subject to a spontaneous decomposition which is generally slow but may be accelerated greatly by the presence of impurities in the sample. For this reason, nitrocellulose and smokeless powder are regularly subjected to *stability tests* for the purpose, not of establishing facts concerning the explosive in question, but rather for determining the quality of the particular sample.[10]

[10] The routine tests which are carried out on military explosives are described in U. S. War Department Technical Manual TM9-2900, "Military Explosives." The testing of explosives for sensitivity, explosive power, etc., is described in the *Bulletins* and *Technical Papers* of the U. S. Bureau of Mines. The student of explosives is advised to secure from the Superintendent of Documents, Washington, D. C., a list of the publications of the Bureau of Mines, and then to supply himself with as many as may be of interest, for they are sold at very moderate prices. The following are especially recommended. Several of these are now no longer procurable from the Superintendent of Documents, but they may be found in many libraries.

Bull. 15. "Investigations of Explosives Used in Coal Mines," by Clarence Hall, W. O. Snelling, and S. P. Howell.

Bull. 17. "A Primer on Explosives for Coal Miners," by Charles E. Munroe and Clarence Hall.

Bull. 48. "The Selection of Explosives Used in Engineering and Mining Operations," by Clarence Hall and Spencer P. Howell.

Bull. 59. "Investigations of Detonators and Electric Detonators," by Clarence Hall and Spencer P. Howell.

Bull. 66. "Tests of Permissible Explosives," by Clarence Hall and Spencer P. Howell.

Tests of Explosive Power and Brisance

For estimating the total energy of an explosive, a test in the manometric bomb probably supplies the most satisfactory single indication. It should be remembered that total energy and actual effectiveness are different matters. The effectiveness of an explosive depends in large part upon the rate at which its energy is liberated.

The high pressures developed by explosions were first measured by means of the Rodman gauge, in which the pressure caused a hardened-steel knife edge to penetrate into a disc of soft copper. The depth of penetration was taken as a measure of the pressure to which the apparatus had been subjected. This gauge was improved by Nobel, who used a copper cylinder placed between a fixed and a movable steel piston. Such *crusher gauges* are at present used widely, both for measuring the maximum pressures produced by explosions within the confined space of the manometric bomb and for determining the pressures which exist in the barrels of guns during the proof firing of powder. The small copper cylinders are purchased in large and uniform lots, their deformations under static pressures are determined and plotted in a chart, and the assumption is made that the sudden pressures resulting from explosions produce the same deformations as static pressures of the same magnitudes. Piezoelectric gauges, in which the pressure on a tourmaline crystal or on discs of quartz produces an electromotive force, have been used in work with manometric bombs and for measuring the pressures which exist in the chambers of guns. Other gauges, which depend

---

Bull. 80. "A Primer on Explosives for Metal Miners and Quarrymen," by Charles E. Munroe and Clarence Hall.

Bull. 346. "Physical Testing of Explosives at the Bureau of Mines Explosives Experiment Station, Bruceton, Pa.," by Charles E. Munroe and J. E. Tiffany.

Tech. Paper 125. "The Sand Test for Determining the Strength of Detonators," by C. G. Storm and W. C. Cope.

Tech. Paper 234. "Sensitiveness of Explosives to Frictional Impact," by S. P. Howell.

On this subject the book "Testing Explosives" by C. E. Bichel, English translation by A. Larnsen, London, 1905, will be found of value, as will also the book of Berthelot, already cited, and many important papers in *Mémorial des poudres* and *Zeitschrift für das gesamte Schiess- und Sprengstoffwesen.*

upon the change of electrical resistance of a conducting wire, are beginning to find application.

The manometric bomb is strongly constructed of steel and has a capacity which is known accurately. In order that the pressure resulting from the explosion may have real significance, the *density of loading*, that is, the number of grams of explosive per cubic centimeter of volume, must also be reported. The pressures produced by the same explosive in the same bomb are in general not directly proportional to the density of loading. The temperatures in the different cases are certainly different, and the compositions of the hot gaseous mixtures depend upon the pressures which exist upon them and determine the conditions of the equilibria between their components. The water in the gases can be determined, their volume and pressure can be measured at ordinary temperature, and the temperature of the explosion can be calculated roughly if the assumptions are made that the gas laws hold and that the composition of the cold gases is the same as that of the hot. If the gases are analyzed, and our best knowledge relative to the equilibria which exist between the components is assumed to be valid for the whole temperature range, then the temperature produced by the explosion can be calculated with better approximation.

Other means of estimating and comparing the capacity of explosives for doing useful work are supplied by the tests with the *ballistic pendulum*[11] and by the Trauzl and small lead block tests. The first of these is useful for comparing a new commercial explosive with one which is standard; the others give indications which are of interest in describing both commercial explosives and pure explosive substances.

In the *Trauzl lead block test* (often called simply the lead block test) 10 grams of the explosive, wrapped in tinfoil and stemmed with sand, is exploded by means of an electric detonator in a cylindrical hole in the middle of a cylindrical block of lead, and the enlargement of the cavity, measured by pouring in water from a graduate and corrected for the enlargement which is ascribable to the detonator alone, is reported. For the standard test, the blocks are cast from chemically pure lead, 200 mm. in height and 200 mm. in diameter, with a central hole made by the mold, 125 mm. deep and 25 mm. in diameter. The test is

[11] *U. S. Bur. Mines Bull.* 15, pp. 79-82.

applicable only to explosives which detonate. Black powder and other explosives which burn produce but little effect, for the gases blow out the stemming and escape. The test is largely one of brisance, but for explosives of substantially equal brisance it gives some indication of their relative power. An explosive of great brisance but little power will create an almost spherical pocket at the bottom of the hole in the block, while one of less brisance and greater power will enlarge the hole throughout its

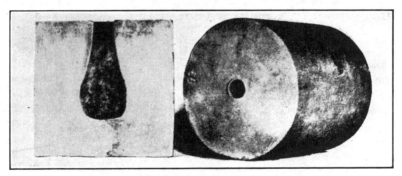

FIGURE 13. Lead Block Tests (above), and Trauzl Tests (below). (Courtesy U. S. Bureau of Mines.)

length and widen its throat at the top of the block. The form of the hole, then, as shown by sectioning the block, is not without significance. The Trauzl test does not give reliable indications with explosives which contain aluminum (such as *ammonal*) or with others which produce very high temperatures, for the hot gases erode the metal, and the results are high.

A small Trauzl block is used for testing commercial detonators.

Another test, known as the *small lead block test*, is entirely a test of brisance. As the test is conducted at the U. S. Bureau of Mines,[12] a lead cylinder 38 mm. in diameter and 64 mm. high is set upright upon a rigid steel support; a disc of annealed steel

[12] *Ibid.,* p. 114.

38 mm. in diameter and 6.4 mm. thick is placed upon it; a strip of manila paper wide enough to extend beyond the top of the composite cylinder and to form a container at its upper end is wrapped and secured around it; 100 grams of explosive is placed

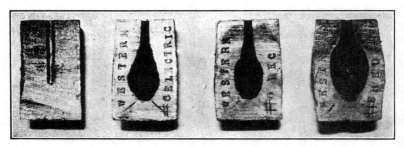

FIGURE 14. Small Trauzl Tests of Detonators. (Courtesy Western Cartridge Company.)

in this container and fired, without tamping, by means of an electric detonator. The result is reported as the compression of the lead block, that is, as the difference between its height before and its height after the explosion. The steel disc receives the force of the explosion and transmits it to the lead cylinder. With-

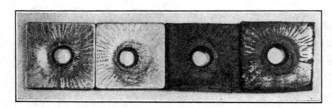

FIGURE 15. Aluminum Plate and Lead Plate Tests of Detonators. (Courtesy Atlas Powder Company.)

out it, the lead cylinder would be so much deformed that its height could not be measured.

In the *lead plate test of detonators*, the detonator is fired while standing upright on a plate of pure lead. Plates 2 to 6 mm. thick are used, most commonly 3 mm. A good detonator makes a clean-cut hole through the lead. The metal of the detonator case is blown into small fragments which make fine and characteristic markings on the lead plate radiating away from the place where the detonator stood. With a good detonator, the surface of the lead plate ought to show no places where it has been torn roughly

FIGURE 16. Effect of Detonator on Lead Plate 10 cm. Distant from Its End. The diameter of the hole is about 1½ times the diameter of the detonator which was fired. The lead has splashed up around the hole in much the same fashion as placid water splashes when a pebble is dropped into it. Note the numerous small splashes on the lead plate where it was struck by fragments of the detonator casing.

by large fragments of the case. Similar tests are often carried out with plates of aluminum.

CHAPTER II

BLACK POWDER

The discovery that a mixture of potassium nitrate, charcoal, and sulfur is capable of doing useful mechanical work is one of the most important chemical discoveries or inventions of all time. It is to be classed with the discovery or invention of pottery, which occurred before the remote beginning of history, and with that of the fixation of nitrogen by reason of which the ecology of the human race will be different in the future from what it has been throughout the time that is past. Three great discoveries signalized the break-up of the Middle Ages: the discovery of America, which made available new foods and drugs, new natural resources, and new land in which people might multiply, prosper, and develop new cultures; the discovery of printing, which made possible the rapid and cheap diffusion of knowledge; and the discovery of the controllable force of gunpowder, which made huge engineering achievements possible, gave access to coal and to minerals within the earth, and brought on directly the age of iron and steel and with it the era of machines and of rapid transportation and communication. It is difficult to judge which of these three inventions has made the greatest difference to mankind.

Black powder and similar mixtures were used in incendiary compositions, and in pyrotechnic devices for amusement and for war, long before there was any thought of applying their energy usefully for the production of mechanical work. The invention of guns—and it seems to be this invention which is meant when "the discovery of gunpowder" is mentioned—did not follow immediately upon the discovery of the composition of black powder. It is possible that other applications antedated it, that black powder was used in petards for blowing down gateways, drawbridges, etc., or in simple operations of blasting, before it was used for its ballistic effect.

Berthold Schwarz

The tradition that the composition of black powder was discovered and that guns were invented about 1250 (or 1350 or even later) by Berthold Schwarz, a monk of Freiburg i. Br., in Germany, is perpetuated by a monument at that place. Constantin Anklitzen assumed the name of Berthold when he joined the Franciscan order, and was known by his confreres as *der schwarzer Berthold* because of his interest in black magic. The records of the Franciscan chapter in Freiburg were destroyed or scattered before the Reformation, and there are no contemporaneous accounts of the alleged discovery. Concerning the absence of documents, Oesper[1] says:

If he is a purely legendary inventor the answer is obvious. However, history may have taken no interest in his doings because guns were said to be execrable inventions and their employment (except against the unbelievers) was decried as destructive of manly valor and unworthy of an honorable warrior. Berthold was reputed to have compounded powder with Satan's blessing, and the clergy preached that as a co-worker of the Evil One he was a renegade to his profession and his name should be forgotten. There is a tradition that he was imprisoned by his fellow monks, and some say he made his diabolic invention while in prison. According to another legend, Berthold blew himself up while demonstrating the power of his discovery; another states that he was executed.

The lovers of fine points may argue over Berthold's existence, but it can be historically established that Freiburg in the fourteenth and fifteenth centuries was a flourishing center for the casting of cannon and the training of gunners.

Boerhaave on Black Powder

Although black powder has done immeasurable good through its civil uses, it has nevertheless been regarded as an evil discovery because of the easy and unsportsmanlike means which it provides for the destruction of life. Boerhaave, more than two centuries ago, wrote in the modern spirit on the importance of chemistry in war and condemned black powder[2] in a manner

[1] Oesper, *J. Chem. Education,* **16**, 305-306 (1939).

[2] Boerhaave, "A New Method of Chemistry," etc., trans. Peter Shaw, London, 1753, Vol. I, pp. 189-191. The quoted passage corresponds to the Latin of Vol. I, pp. 99-101, of the first edition of Boerhaave's "Elementa Chemiae," Leiden, 1732.

similar to that in which some of our latest devices of warfare
have been decried in public print.

It were indeed to be wish'd that our art had been less ingenious, in contriving means destructive to mankind; we
mean those instruments of war, which were unknown to the
ancients, and have made such havoc among the moderns.
But as men have always been bent on seeking each other's
destruction by continual wars; and as force, when brought
against us, can only be repelled by force; the chief support
of war, must, after money, be now sought in chemistry.

Roger Bacon, as early as the twelfth century,[3] had found
out gunpowder, wherewith he imitated thunder and lightning; but that age was so happy as not to apply so extraordinary a discovery to the destruction of mankind. But two
ages afterwards, *Barthol. Schwartz,*[4] a *German* monk and
chemist, happening by some accident to discover a prodigious
power of expanding in some of this powder which he had
made for medicinal uses, he apply'd it first in an iron barrel,
and soon after to the military art, and taught it to the
Venetians. The effect is, that the art of war has since that
time turned entirely on this one chemical invention; so that
the feeble boy may now kill the stoutest hero: Nor is there
anything, how vast and solid soever, can withstand it. By a
thorough acquaintance with the power of this powder, that
intelligent *Dutch* General *Cohorn* quite alter'd the whole art
of fighting; making such changes in the manner of fortifica-
tion, that places formerly held impregnable, now want de-
fenders. In effect, the power of gun-powder is still more to be
fear'd. I tremble to mention the stupendous force of another

[3] Bacon lived in the thirteenth century; we quote the passage as it is
printed.

[4] Shaw's footnote (*op. cit.,* p. 190) states:

What evidently shews the ordinary account of its invention false, is,
that *Schwartz* is held to have first taught it to the *Venetians* in the year
1380; and that they first used it in the war against the *Genoese,* in a
place antiently called *Fossa Caudeana,* now *Chioggia.* For we find mention of fire arms much earlier: *Peter Messius,* in his *variae lectiones,*
relates that *Alphonsus XI.* king of Castile used mortars against the
Moors, in a siege in 1348; and *Don Pedro,* bishop of *Leon,* in his chron-
icle, mentions the same to have been used above four hundred years
ago, by the people of *Tunis,* in a sea fight against the *Moorish* king of
Sevil. Du Cange adds, that there is mention made of this powder in
the registers of the chambers of accounts in *France,* as early as the
year 1338.

powder, prepar'd of sulfur, nitre, and burnt lees of wine;[5] to
say nothing of the well-known power of *aurum fulminans*.
Some person taking a quantity of fragrant oil, chemically
procured from spices, and mixing it with a liquor procured
from salt-petre, discover'd a thing far more powerful than
gun-powder itself; and which spontaneously kindles and

[5] This is *fulminating powder*, made, according to Ure's "Dictionary of
Chemistry," first American edition, Philadelphia, 1821:

by triturating in a warm mortar, three parts by weight of nitre, two
of carbonate of potash, and one of flowers of sulfur. Its effects, when
fused in a ladle, and then set on fire, are very great. The whole of the
melted fluid explodes with an intolerable noise, and the ladle is commonly disfigured, as if it had received a strong blow downwards.

Samuel Guthrie, Jr. (cf. *Archeion*, 13, 11 ff. [1931]), manufactured and
sold in this country large quantities of a similar material. In a letter to
Benjamin Silliman dated September 12, 1831 (*Am. J. Sci. Arts,* 21, 288 ff.
[1832]), he says:

I send you two small phials of nitrated sulphuret of potash, or yellow
powder, as it is usually called in this country. . . . I have made some
hundred pounds of it, which were eagerly bought up by hunters and
sportsmen for *priming* fire arms, a purpose which it answered most
admirably; and, but for the happy introduction of powder for priming,
which is ignited by percussion, it would long since have gone into
extensive use.

With this preparation I have had much to do, and I doubt whether,
in the whole circle of experimental philosophy, many cases can be found
involving dangers more appalling, or more difficult to be overcome,
than melting fulminating powder and saving the product, and reducing
the process to a business operation. I have had with it some eight or
ten tremendous explosions, and in one of them I received, full in my
face and eyes, the flame of a quarter of a pound of the composition,
just as it had become thoroughly melted.

The common proportions of 3 parts of nitre, 2 parts of carbonate of
potash and 1 part of sulphur, gave a powder three times quicker than
common black powder; but, by *melting together* 2 parts of nitre and 1
of carbonate of potash, and when the mass was cold adding to 4½ parts
of it, 1 part of sulphur—equal in the 100, to 54.54 dry nitre, 27.27 dry
carbonate of potash and 18.19 sulphur—a greatly superior composition
was produced, burning no less than eight and one half times quicker
than the best common powder. The substances were intimately ground
together, and then melted to a *waxy* consistence, upon an iron plate of
one inch in thickness, heated over a muffled furnace, taking care to
knead the mass assiduously, and remove the plate as often as the bottom of the mass became pretty slippery.

By the previously melting together of the nitre and carbonate of
potash, a more intimate union of these substances was effected than
could possibly be made by mechanical means, or by the slight melting
which was admissible in the after process; and by the slight melting of
the whole upon a *thick* iron plate, I was able to conduct the business
with facility and safety.

The melted mass, after being cold, is as hard and porous as pumice
stone, and is grained with difficulty; but there is a stage when it is
cooling in which it is very crumbly, and it should then be powdered
upon a board, with a small wooden cylinder, and put up hot, without
sorting the grains or even sifting out the flour.

burns with great fierceness, without any application of fire.[6]
I shall but just mention a fatal event which lately happen'd
in *Germany*, from an experiment made with balsam of sulphur terebinthinated, and confined in a close chemical vessel,
and thus exploded by fire; God grant that mortal men may
not be so ingenious at their own cost, as to pervert a profitable science any longer to such horrible uses. For this reason
I forbear to mention several other matters far more horrible
and destructive, than any of those above rehearsed.

Greek Fire

Fire and the sword have been associated with each other from
earliest times. The invention of Greek fire appears to have consisted of the addition of saltpeter to the combustible mixtures
already in use, and Greek fire is thus seen as the direct ancestor
both of black gunpowder and of pyrotechnic compositions.

The Byzantine historian, Theophanes the Confessor, narrates
that "Constantine [Constantine IV, surnamed Pogonatus, the
Bearded], being apprised of the designs of the unbelievers against
Constantinople, commanded large boats equipped with cauldrons
of fire (tubs or vats of fire) and fast-sailing galleys equipped
with siphons." The narrative refers to events which occurred in
the year 670, or possibly 672. It says for the next year: "At this
time Kallinikos, an architect (engineer) from Heliopolis of Syria,
came to the Byzantines and having prepared a sea fire (or marine
fire) set fire to the boats of the Arabs, and burned these with
their men aboard, and in this manner the Byzantines were
victorious and found (discovered) the marine fire."[7] The Mos-
lem fleet was destroyed at Cyzicus by the use of this fire which
for several centuries afterwards continued to bring victory
to the Byzantines in their naval battles with the Moslems and
Russians.

Leo's *Tactica*, written about A.D. 900 for the generals of the
empire, tells something of the manner in which the Greek fire was
to be used in combat.

[6] Shaw's footnote (*op. cit.*, p. 191): "A drachm of compound spirit of
nitre being poured on half a drachm of oil of carraway seeds *in vacuo*;
the mixture immediately made a flash like gun-powder, and burst the exhausted receiver, which was a glass six inches wide, and eight inches deep."

[7] Quoted by N. D. Cheronis, article entitled "Chemical Warfare in the
Middle Ages. Kallinikos's Prepared Fire," *J. Chem. Education*, **14,** 360
(1937).

And of the last two oarsmen in the bow, let the one be the *siphonator*, and the other to cast the anchor into the sea. . . . In any case, let him have in the bow the *siphon* covered with copper, as usual, by means of which he shall shoot the *prepared fire* upon the enemy. And above such siphon (let there be) a false bottom of planks also surrounded by boards, in which the warriors shall stand to meet the oncoming foes. . . . On occasion [let there be] formations immediately to the front [without maneuvers] so, whenever there is need, to fall upon the enemy at the bow and set fire to the ships by means of the *fire of the siphons*. . . . Many very suitable contrivances were invented by the ancients and moderns, with regard to both the enemy's ships and the warriors on them—such as at that time the *prepared fire* which is ejected (thrown) by means of *siphons* with a roar and a lurid (burning) smoke and filling them [the ships] with smoke. . . . They shall use also the other method of small *siphons* thrown (i.e., directed) by hand from behind iron shields and held [by the soldiers], which are called *hand siphons* and have been recently manufactured by our state. For these can also throw (shoot) the *prepared fire* into the faces of the enemy.[8]

Leo also described the use of *strepta*, by which a liquid fire was ejected, but he seemed to have been vague upon the details of construction of the pieces and upon the force which propelled the flame, and, like the majority of the Byzantine writers, he failed to mention the secret ingredient, the saltpeter, upon which the functioning of the fires undoubtedly depended, for their flames could be directed downward as well as upward.

The Byzantines kept their secret well and for a long time, but the Moslems finally learned about it and used the fire against the Christians at the time of the Fifth Crusade. In the Sixth Crusade the army of Saint Louis in Egypt was assailed with incendiaries thrown from ballistae, with fire from tubes, and with grenades of glass and metal, thrown by hand, which scattered fire on bursting. Brock[9] thinks that the fire from tubes operated in the manner of Roman candles. The charge, presumed to be a non-homogeneous mixture of combustible materials with saltpeter, "will, in certain proportions, if charged into a strong tube, give intermittent bursts, projecting blazing·masses of the mixture to a

[8] Cheronis, *op. cit.*, p. 362.

[9] A. St. H. Brock, "Pyrotechnics: the History and Art of Firework Making," London, 1922, p. 15.

considerable distance. The writer has seen this effect produced in a steel mortar of 5½ inches diameter, the masses of composition being thrown a distance of upwards of a hundred yards, a considerable range in the days of close warfare." There is no reason to believe that the fire tubes were guns.

Marcus Graecus

In the celebrated book of Marcus Graecus, *Liber ignium ad comburendos hostes*,[10] Greek fire and other incendiaries are described fully, as is also black powder and its use in rockets and crackers. This work was quoted by the Arabian physician, Mesue, in the ninth century, and was probably written during the eighth.

Greek fire is made as follows: take sulfur, tartar, sarcocolla, pitch, melted saltpeter, petroleum oil, and oil of gum, boil all these together, impregnate tow with the mixture, and the material is ready to be set on fire. This fire cannot be extinguished by urine, or by vinegar, or by sand. . . .

Flying fire (rockets) may be obtained in the following manner: take one part of colophony, the same of sulfur, and two parts of saltpeter. Dissolve the pulverized mixture in linseed oil, or better in oil of lamium. Finally, the mixture is placed in a reed or in a piece of wood which has been hollowed out. When it is set on fire, it will fly in whatever direction one wishes, there to set everything on fire.

Another mixture corresponds more closely to the composition of black powder. The author even specifies grapevine or willow charcoal which, with the charcoal of black alder, are still the preferred charcoals for making fuze powders and other grades where slow burning is desired.

Take one pound of pure sulfur, two pounds of grapevine or willow charcoal, and six pounds of saltpeter. Grind these three substances in a marble mortar in such manner as to reduce them to a most subtle powder. After that, the powder in desired quantity is put into an envelope for flying (a rocket) or for making thunder (a cracker). Note that the envelope for flying ought to be thin and long and well-filled with the above-described powder tightly packed, while the envelope for making thunder ought to be short and thick,

---

[10] Book of fires for burning the enemy, reprinted in full by Hoefer, "Histoire de la chimie," second edition, Paris, 1866, Vol. 1, pp. 517-524, and discussed *ibid.*, Vol. 1, p. 309.

only half filled with powder, and tightly tied up at both ends with an iron wire. Note that a small hole ought to be made in each envelope for the introduction of the match. The match ought to be thin at both ends, thick in the middle, and filled with the above-described powder. The envelope intended to fly in the air has as many thicknesses (ply) as one pleases; that for making thunder, however, has a great many.

Toward the end of the *Liber ignium* the author gives a slightly different formula for the black powder to be used in rockets.

> The composition of flying fire is threefold. The first composition may be made from saltpeter, sulfur, and linseed oil. These ground up together and packed into a reed, and lighted, will make it ascend in the air. Another flying fire may be made from saltpeter, sulfur, and grapevine or willow charcoal. These materials, mixed and introduced into a papyrus tube, and ignited, will make it fly rapidly. And note that one ought to take three times as much charcoal as sulfur and three times as much saltpeter as charcoal.

Roger Bacon

Roger Bacon appears to have been the first scholar in northern Europe who was acquainted with the use of saltpeter in incendiary and explosive mixtures. Yet the passage in which he makes specific mention of this important ingredient indicates that toy firecrackers were already in use by the children of his day. In the "Opus Majus," Sixth Part, On Experimental Science, he writes:

> For malta, which is a kind of bitumen and is plentiful in this world, when cast upon an armed man burns him up. The Romans suffered severe loss of life from this in their conquests, as Pliny states in the second book of the Natural History, and as the histories attest. Similarly yellow petroleum, that is, oil springing from the rock, burns up whatever it meets if it is properly prepared. For a consuming fire is produced by this which can be extinguished with difficulty; for water cannot put it out. Certain inventions disturb the hearing to such a degree that, if they are set off suddenly at night with sufficient skill, neither city nor army can endure them. No clap of thunder could compare with such noises. Certain of these strike such terror to the sight that the coruscations of the clouds disturb it incomparably less. . . . We have an example of this in that toy of children which is made in many parts of the world, namely an instrument as

FIGURE 17. Roger Bacon (c. 1214-1292). Probably the first man in Latin Europe to publish a description of black powder. He was acquainted with rockets and firecrackers, but not with guns.

large as the human thumb. From the force of the salt called saltpeter so horrible a sound is produced at the bursting of so small a thing, namely a small piece of parchment, that we perceive it exceeds the roar of sharp thunder, and the flash exceeds the greatest brilliancy of the lightning accompanying the thunder.[11]

A description in cypher of the composition of black powder in the treatise "De nullitate magiae"[12] which is ascribed to Roger Bacon has attracted considerable attention. Whether Bacon wrote the treatise or not, it is certain at any rate that the treatise dates from about his time and certain, too, that much of the material which it contains is to be found in the "Opus Majus." The author describes many of the wonders of nature, mechanical, optical, medicinal, etc., among them incendiary compositions and firecrackers.

> We can prepare from saltpeter and other materials an artificial fire which will burn at whatever distance we please. The same may be made from red petroleum and other things, and from amber, and naphtha, and white petroleum, and from similar materials. . . . Greek fire and many other combustibles are closely akin to these mixtures. . . . For the sound of thunder may be artificially produced in the air with greater resulting horror than if it had been produced by natural causes. A moderate amount of proper material, of the size of the thumb, will make a horrible sound and violent coruscation.

Toward the end of the treatise the author announces his intention of writing obscurely upon a secret of the greatest importance, and then proceeds to a seemingly incoherent discussion of something which he calls "the philosopher's egg." Yet a thoughtful reading between the lines shows that the author is describing the purification of "the stone of Tagus" (saltpeter), and that this material is somehow to be used in conjunction with "certain parts of burned shrubs or of willow" (charcoal) and with the "vapor of pearl" (which is evidently sulfur in the language of the medieval

[11] "The Opus Majus of Roger Bacon," trans. Robert Belle Burke, University of Pennsylvania Press, Philadelphia, 1928, Vol. 2, p. 629.

[12] Cf. "Roger Bacon's Letter Concerning the Marvelous Power of Art and of Nature and Concerning the Nullity of Magic," trans. Tenney L. Davis, Easton, Pennsylvania, 1922.

chemists). The often-discussed passage which contains the black powder anagram is as follows:

Sed tamen salis petrae LVRV VO .PO VIR CAN VTRIET sulphuris, et sic facies tonitruum et coruscationem: sic facies artificium.

A few lines above the anagram, the author sets down the composition of black powder in another manner. "Take then of the bones of Adam (charcoal) and of the Calx (sulfur), the same weight of each; and there are six of the Petral Stone (saltpeter) and five of the Stone of Union." The Stone of Union is either sulfur or charcoal, probably sulfur, but it doesn't matter for the context has made it evident that only three components enter into the composition. Of these, six parts of saltpeter are to be taken, five each of the other two. The little problem in algebra supplies a means of checking the solution of the anagram, and it is evident that the passage ought to be read as follows:

Sed tamen salis petrae R. VI. PART. V. NOV. CORVLI. ET V. sulphuris, et sic facies tonitruum et coruscationem: sic facies artificium.

But, however, of saltpeter take six parts, five of young willow (charcoal), and five of sulfur, and so you will make thunder and lightning, and so you will turn the trick.

The 6:5:5 formula is not a very good one for the composition of black powder for use in guns, but it probably gave a mixture which produced astonishing results in rockets and firecrackers, and it is not unlike the formulas of mixtures which are used in certain pyrotechnic pieces at the present time.

Although Roger Bacon was not acquainted with guns or with the use of black powder for accomplishing mechanical work, yet he seems to have recognized the possibilities in the mixture, for the treatise "On the Nullity of Magic" comes to an end with the statement: "Whoever will rewrite this will have a key which opens and no man shuts, and when he will shut, no man opens."[13]

[13] Compare *Revelations*, 3: 7 and 8. "And to the angel of the church in Philadelphia write: These things saith he that is holy, he that is true, he that hath the key of David, he that openeth, and no man shutteth; and shutteth, and no man openeth; I know thy works: behold, I have set before thee an open door, and no man can shut it."

Development of Black Powder[14]

Guns apparently first came into use shortly after the death of Roger Bacon. A manuscript in the Asiatic Museum at Leningrad, probably compiled about 1320 by Shems ed Din Mohammed, shows tubes for shooting arrows and balls by means of powder. In the library of Christ Church, Oxford, there is a manuscript entitled "De officiis regum," written by Walter de Millemete in 1325, in which a drawing pictures a man applying a light to the touch-hole of a bottle-shaped gun for firing a dart. On February 11, 1326, the Republic of Venice ordered iron bullets and metal cannon for the defense of its castles and villages, and in 1338 cannon and powder were provided for the protection of the ports of Harfleur and l'Heure against Edward III. Cannon were used in 1342 by the Moors in the defense of Algeciras against Alphonso XI of Castile, and in 1346 by the English at the battle of Crécy.

When guns began to be used, experiments were carried out for determining the precise composition of the mixture which would produce the best effect. One notable study, made at Bruxelles about 1560, led to the selection of a mixture containing saltpeter 75 per cent, charcoal 15.62 per cent, and sulfur 9.38 per cent. A few of the formulas for black powder which have been used at various times are calculated to a percentage basis and tabulated below: ·

	SALTPETER	CHARCOAL	SULFUR
8th century, Marcus Graecus...........	66.66	22.22	11.11
8th century, Marcus Graecus...........	69.22	23.07	7.69
c. 1252, Roger Bacon.................	37.50	31.25	31.25
1350, Arderne (laboratory recipe)........	66.6	22.2	11.1
1560, Whitehorne....................	50.0	33.3	16.6
1560, Bruxelles studies................	75.0	15.62	9.38
1635, British Government contract......	75.0	12.5	12.5
1781, Bishop Watson.................	75.0	15.0	10.0

It is a remarkable fact, and one which indicates that the improvements in black powder have been largely in the methods of manufacture, that the last three of these formulas correspond very closely to the composition of all potassium nitrate black powder for military and sporting purposes which is used today. Any considerable deviation from the 6:1:1 or 6:1.2:0.8 formulas

[14] An interesting and well-documented account of the history of black powder and of other explosives may be found in Molinari and Quartieri's "Notizie sugli esplodenti in Italia," Milano, 1913.

produces a powder which burns more slowly or produces less vigorous effects, and different formulas are used for the compounding of powders for blasting and for other special purposes. In this country blasting powder is generally made from sodium nitrate.

John Bate early in the seventeenth century understood the individual functions of the three components of black powder

Figure 18. Gunpowder Manufacture, Lorrain, 1630. After the materials had been intimately ground together in the mortar, the mixture was moistened with water, or with a solution of camphor in brandy, or with other material, and formed into grains by rubbing through a sieve.

when he wrote: "The Saltpeter is the Soule, the Sulphur the Life, and the Coales the Body of it."[15] The saltpeter supplies the oxygen for the combustion of the charcoal, but the sulfur is the life, for this inflammable element catches the first fire, communicates it throughout the mass, makes the powder quick, and gives it vivacity.

Hard, compressed grains of black powder are not porous—the sulfur appears to have colloidal properties and to fill completely

[15] John Bate, "The Mysteries of Nature and Art," second edition, London, 1635, p. 95.

the spaces between the small particles of the other components—
and the grains are poor conductors of heat. When they are lighted,
they burn progressively from the surface. The area of the surface
of an ordinary grain decreases as the burning advances, the grain
becomes smaller and smaller, the rate of production of gas decreases, and the duration of the whole burning depends upon the
dimension of the original grain. Large powder grains which required more time for their burning were used in the larger guns.
Napoleon's army used roughly cubical grains 8 mm. thick in its
smaller field guns, and cubical or lozenge-shaped grains twice as
thick in some of its larger guns. Grains in the form of hexagonal
prisms were used later, and the further improvement was introduced of a central hole through the grain in a direction parallel
to the sides of the prism. When these single-perforated hexagonal
prisms were lighted, the area of the outer surfaces decreased as
the burning advanced, but the area of the inner surfaces of the
holes actually increased, and a higher rate of production of gas
was maintained. Such powder, used in rifled guns, gave higher
velocities and greater range than had ever before been possible.
Two further important improvements were made: one, the use of
multiple perforations in the prismatic grain by means of which
the burning surface was made actually to increase as the burning
progressed, with a resultant acceleration in the rate of production
of the gases; and the other, the use of the slower-burning *cocoa
powder* which permitted improvements in gun design. These,
however, are purely of historical interest, for smokeless powder
has now entirely superseded black powder for use in guns.

If a propellent powder starts to burn slowly, the initial rise of
pressure in the gun is less and the construction of the breech end
of the gun need not be so strong and so heavy. If the powder
later produces gas at an accelerated rate, as it will do if its
burning surface is increasing, then the projectile, already moving
in the barrel, is able to take up the energy of the powder gases
more advantageously and a greater velocity is imparted to it.
The desired result is now secured by the use of progressive-
burning colloided smokeless powder. Cocoa powder was the most
successful form of black powder for use in rifled guns of long
range.

Cocoa powder or brown powder was made in single-perforated

hexagonal or octagonal prisms which resembled pieces of milk chocolate. A partially burned brown charcoal made from rye straw was used. This had colloidal properties and flowed under pressure, cementing the grains together, and made it possible to manufacture powders which were slow burning because they contained little sulfur or sometimes even none. The compositions of several typical cocoa powders are tabulated below:

	SALTPETER	BROWN CHARCOAL	SULFUR
England.........	79	18	3
England.........	77.4	17.6	5
Germany.........	78	19	3
Germany.........	80	20	0
France..........	78	19	3

Cocoa powder was more sensitive to friction than ordinary black powder. Samples were reported to have inflamed from shaking in a canvas bag. Cocoa powder was used in the Spanish-American war, 1898. When its use was discontinued, existing stocks were destroyed, and single grains of the powder are now generally to be seen only in museums.

Burning of Black Powder

Black powder burns to produce a white smoke. This, of course, consists of extremely small particles of solid matter held temporarily in suspension by the hot gases from the combustion. Since the weight of these solids is equal to more than half of the weight of the original powder, the superiority of smokeless powder, which produces practically no smoke and practically 100 per cent of its weight of hot gas, is immediately apparent. The products of the burning of black powder have been studied by a number of investigators, particularly by Noble and Abel,[16] who showed that the burning does not correspond to any simple chemical reaction between stoichiometrical proportions of the ingredients. Their experiments with RLG powder having the percentage composition indicated below showed that this powder burned to produce (average results) 42.98 per cent of its weight of gases, 55.91 per cent solids, and 1.11 per cent water.

[16] Noble and Abel, *Phil. Trans.*, 1875, 49; 1880, 203; *Mém. poudres,* **1,** 193 (1882). See also Debus, *Proc. Roy. Soc.,* **30,** 198 (1880).

```
Potassium nitrate...................... 74.430
Potassium sulfate...................... 0.133
Sulfur................................. 10.093
           ⎡ Carbon...... 12.398 ⎤
Charcoal   ⎢ Hydrogen.... 0.401  ⎥ ..... 14.286
           ⎢ Oxygen...... 1.272  ⎥
           ⎣ Ash......... 0.215  ⎦
Moisture............................. 1.058
```

Their mean results from the analysis of the gaseous products (percentage by volume) and of the solid products (percentage by weight) are shown in the following tables.

Carbon dioxide.............	49.29	Potassium carbonate........	61.03
Carbon monoxide...........	12.47	Potassium sulfate..........	15.10
Nitrogen...................	32.91	Potassium sulfide..........	14.45
Hydrogen sulfide...........	2.65	Potassium thiocyanate.......	0.22
Methane...................	0.43	Potassium nitrate..........	0.27
Hydrogen..................	2.19	Ammonium carbonate.......	0.08
		Sulfur....................	8.74
		Carbon...................	0.08

One gram of the powder in the state in which it was normally used, that is, while containing 1.058 per cent of moisture, produced 718.1 calories and 271.3 cc. of permanent gas measured at 0° and 760 mm. One gram of the completely desiccated powder gave 725.7 calories and 274.2 cc. These results indicate by calculation that the explosion of the powder produces a temperature of about 3880°.

Uses of Black Powder

Where smoke is no objection, black powder is probably the best substance that we have for communicating fire and for producing a quick hot flame, and it is for these purposes that it is now principally used in the military art. Indeed, the fact that its flame is filled with finely divided solid material makes it more efficient as an igniter for smokeless powder than smokeless powder itself. Standard black powder (made approximately in accordance with the 6:1:1 or the 6:1.2:0.8 formula) is used in *pétards*, as a base charge or expelling charge for shrapnel shells, in saluting and blank fire charges, as the bursting charge of practice shells and bombs, as a propelling charge in certain pyrotechnic pieces, and, either with or without the admixture of other substances which modify the rate of burning, in the time-train

FIGURE 19. Stamp Mill for Making Black Powder. (Courtesy National Fireworks Company and the *Boston Globe*.) This mill, which makes powder for use in the manufacture of fireworks, consists of a single block of granite in which three deep cup-shaped cavities have been cut. The stamps which operate in these cups are supplied at their lower ends with cylindrical blocks of wood, sections cut from the trunk of a hornbeam tree. These are replaced when worn out. The powder from the mill is called "meal powder" and is used as such in the manufacture of fireworks. Also it is moistened slightly with water and rubbed through sieves to form granular gunpowder for use in making rockets, Roman candles, aerial bombshells, and other artifices.

rings and in other parts of fuzes. Modified black powders, in which the proportion of the ingredients does not approximate to the standard formulas just mentioned, have been used for blasting, especially in Europe, and have been adapted to special uses in pyrotechny. Sodium nitrate powder, *ammonpulver*, and other more remote modifications are discussed later in this chapter or in the chapter on pyrotechnics.

Manufacture

During the eighteenth century, stamp mills (Figure 19) for incorporating the ingredients of black powder largely superseded the more primitive mortars operated by hand. The meal powder, or *pulverin* as the French call it, was made into gunpowder by moistening slightly and then pressing through sieves.[17] The powder grains were not uniform with one another either in their composition or their density, and could not be expected to give very uniform ballistic results. The use of a heavy wheel mill for grinding and pressing the materials together, and the subsequent pressing of the material into a hard cake which is broken up into grains, represent a great advance in the art and produce hard grains which are physically and ballistically uniform.[18] The operations in the manufacture of black powder as it is carried out at present are briefly as follows:

1. *Mixing* of the powdered materials is accomplished by hand or mechanical blending while they are dampened with enough water to prevent the formation of dust, or the powdered sulfur and charcoal are stirred into a saturated solution of the requisite amount of potassium nitrate at a temperature of about 130°, the hot mass is spread out on the floor to cool, and the lumps are broken up.

2. *Incorporating or Milling.* The usual wheel mill has wheels which weigh 8 or 10 tons each. It takes a charge of 300 pounds of

[17] The French still make *pulverin*, for the preparation of black match and for use in pyrotechnics, by rolling the materials with balls, some of lead and some of lignum vitae, in a barrel of hardwood. They also sometimes use this method for mixing the ingredients before they are incorporated more thoroughly in the wheel mill.

[18] The black powder wheel mill is also used for reducing (under water) deteriorated smokeless powder to a fine meal in order that it may be reworked or used in the compounding of commercial explosives, and for the intimate incorporation of such explosives as the French *schneiderite*.

the mixture. The wheels rotate for about 3 hours at a rate of about 10 turns per minute. Edge runners turn back under the tread of the wheels material which would otherwise work away from the center of the mill. Considerable heat is produced during the milling, and more water is added from time to time to replace that which is lost by evaporation in order that the material may always be moist. The "wheel cake" and "clinker" which result from the milling are broken up into small pieces for the pressing.

Figure 20. Modern Wheel Mill for Making Black Powder. (Courtesy Atlas Powder Company.) The large wheels weigh 10 tons each.

3. *Pressing* is done in a horizontal hydraulic press. Layers of powder are built up by hand between plates of aluminum, and the whole series of plates is pressed in one operation. The apparatus is so designed that fragments of powder are free to fall out at the edges of the plates, and only as much of the material remains between them as will conveniently fill the space. An effective pressure of about 1200 pounds per square inch is applied, and the resulting press cakes are about ¾ inch thick and 2 feet square.

4. *Corning or granulating* is the most dangerous of the operations in the manufacture of black powder. The corning mill is usually situated at a distance from the other buildings, is barri-

caded, and is never approached while the machinery, controlled from a distance, is in operation. The press cake is cracked or granulated between crusher rolls. Screens, shaken mechanically, separate the dust and the coarse pieces from the grains which are of the right size for use. The coarse pieces pass between other crusher rolls and over other screens, four sets of crusher rolls being used. Corning mill dust is used in fuse powder and by the makers of fireworks, who find it superior for certain purposes to other kinds of meal powder.

5. *Finishing.* The granulated powder from the corning mill is rounded or polished and made ":bright" by tumbling in a revolving wooden cylinder or barrel. Sometimes it is dried at the same time by forcing a stream of warm air through the barrel. Or the polished powder is dried in wooden trays in a dry-house at 40°. If a glazed powder is desired, the glaze is usually applied before the final drying. To the polished powder, still warm from the tumbling, a small amount of graphite is added, and the tumbling is continued for a short time. Black powder of commerce usually contains about 1 or 1.5 per cent moisture. If it contains less than this, it has a tendency to take up moisture from the air; if it contains much more, its efficiency is affected.

6. *Grading.* The powder is finally rescreened and separated into the different grain sizes, C (coarse), CC, CCC, F (fine), FF or 2F, 3F, 4F, etc. The word *grade* applied to black powder, refers to the grain size, not to the quality.

Analysis[19]

A powdered sample for analysis may be prepared safely by grinding granulated black powder, in small portions at a time, in a porcelain mortar. The powder may be passed through a 60-mesh sieve and transferred quickly to a weighing bottle without taking up an appreciable amount of moisture.

[19] A test which from ancient times has been applied to black powder is carried out by pouring a small sample onto a cold flat surface and setting fire to it. A good powder ought to burn in a flash and leave no "pearls" or residue of globules of fused salt. A solid residue indicates either that the ingredients have not been well incorporated, or that the powder at some time in its history has been wet (resulting in larger particles of saltpeter than would be present in good powder, the same result as poor incorporation), or that the powder at the time of the test contains an undue amount of moisture.

Moisture is determined by drying in a desiccator over sulfuric acid for 3 days, or by drying to constant weight at 60° or 70°, at which temperature 2 hours is usually long enough.

For determining *potassium nitrate*, the weighed sample in a Gooch crucible is washed with hot water until the washings no longer give any test for nitrate,[20] and the crucible with its contents is dried to constant weight at 70°. The loss of weight is equal to potassium nitrate *plus* moisture. In this determination, as in the determination of moisture, care must be taken not to dry the sample too long, for there is danger that some of the sulfur may be lost by volatilization.

Sulfur is determined as the further loss of weight on extraction with carbon disulfide in a Wiley extractor or other suitable apparatus. After the extraction, the crucible ought to be allowed to dry in the air away from flames until all the inflammable carbon disulfide has escaped. It is then dried in the oven to constancy of weight, and the residue is taken as *charcoal*. *Ash* is determined by igniting the residue in the crucible until all carbon has burned away. A high result for ash may indicate that the water extraction during the determination of potassium nitrate was not complete. The analytical results may be calculated on a moisture-free basis for a closer approximation to the formula by which the manufacturer prepared the powder.

Blasting Powder

The 6:1:1 and 6:1.2:08 formulas correspond to the quickest and most vigorous of the black-powder compositions. A slower and cheaper powder is desirable for blasting, and both these desiderata are secured by a reduction in the amount of potassium nitrate. For many years the French government has manufactured and sold three kinds of blasting or mining powder, as follows:

	SALTPETER	CHARCOAL	SULFUR
Forte	72	15	13
Lente	40	30	30
Ordinaire	62	18	20

In the United States a large part of all black powder for blast-

[20] A few drops, added to a few cubic centimeters of a solution of 1 gram of diphenylamine in 100 cc. of concentrated sulfuric acid, give a blue color if a trace of nitrate is present.

ing is made from sodium nitrate. This salt is hygroscopic, but a heavy graphite glaze produces a powder from it which is satisfactory under a variety of climatic conditions. Analyses of samples of granulated American blasting powder have shown that the compositions vary widely, sodium nitrate from 67.3 to 77.1 per cent, charcoal from 9.4 to 14.3 per cent, and sulfur from 22.9 to 8.6 per cent. Perhaps sodium nitrate 73, charcoal 11, and sulfur 16 may be taken as average values.

Pellet powders, made from sodium nitrate, are finding extensive use. These consist of cylindrical "pellets," 2 inches long, wrapped in paraffined paper cartridges, $1\frac{1}{4}$, $1\frac{3}{8}$, $1\frac{1}{2}$, $1\frac{3}{4}$, and 2 inches in diameter, which resemble cartridges of dynamite. The cartridges contain 2, 3, or 4 pellets which are perforated in the direction of their axis with a $\frac{3}{8}$-inch hole for the insertion of a squib or fuse for firing.

Ammonpulver

Propellent powder made from ammonium nitrate is about as powerful as smokeless powder and has long had a limited use for military purposes, particularly in Germany and Austria. The Austrian army used Ammonpulver, among others, during the first World War, and it is possible that the powder is now, or may be at any time, in use.

Gäns of Hamburg in 1885 patented[21] a powder which contained no sulfur and was made from 40 to 45 per cent potassium nitrate, 35 to 38 per cent ammonium nitrate, and 14 to 22 per cent charcoal. This soon came into use under the name of *Amidpulver,* and was later improved by decreasing the proportion of potassium nitrate. A typical improved Amidpulver, made from potassium nitrate 14 per cent, ammonium nitrate 37 per cent, and charcoal 49 per cent, gives a flashless discharge when fired in a gun and only a moderate amount of smoke. Ammonpulver which contains no potassium nitrate—in a typical example ammonium nitrate 85 per cent and charcoal 15 per cent, or a similar mixture containing in addition a small amount of aromatic nitro compound—is flashless and gives at most only a thin bluish-gray smoke which disappears rapidly. Rusch[22] has published data

[21] Ger. Pat. 37,631.
[22] *Seewesen,* January, 1909, cited by Escales, "Ammonsalpetersprengstoffe," Leipzig, 1909, p. 217.

which show that the temperature of the gases from the burning of ammonpulver (ammonium nitrate 80 to 90 per cent, charcoal 20 to 10 per cent) is below 900°, and that the ballistic effect is approximately equal to that of ballistite containing one-third of its weight of nitroglycerin.

Ammonpulver has the advantages of being cheap, powerful, flashless, and smokeless. It is insensitive to shock and to friction, and is more difficult to ignite than black powder. In use it requires a strong igniter charge. It burns rapidly, and in gunnery issused in the form of single-perforated cylindrical grains usually of a diameter nearly equal to that of the space within the cartridge. It has the disadvantages that it is extremely hygroscopic and that it will not tolerate wide changes of temperature without injury. The charges must be enclosed in cartridges which are effectively sealed against the ingress of moisture from the air. Ammonium nitrate has a transition point at 32.1°. If Ammonpulver is warmed above this temperature, the ammonium nitrate which it contains undergoes a change of crystalline state; this results in the crumbling of the large powder grains and consequent high pressures and, perhaps, bursting of the gun if the charge is fired. At the present time Ammonpulver appears to be the only modification of black powder which has interesting possibilities as a military propellant.

Other Related Propellent Explosives

Guanidine nitrate powders have not been exploited, but the present availability of guanidine derivatives from the cyanamide industry suggests possibilities. The salt is stable and non-hygroscopic, and is a flashless explosive—cooler indeed than ammonium nitrate. Escales[23] cites a German patent to Gäns[24] for a blasting powder made from potassium nitrate 40 to 60 per cent, guanidine nitrate 48 to 24 per cent, and charcoal 12 to 16 per cent.

Two other powders, now no longer used, are mentioned here as historically interesting examples of propellants made up in accordance with the same principle as black powder, namely, the principle of mixing an oxidizing salt with a combustible material.

Raschig's white blasting powder was made by dissolving 65 parts of sodium nitrate and 35 parts of sodium cresol sulfonate

[23] Escales, *op. cit.*, p. 225.
[24] Ger. Pat. 54,429.

together in water, running the solution in a thin stream onto a rotating and heated steel drum whereby the water was evaporated, and scraping the finished powder off from the other side of the drum. It was cheap, and easy and safe to make, but was hygroscopic. For use in mining, it was sold in waterproof paper cartridges.

Poudre Brugère was made by grinding together 54 parts of ammonium picrate and 46 parts of potassium nitrate in a black powder wheel mill, and pressing and granulating, etc., as in the manufacture of black powder. The hard grains were stable and non-hygroscopic. The powder was used at one time in military weapons. It was more powerful than black powder and gave less smoke.

CHAPTER III

PYROTECHNICS

The early history of pyrotechnics and the early history of black powder are the same narrative. Incendiary compositions containing saltpeter, and generally sulfur, mixed with combustible materials were used both for amusement and for purposes of war. They developed on the one hand into black powder, first used in crackers for making a noise and later in guns for throwing a projectile, and on the other into pyrotechnic devices. The available evidence indicates that fireworks probably developed first in the Far East, possibly in India earlier than in China, and that they were based upon various compositions of potassium nitrate, sulfur, and charcoal, with the addition of iron filings, coarse charcoal, and realgar (As_2S_2) to produce different visual effects. The nature of the composition and the state of subdivision of its ingredients determine the rate of burning and the appearance of the flame. In Chinese fire, coarse particles of hard-wood charcoal produce soft and lasting sparks; filings of cast iron produce bright and scintillating ones. The original Bengal lights were probably made more brilliant by the addition of realgar.

The manufacture of pyrotechnics from the Renaissance onward has been conducted, and still is practiced in certain places, as a household art or familiar craft. The artificer[1] needs patience and skill and ingenuity for his work. For large-scale factory production, the pyrotechnist has few problems in chemical engineering but many in the control of craftsmanship. His work, like that of the wood-carver or bookbinder, requires manual dexterity but transcends artistry and becomes art by the free play of the imagination for the production of beauty. He knows the kinds of effects, audible and visible, which he can get from his materials. He knows this as the graphic artist knows the appearance of his

[1] In the French language the word *artificier* means fireworks maker, and *artifice* means a pyrotechnic device.

52

colors. His problem is twofold: the esthetic one of combining these effects in a manner to produce a result which is pleasing, and the wholly practical one of contriving devices—and the means for the construction of devices—which shall produce these results. Like the graphic artist, he had but few colors at first, and he created designs with those which he had—lights, fountains, showers, Roman candles, rockets, etc. As new colors were discovered, he applied them to the production of better examples of the same or slightly modified designs. At the same time he introduced factory methods, devised improvements in the construction of his devices, better tools, faster and more powerful machinery, and learned to conduct his operations with greater safety and with vastly greater output, but the essential improvements in his products since the beginning of the seventeenth century have been largely because of the availability of new chemical materials.

Development of Pyrotechnic Mixtures

The use of antimony sulfide, Sb_2S_3, designated in the early writings simply as antimony, along with the saltpeter, sulfur, and charcoal, which were the standard ingredients of all pyrotechnic compositions, appears to have been introduced in the early part of the seventeenth century. John Bate's "Book of Fireworks," 1635, containing information derived from "the noted Professors, as Mr. Malthus, Mr. Norton, and the French Authour, Des Récreations Mathématiques,"[2] mentions no mixtures which contain antimony. Typical of his mixtures are the following.

> *Compositions for Starres.* Take saltpeter one pound, brimstone half a pound, gunpowder foure ounces, this must be bound up in paper or little ragges, and afterwards primed.
> *Another receipt for Starres.* Take of saltpeter one pound, gunpowder and brimston of each halfe a pound; these must be mixed together, and of them make a paste, with a sufficient quantity of oil of peter (petroleum), or else of faire water; of this paste you shal make little balles, and roll them in drie gunpowder dust; then dry them, and keepe them for your occasions.[3]

The iron scale which John Bate used in certain of his rocket

[2] F. Malthus (François de Malthe), "Treatise of Artificial Fireworks," 1629; Robert Norton, "The Gunner," 1628.

[3] John Bate, "The Mysteries of Nature and Art," London, 1635, Second Part, p. 101.

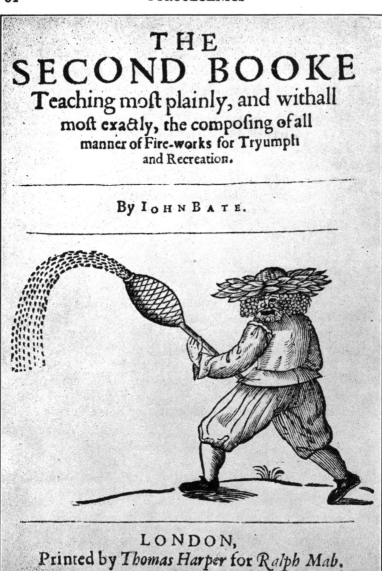

THE
SECOND BOOKE
Teaching moſt plainly, and withall
moſt exactly, the compoſing of all
manner of Fire-works for Tryumph
and Recreation.

By IOHN BATE.

LONDON,
Printed by *Thomas Harper* for *Ralph Mab.*
1635.

FIGURE 21. Title Page of John Bate's "Book of Fireworks." A "green man," such as might walk at the head of a procession, is shown scattering sparks from a fire club. The construction of this device is described as follows: "To make . . . you must fill diverse canes open at both ends (and

compositions probably produced no brilliant sparks but only glowing globules of molten slag which gave the rocket a more luminous tail. Hanzelet Lorrain[4] in 1630 showed a more advanced knowledge of the art and gave every evidence of being acquainted with it by his own experience. He described several mixtures containing antimony sulfide and compositions, for balls of brilliant fire to be thrown from the hand, which contain orpiment (As_2S_3) and verdigris.

Stars of the only two compositions which are well approved. Take of powder (gunpowder) four ounces, of saltpeter two ounces, of sulfur two ounces, of camphor half an ounce, of steel filings two *treseaux*, of white amber half an ounce, of antimony (sulfide) half an ounce, of (corrosive) sublimate half an ounce. For double the efficacy it is necessary to temper all these powders with gum *agragante* dissolved in brandy over hot cinders. When you see that the gum is well swollen and fully ready to mix with the said brandy, it is necessary forthwith to mix them in a mortar with the powder, the quicker the better, and then to cut up the resulting paste into pieces. These stars are very beautiful and very flowery. Note that it is necessary to put them to dry in a pastry or baking oven after the bread has been taken off of the hearth.

Second star composition. Take of saltpeter in fine and dry flour ten ounces, of charcoal, of sulfur, of powder (gunpowder), of antimony (sulfide), and of camphor each two *treseaux*. Temper the whole with oil of turpentine, and make it into a powdery (mealy) paste which you will put into little cartridges; and you will load them in the same manner as rockets [that is, by pounding in the charge]. When you wish to use them, it is necessary to remove the paper wrapper and to cut them into pieces setting a little black match (*mèche d'estoupin*) in the middle (of each piece) through a little hole which you will pierce there.

How fire balls are made so white that one can scarcely look at them without being dazzled. Take a pound of sulfur, three pounds of saltpeter, half a pound of gum arabic, four ounces of orpiment: grind all together, and mix well by hand,

[4] Hanzelet Lorrain, "La pyrotechnie," Pont à Mousson, 1630. The author's name is a pseudonym of Jean Appier.

---

of a foot long, or more, or lesse, as you think fit) with a slow composition, and binde them upon a staffe of four or five foot long; prime them so that one being ended, another may begin: you may prime them with a stouple or match (prepared as before). Make an osier basket about it with a hole in the very top to fire it by, and it is done."

FIGURE 22. Seventeenth-Century Fireworks Display, Lorrain, 1630. Flaming swords, shields and pikes, wheel of fire, rockets, stars, candles, serpents, water fireworks. The sun and the moon which are pictured are presumably aerial bombs, and the dragons are probably dragon rockets running on ropes but may possibly be imaginative representations of serpents of fire. The picture is convincing evidence that many of the varieties of fireworks which are now used (in improved form) for display purposes were already in use three centuries ago.

and moisten with brandy and make into a stiff paste into which you will mix half a pound of ground glass, or of crystal in small grains, not in powder, which you will pass through a screen or sieve. Then, mixing well with the said paste, you will form balls of it, of whatever size you please and as round as you can make them, and then you will let them dry. If you wish to have green fire, it is necessary merely to add a little verdigris to the composition. This is a very beautiful fire and thoroughly tested, and it needs no other primer to fire it than the end of a lighted match, for, as soon as the fire touches it, it inflames forthwith. It is beautiful in saluting a prince or nobleman to have such agreeable hand fire balls before setting off any other fireworks.[5]

Audot, whose little book[6] we take to be representative of the state of the art at the beginning of the nineteenth century, had a slightly larger arsenal of materials.

Iron and steel filings. "They give white and red sparks. It is necessary to choose those which are long and not rolled up, and to separate them from any dirt. They are passed through two sieves, in order to have two sizes, fine filings and coarse filings. Those of steel are in all respects to be preferred. It is easy to procure them from the artisans who work in iron and steel."

Ground and filed cast iron. "Cast iron is used in the fires which are designated by the name of *Chinese fire.* Two kinds, fine and coarse. The cast iron is ground in a cast iron mortar with a cast iron or steel pestle, and then sifted."

Red copper filings. "This gives greenish sparks."

Zinc filings "produce a beautiful blue color; it is a substance very difficult to file."

Antimony (sulfide) "gives a blue flame. It is ground up and passed through a screen of very fine silk."

Yellow amber. "Its color, when it burns, is yellow. It is used only for the fire of lances. It is very common in the drug trade. It ought to be ground and passed through a sieve."

Lampblack. "It gives a very red color to fire, and it gives rose in certain compositions."

Yellow sand or gold powder. "It is used in suns where it produces golden yellow rays. It is a reddish yellow sand mixed with

[5] Hanzelet Lorrain, *op. cit.,* pp. 256-258.

[6] Anon. (L.-E. Audot), "L'art de faire, à peu de frais, les feux d'artifice," Paris, 1818.

little brilliant scales. The paperers sell it under the name of gold powder. It is very common in Paris."[7]

Some of Audot's compositions are as follows:

Common fire: meal powder 16 parts, coarse and fine charcoal 5 parts.

Chinese fire: meal powder 16 parts, cast iron 6 parts.

Brilliant fire: meal powder 16 parts, steel filings 4 parts.

Blue fire for cascades: meal powder 16 parts, saltpeter 8, sulfur 12, and zinc filings 12 parts.

Fixed star: saltpeter 16 parts, sulfur 4, meal powder 4, and antimony (sulfide) 2 parts.

Silver rain for a turning sun or fire wheel: meal powder 16 parts, saltpeter 1, sulfur 1, steel filings 5 parts.

Green fire for the same: meal powder 16 parts, copper filings 3 parts.

Chinese fire for the same: meal powder 16 parts, saltpeter 8, fine charcoal 3, sulfur 3, fine and coarse cast iron 10 parts.

Composition for lances. Yellow: saltpeter 16 parts, meal powder 16, sulfur 4, amber 4, and colophony 3 parts. *Rose:* saltpeter 16 parts, lampblack 1, meal powder 3. *White:* saltpeter 16 parts, sulfur 8, meal powder 4. *Blue:* saltpeter 16 parts, antimony (sulfide) 8, very fine zinc filings 4. *Green:* saltpeter 16 parts, sulfur 6, verdigris 16, and antimony (sulfide) 6 parts.

Bengal flame: saltpeter 16 parts, sulfur 4, and antimony (sulfide) 2 parts. This mixture was to be lighted by quickmatch and burned in small earthenware pots for general illumination.[8]

The Ruggieri, father and son, contributed greatly to the development of fireworks by introducing new, and often very elaborate, pieces for public display and by introducing new materials into the compositions. They appear to have been among the first who attempted to modify the colors of flames by the addition of salts. The compositions which we have cited from Audot are similar to some of those which the elder Ruggieri undoubtedly used at an earlier time, and the younger Ruggieri, earlier than Audot's book, was using materials which Audot does not mention, in particular, copper sulfate and ammonium chloride for the green fire of the palm-tree set piece. The use of ammonium chloride was a definite advance, for the chloride helps to volatilize the copper and to produce a brighter color. But ammonium

[7] Audot, *op. cit.,* pp. 15-19.

[8] Audot, *op. cit.,* pp. 48, 49, 50, 52, 63, 64, 67.

chloride is somewhat hygroscopic and tends to cake, and it is now no longer used; indeed, the chloride is unnecessary in compositions which contain chlorate or perchlorate. In the Ruggieri "we have two pyrotechnists who can be considered to represent the best skill of France and Italy; in fact, it was Ruggieri whose arrival in France from Italy in or about 1735 marked the great advance in pyrotechny in the former country."[9] The elder Ruggieri conducted a fireworks display at Versailles in 1739. In 1743 he exhibited for the first time, at the Théâtre de la Comédie Italienne and before the King, the passage of fire from a moving to a fixed piece. "This ingenious contrivance at first astonished the scientists of the day, who said when it was explained to them that nothing could be more simple and that any one could have done it at once."[10] In 1749 he visited England to conduct, with Sarti, a fireworks display in Green Park in celebration of the peace of Aix-la-Chapelle. The younger Ruggieri conducted many public pyrotechnic exhibitions in France during the years 1800-1820, and wrote a treatise on fireworks which was published both in French and in German.

Potassium chlorate had been discovered, or at least prepared in a state of purity, by Berthollet in 1786. It had been tried unsuccessfully and with disastrous results in gunpowder. Forty years elapsed before it began to be used in pyrotechnic mixtures, where, with appropriate salts to color the flame, it yields the brilliant and many-colored lights which are now familiar to us. At present it is being superseded for certain purposes by the safer perchlorate.

James Cutbush, acting professor of chemistry and mineralogy at West Point, in his posthumous "System of Pyrotechny," 1825, tells[11] of the detonation of various chlorate mixtures and of their use for the artificial production of fire. "Besides the use of nitre in pyrotechnical compositions, as it forms an essential part of all

[9] A. St. H. Brock, "Pyrotechnics: The History and Art of Fireworks Making," London, 1922. This is a scholarly and handsome book, bountifully illustrated, which contains excellent accounts both of the history of fireworks and of present manufacturing practice. The author comes from several generations of fireworks makers.

[10] Quoted by Brock, *op. cit.*, p. 124.

[11] James Cutbush, "A System of Pyrotechny, Comprehending the Theory and Practice, with the Application of Chemistry; Designed for Exhibition and for War," Philadelphia, 1825, p. 22.

of them, there is another salt . . . that affords a variety of amusing experiments. This salt is the hyperoxymuriate or chlorate of potassa. Although it has neither been used for fire-works on an extensive scale, nor does it enter into any of the compositions usually made for exhibition, yet its effect is not the less amusing." At a later place Cutbush says: "M. Ruggieri is of opinion, that chlorate, or hyperoxymuriate of potassa may be employed with advantage in the composition of rockets, but we have not heard that it has been used. It is more powerful in its effects, and probably for this reason he recommended it. This salt, mixed with other substances, will produce the *green fire* of the palm-tree, in imitation of the Russian fire."[12]

Ruggieri's Russian fire, as his son later described it, consisted of crystallized copper acetate 4 parts, copper sulfate 2 parts, and ammonium chloride 1 part,[13] all finely pulverized and mixed with alcohol, and placed upon cotton wick attached to spikes upon the thin metal pieces which were the leaves of the palm tree. The resulting display would not be impressive according to modern standards.

Cutbush also knew how to color the flame, for he says:

We are of opinion, that many of the nitrates might be advantageously employed in the manufacture of fire works. Some, as nitrate of strontian, communicate a red color to flame, as the flame of alcohol. Nitrate of lime also might be used. . . . Muriate of strontian, mixed with alcohol, or spirit of wine, will give a carmine-red flame. For this experiment, one part of the muriate is added to three or four parts of alcohol. Muriate of lime produces, with alcohol, an orange-coloured flame. Nitrate of copper produces an emerald-green flame. Common salt and nitre, with alcohol, give a yellow flame.[14]

According to Brock, the use of chlorate in pyrotechnic mixtures, initiating the modern epoch in the art, first occurred about 1830. Lieut. Hippert of the Belgian artillery published at Bruxelles in 1836 a French translation, "Pyrotechnie raisonnée," of a work by Prussian artillery Captain Moritz Meyer in which one chapter is devoted to colored fires, and listed several com-

[12] Cutbush, *op. cit.*, p. 77.
[13] Ruggieri, "Handbüchlein der Lustfeuerwerkerei," second edition, Quedlinburg and Leipzig, 1845, p. 142.
[14] Cutbush, *op. cit.*, pp. 8 and 20.

positions which contain potassium chlorate. Meyer states, incidentally, that the English at that time used colored rockets for signaling at sea and were able to produce ten distinguishable shades. His descriptions of his compositions give one reason to suspect that he had had little experience with them himself. The first, a mixture of potassium chlorate and sugar, burns, he says, with a red light; but the color is actually a bluish white.

A powder which burns with a green flame is obtained by the addition of nitrate of baryta to chlorate of potash, nitrate of potash, acetate of copper. A white flame is made by the addition of sulfide of antimony, sulfide of arsenic, camphor. Red by the mixture of lampblack, coal, bone ash, mineral oxide of iron, nitrate of strontia, pumice stone, mica, oxide of cobalt. Blue with ivory, bismuth, alum, zinc, copper sulfate purified of its sea water [sic]. Yellow by amber, carbonate of soda, sulfate of soda, cinnabar. It is necessary in order to make the colors come out well to animate the combustion by adding chlorate of potash.[15]

Although Meyer's formulas are somewhat incoherent, they represent a definite advance. Equally significant with the use of chlorate is his use of the nitrates of strontium and barium.

The second German edition of Ruggieri's book (we have not seen the first) contains a *Nachtrag* or supplement which lists nine compositions,[16] of which four contain *Kali oxym.* or potassium chlorate. These are: (1) for red fire, strontium nitrate 24 parts, sulfur 3, fine charcoal 1, and potassium chlorate 5; (2) for green fire, barium carbonate 20 parts, sulfur 5, and potassium chlorate 8 parts; (3) for green stars, barium carbonate 20 parts, sulfur 5, and potassium chlorate 9 parts; and (4) for red lances, strontium carbonate 24 parts, sulfur 4, charcoal 1, and potassium chlorate 4 parts. Ruggieri says:

The most important factor in the preparation of these compositions is the fine grinding and careful mixing of the several materials. Only when this is done is a beautiful flame to be expected. And it is further to be noted that the potassium chlorate, which occurs in certain of the compositions, is to be wetted with spirit for the grinding in order to avoid an explosion.

[15] Brock, *op. cit.,* pp. 145, 146.
[16] Ruggieri, *op. cit.,* pp. 147, 148.

The chlorate compositions recommended by Ruggieri would undoubtedly give good colors, but are not altogether safe and would probably explode if pounded into their cases. They could be loaded with safety in an hydraulic press, and would probably not explode if tamped carefully by hand.

F. M. Chertier, whose book "Nouvelles recherches sur les feux d'artifice" was published at Paris in 1854, devotes most of his attention to the subject of color, so successfully that, although new materials have come into use since his time, Brock says that "there can be no doubt that Chertier stands alone in the literature of pyrotechny and as a pioneer in the modern development of the art."[17] Tessier, in his "Chimie pyrotechnique ou traité pratique des feux colores," first edition, Paris, 1859, second edition 1883, discusses the effect of individual chemicals upon the colors of flames and gives excellent formulas for chlorate and for non-chlorate compositions which correspond closely to present practice. He used sulfur in many but not in all of his chlorate mixtures. Pyrotechnists in France, with whom the present writer talked during the first World War, considered Tessier's book at that time to be the best existing work on the subject of colored fires—and this in spite of the fact that its author knew nothing of the use of magnesium and aluminum. The spectroscopic study of the colors produced by pure chemicals, and of the colors of pyrotechnic devices which are best suited for particular effects, is the latest of current developments.

Chlorate mixtures which contain sulfur give brighter flames than those which lack it, and such mixtures are still used occasionally in spite of their dangerous properties. The present tendency, however, is toward chlorate mixtures which contain no sulfur, or toward potassium nitrate mixtures (for stars, etc.) which contain sulfur but no chlorate, or toward nitrates, such as those of strontium and barium, which supply both color for the flame and oxygen for the combustion and are used with magnesium or aluminum to impart brilliancy. Magnesium was first used for pyrotechnic purposes about 1865 and aluminum about 1894, both of them for the production of dazzling white light. These metals were used in the compositions of colored airplane

---

[17] Brock, *op. cit.*, p. 147. Chertier also published a pamphlet on colored fires nearly thirty years earlier than the above-mentioned book.

flares during the first World War, but their use in the colored fires of general pyrotechny is largely a later development.

Tessier introduced the use of cryolite ($AlNa_3F_6$) for the yellow coloring of stars, lances, and Bengal lights. In his second edition he includes a chapter on the small pyrotechnic pieces which are known as Japanese fireworks, giving formulas for them, and another on the picrates, which he studied extensively. The picrates of sodium, potassium, and ammonium crystallize in the anhydrous condition. Those of barium, strontium, calcium, magnesium, zinc, iron, and copper are hygroscopic and contain considerable water of crystallization which makes them unfit for use in pyrotechnic compositions. Lead picrate, with 1 H_2O, detonates from fire and from shock, and its use in caps and primers was patented in France in 1872. Potassium and sodium picrate deflagrate from flame, retaining that property when mixed with other substances. Ammonium picrate detonates from fire and from shock when in contact with potassium chlorate or lead nitrate, but in the absence of these substances it has the special advantage for colored fires that the mixtures give but little smoke and this without offensive odor. Tessier recommends ammonium picrate compositions for producing colored lights in the theater and in other places where smoke might be objectionable. "Indoor fireworks" have been displaced in the theater by electric lighting devices, but are still used for certain purposes. Tessier's formulas, which are excellent, are described later in the section on picrate compositions.

Colored Lights

Colored light compositions are used in the form of a loose powder, or are tamped into paper tubes in torches for political parades, for highway warnings, and for railway and marine signals, in Bengal lights, in airplane flares, and in lances for set pieces, or are prepared in the form of compact pellets as stars for Roman candles, rockets, and aerial bombs, or as stars to be shot from a special pistol for signaling.

Colored fire compositions intended for burning in conical heaps or in trains are sometimes sold in paper bags but more commonly in boxes, usually cylindrical, of pasteboard, turned wood, or tinned iron. The mixtures are frequently burned in the boxes in which they are sold. Compositions which contain no chlorate (or perchlorate) are the oldest, and are still used where the most brilliant colors are not necessary.

	White					Red	Pink		Yellow
Potassium nitrate..........	5	3	32	8	14	..	12	14	..
Sulfur....................	2	1	15	2	4	5	5	4	3
Strontium nitrate..........	..	..	..	..	..	18	48	36	..
Barium nitrate.............	..	..	..	..	..	..	..	..	36
Sodium oxalate.............	..	..	..	..	..	..	..	..	6
Antimony metal.............	..	1	12	..	..	..	..	..	..
Antimony sulfide...........	1	1	..	..	..	..	..	..	..
Realgar...................	..	..	..	1	5	..	..	..	..
Minium...................	..	..	10	..	..	..	..	..	..
Lampblack.................	..	..	..	..	..	1	..	..	..
Charcoal..................	..	..	..	..	..	..	4	1	..
Red gum..................	..	..	..	..	..	..	..	4	5
Dextrin...................	..	..	1	..	..	..	..	1	..

The chlorate compositions listed below, which contain no sulfur, burn rapidly with brilliant colors and have been recommended for indoor and theatrical uses.

	White	Red	Yellow	Green
Potassium chlorate......	12	1	6	2
Potassium nitrate.......	4	..	6	..
Strontium nitrate.......	..	4	..	..
Barium nitrate.........	..	..	..	1
Barium carbonate.......	1	..	..	..
Sodium oxalate.........	..	..	5	..
Cane sugar............	4	..	..	1
Stearine..............	1	..	..	..
Shellac...............	..	1	3	..

The following are brilliant, somewhat slower burning, and suitable for outdoor use and general illumination. The smokes from the compositions which contain calomel and Paris green are poisonous. In mixing Paris green, care must be exercised not to inhale the dust.

	Red			Green		Blue		
Potassium chlorate.................	10	4	8	4	4	6	8	16
Strontium nitrate..................	40	10	16	..	..	..	..	..
Barium nitrate.....................	..	..	..	8	8	4	..	14
Paris green........................	..	..	..	..	..	4	..	12
Shellac............................	..	..	3	..	3	..	..	1
Stearine...........................	..	..	..	..	..	1	..	2
Red gum..........................	6	3	..	2	..	..	..	..
Calomel............................	..	..	..	..	..	..	6	2
Sal ammoniac......................	..	..	..	1	..	1	..	..
Copper ammonium chloride.........	..	..	..	..	..	..	2	..
Fine sawdust......................	6	..	..	..	..	..	..	..
Rosin..............................	..	1	..	..	..	..	..	..
Lampblack........................	1	1	..	..	..	..	..	..
Milk sugar........................	..	..	..	..	..	..	3	..

Railway Fusees (Truck Signal Lights)

Motor trucks are required by law to be equipped with red signal lights for use as a warning in case an accident causes them to be stopped on the road at night without the use of their electric lights. Similar lights are used for signaling on the railways. The obvious requirement is that the signal should burn conspicuously and for a long time. A. F. Clark recommends a mixture of:

	PARTS
Strontium nitrate (100 mesh)...........	132
Potassium perchlorate (200 mesh).......	15
Prepared maple sawdust (20 mesh)......	20
Wood flour (200 mesh).....	1
Sulfur (200 mesh)....................	25

The prepared maple sawdust is made by cooking with miner's wax, 10 pounds of sawdust to 1 ounce of wax, in a steam-jacketed kettle. The mixture is tamped dry into a paper tube, $7/8$ inch in external diameter, 1/32 inch wall, and burns at the rate of about 1 inch per minute. The fusee is supplied at its base with a pointed piece of wood or iron for setting it up in the ground, and it burns best when set at an angle of about 45°. In order to insure certain ignition, the top of the charge is covered with a primer or *starting fire,* loaded while moistened with alcohol, which consists of potassium chlorate 16 parts, barium chlorate 8, red gum (gum yacca) 4, and powdered charcoal 1. This is covered with a piece of paper on which is painted a *scratch mixture* similar to that which composes the head of a safety match. The top of the fusee is supplied with a cylindrical paper cap, the end of which is coated with a material similar to that with which the striking surface on the sides of a box of safety matches is coated. To light the fusee, the cap is removed and inverted, and its end or bottom is scratched against the mixture on the top of the fusee.

Weingart[18] recommends the first four of the following compositions for railway fusees; Faber[19] reports the fifth. Weingart's mixtures are to be moistened with kerosene before they are tamped into the tubes.

Potassium chlorate					12
Potassium perchlorate				5	
Strontium nitrate	48	36	16	36	72
Saltpeter	12	14	4		
Sulfur	5	4	5	5	10
Fine charcoal	4	1	1		
Red gum	10	4			4
Dextrin		1			
Sawdust				2	
Sawdust and grease					4
Calcium carbonate					1

Scratch Mixture

Typical scratch mixtures are the pair: (A) potassium chlorate 6, antimony sulfide 2, glue 1; and (B) powdered pyrolusite (MnO_2) 8, red phosphorus 10, glue 3, recommended by Weingart; and the pair: (A) potassium chlorate 86, antimony sulfide 52, dextrin 35; and (B) red phosphorus 9, fine sand 5, dextrin 4, used with gum arabic as a binder, and recommended by A. F. Clark.

Marine Signals

Other interesting signal lights, reported by Faber,[20] are as follows.

---

[18] Weingart, "Dictionary and Manual of Fireworks," Boston, 1937, p. 61.
[19] Faber, "Military Pyrotechnics," 3 vols., Washington, 1919, Vol. I, p. 189.
[20] *Loc. cit.*

	Marine Flare Torch	Pilot's Blue Light
Barium nitrate......	16	..
Potassium nitrate....	8	..
Potassium chlorate ..	..	46
Strontium carbonate.	1	..
Copper oxychloride..	..	32
Sulfur..............	2	28
Red gum...........	2	..
Shellac.............	..	48
Calomel...........	..	3

Parade Torches

Parade torches are made in various colors; they are of better quality than railway fusees, burn with a deeper color and a brighter light, and are generally made with more expensive compositions. Below are a few typical examples. Parade torches are

	Red			Green		Purple	Amber	Blue
Strontium nitrate.......	16	5	9	..	..	7	36	..
Barium nitrate..........	..	..	..	40	30	..	..	..
Potassium chlorate......	8	1	..	11	..	..	..	..
Potassium perchlorate....	..	..	2	..	6	9	10	5
Sodium oxalate..........	..	..	..	..	..	..	8	..
Cupric oxide............	..	..	..	..	..	6	..	..
Paris green.............	..	..	..	..	..	..	..	2
Sal ammoniac...........	..	..	..	1	..	..	..	..
Calomel................	..	..	..	..	..	3	..	1
Sulfur.................	..	..	2	..	3	5	3	..
K.D. gum..............	..	..	..	6	2	..	..	..
Shellac................	3	..	..	..	..	..	5	..
Red gum...............	..	1	1	..	..	..	..	..
Dextrin................	..	..	..	..	..	..	..	1

equipped with wooden handles at the lower ends, and are sealed at their upper ends with a piece of cloth or paper, pasted on, through which a hole has been punched into the composition to a depth of about 1 inch—and through this a piece of *black match*[21]

[21] The match, prepared by dipping a few strands of cotton twine, twisted together, into a paste of meal powder and allowing to dry while stretched on a frame, is called *black match* by the pyrotechnists. When this is enclosed in a paper tube, it burns almost instantaneously and is then known as *quickmatch*. Such *quickmatch* is used for communicating fire in set pieces, Catherine wheels, etc.

has been inserted and fixed in place by a blob of paste of meal powder with gum-arabic water.

Aluminum and Magnesium Flares

When barium and strontium nitrates are used in colored lights, these substances serve the twofold purpose of coloring the flame and of supplying oxygen for its maintenance. The materials which combine with the oxygen to yield the flame, in the compositions which have been described, have been sulfur and carbonaceous matter. If, now, part or all of these materials is substituted by magnesium or aluminum powder or flakes, the resulting composition is one which burns with an intensely bright light. A mixture of potassium perchlorate 7 parts, mixed aluminum powder and flakes 5 parts, and powdered sulfur 2 parts burns with a brilliant light having a lilac cast. A balanced mixture of barium and strontium nitrates, that is, of green and red, gives a light which is practically white. Such lights are used in parade torches and signals, but are so bright as to be trying to the eyes. They find important use in aviation for signaling and for illuminating landing fields and military objectives.

Magnesium is attacked fairly rapidly by moisture, and pyrotechnic mixtures containing this metal do not keep well unless the particles of magnesium are first coated with a protecting layer of linseed oil or similar material. Aluminum does not have the same defect and is more widely used. An excellent magnesium light, suitable for illumination, is described in a patent recently granted to George J. Schladt.[22] It consists of a mixture of 36 to 40 per cent barium nitrate, 6 to 8 per cent strontium nitrate, 50 to 54 per cent flake magnesium coated with linseed oil, and 1 to 4 per cent of a mixture of linseed and castor oils.

The airplane wing-tip flares which were used for signaling during the first World War are good examples[23] of aluminum compositions. They were loaded in cylindrical paper cases 4¼ inches in length and 1⅝ inches in internal diameter. The white light composition consisted of 77 parts of barium nitrate, 13 of flake aluminum, and 5 of sulfur intimately mixed and secured by a binder of shellac, and burned in the cases mentioned, for 1

[22] U. S. Pat. 2,149,314, March 7, 1939.
[23] Faber, *op. cit.*, Vol. 2, pp. 223, 225-227.

minute with an illumination of 22,000 candlepower. The red light was made from 24 parts of strontium nitrate, 6 of flake aluminum, and 6 of sulfur with a shellac binder and burned for 1 minute with an illumination of 12,000 to 15,000 candlepower. The compositions were loaded into the cases by means of a pneumatic press, and filled them to within 5/16 inch of the top. The charge was then covered with a ⅛-inch layer of starting fire or *first fire composition*, made from saltpeter 6 parts, sulfur 4, and charcoal 1, dampened with a solution of shellac in alcohol, and this, when the device was used, was fired by an electric squib.

Lances

Lances are paper tubes, generally thin and of light construction, say, ¼ to ⅜ inch in diameter and 2 to 3½ inches long, filled with colored fire composition, loaded by tamping, not by ramming, and are used in set pieces, attached to wooden frameworks, to outline the figure of a temple or palace, to represent a flag, to spell words, etc. When set up, they are connected by quickmatch (black match in a paper tube) and are thus lighted as nearly simultaneously as may be. They are often charged in such manner as to burn with a succession of color, in which event the order of loading the various colors becomes important. Green should not be next to white, for there is not sufficient contrast. And green should not burn after red, for the color of the barium flame appears to one who has been watching the flame of strontium to be a light and uninteresting blue. The order of loading (the reverse of the order of burning) is generally white, blue (or yellow or violet, green, red, white. In the tables on page 70 a number of lance compositions are listed, illustrative of the various types and corresponding to considerable differences in cost of manufacture.

Picrate Compositions

Ammonium picrate is used in the so-called indoor fireworks which burn with but little smoke and without the production of objectionable odor. On page 71 some of the compositions recommended by Tessier[24] for Bengal lights are tabulated.

[24] *Op. cit.*, second edition, pp. 383-396.

White	Potassium nitrate	33	5	9	8	11
	Antimony sulfide	5		2		1
	Antimony metal		1			3
	Realgar				1	
	Sulfur	11	2	1	2	3
	Meal powder	2	1			
Red	Potassium chlorate			10	6	36
	Strontium nitrate					54
	Strontium carbonate			3	2	
	Sulfur					13
	Lampblack					2
	Shellac			2		12
	Paraffin				1	
Yellow	Potassium perchlorate					24
	Potassium chlorate	8	4	4		
	Barium nitrate	1		22		
	Sodium oxalate		2			8
	Sodium bicarbonate	2				
	Cryolite			2		
	Sulfur	4		5		
	Lampblack			1		
	Shellac		1			3
Green	Potassium chlorate			7		
	Barium nitrate	12	4	7		
	Barium chlorate	9	5			6
	Lampblack	1				
	Shellac	10	1	2		1
Blue	Potassium perchlorate			16		
	Potassium chlorate				32	5
	Copper oxychloride					2
	Paris green			6	10	
	Calomel			1	6	
	Shellac			1		1
	Stearine				3	
Lilac	Potassium chlorate					26
	Strontium sulfate					10
	Basic copper sulfate					6
	Lead nitrate					5
	Sulfur					4
	Shellac					1
	Stearine					1
Violet	Potassium chlorate					25
	Strontium sulfate					20
	Basic copper sulfate					1
	Sulfur					20

To be burned without compression, in the open, in trains or in heaps:

	Red		Green		Aurora	Yellow	White
Ammonium picrate..........	5	10	8	5	10	20	5
Strontium nitrate...........	25	40	..	..	31	12	..
Barium nitrate...,..........	..	..	32	25	..	58	30
Cryolite...................	..	..	..	..	3	7	..
Antimony metal............	..	..	..	..	..	..	5
Lampblack.................	1	2	1	..	2	4	1
Paraffin...................	1	1	1	1	1	2	1

To be compressed in paper cartridges, 25-30 mm. internal diameter, to be burned in a horizontal position in order that the residue may not interfere with the burning:

	Red		Green			Aurora	Yellow		White	
Ammonium picrate.......	1	20	5	5	11	20	20	24	6	5
Strontium nitrate........	1	60	..	..	..	60	10	3	..	..
Barium nitrate..........	..	..	6	28	36	..	60	20	4	30
Cryolite................	..	..	..	..	..	7	7	4	..	..
Calcium fluoride..........	..	7	..	..	..	..	..	..	..	..
Antimony metal..........	..	..	..	..	..	..	..	..	..	4
Antimony sulfide........	..	..	..	..	..	..	..	..	3	1
Lampblack..............	..	4	..	1	1	4	4	..	..	1
Paraffin................	..	2	..	1	1	2	2	..	..	1

Picric acid added in small quantities to colors deepens them and increases their brilliancy without making them burn much faster. Stars containing picric acid ought not to be used in aerial shells, for they are likely to detonate either from the shock of setback or later from being ignited in a confined space. Mixtures which contain picric acid along with potassium chlorate or salts of heavy metals are liable to detonate from shock.

Weingart[25] lists two "smokeless tableau" fires which contain picric acid, as follows:

[25] *Op. cit.*, p. 60.

	Red	Green
Strontium nitrate......	8	..
Barium nitrate.........	..	4
Picric acid............	5	2
Charcoal..............	2	1
Shellac...............	1	..

The picric acid is to be dissolved in boiling water, the strontium or barium nitrate added, the mixture stirred until cold, and the solid matter collected and dried. The same author[26] gives picric acid compositions for stars, "not suitable for shells," as follows:

	Red		Green
Strontium nitrate.....	8	..	..
Strontium carbonate..	..	3	..
Barium chlorate......	..	..	12
Potassium chlorate ...	4	10	8
Picric acid..........	1.5	1.5	2
Calomel.............	..	..	6
Shellac.............	1.5	0.75	2
Fine charcoal........	1	1	..
Lampblack..........	..	..	1.5
Dextrin............	0.5	0.75	0.5

Picrate Whistles

An intimate mixture of finely powdered dry potassium picrate and potassium nitrate, in the proportion about 60/40, rammed tightly into paper, or better, bamboo tubes from ¼ to ¾ inch in diameter, burns with a loud whistling sound. The mixture is dangerous, exploding from shock, and cannot be used safely in aerial shells. Whistling rockets are made by attaching a tube of the mixture to the outside of the case in such manner that it burns, and whistles, during the flight—or by loading a small tube, say ¼ inch in diameter and 2½ inches long, into the head of the rocket to produce a whistle when the rocket bursts. The mixture

[26] *Op. cit.*, pp. 114, 115.

is used in whistling firecrackers, "musical salutes," "whistling whizzers," "whistling tornados," etc. The effect of a whistle as an accompaniment to a change in the appearance of a burning wheel is amusing. Whistles are perhaps most effective when six or eight of them, varying in size from the small to the large, are fired in series, the smallest caliber and the highest pitch being first.

Non-Picrate Whistles

Non-picrate whistles, made from a mixture of 1 part powdered gallic acid and 3 parts potassium chlorate, are considered to be safer than those which contain picrate. The mixture is charged into a ½-inch case, 5/16 inch in internal diameter. The case is loaded on a 1-inch spindle, and the finished whistle has a 1-inch length of empty tube which is necessary for the production of the sound. Whistles of this sort, with charges of a chlorate or perchlorate explosive at their ends, are used in "chasers," "whizzers," etc., which scoot along the ground while whistling and finally explode with a loud report.

Rockets

The principle of the rocket and the details of its design were worked out at an early date. Improvements have been in the methods of manufacture and in the development of more brilliant and more spectacular devices to load in the rocket head for display purposes. When rockets are made by hand, the present practice is still very much like that which is indicated by Figure 23. The paper casing is mounted on a spindle shaped to form the long conical cavity on the surface of which the propelling charge will start to burn. The composition is rammed into the space surrounding the spindle by means of perforated ram rods or *drifts* pounded by a mallet. The base of the rocket is no longer choked by crimping, but is choked by a perforated plug of clay. The clay, dried from water and moistened lightly with crankcase oil, is pounded or pressed into place, and forms a hard and stable mass. The tubular paper cases of rockets, *gerbs*,[27] etc., are now often made by machinery, and the compositions are loaded into them automatically or semi-automatically and pressed by hydraulic presses.

[27] Pronounced *jurbs*.

John Bate and Hanzelet Lorrain understood that the heavier rockets require compositions which burn more slowly.

> It is necessary to have compositions according to the greatness or the littleness of the rockets, for that which is proper for the little ones is too violent for the large—because the fire, being lighted in a large tube, lights a composition of great amplitude, and burns a great quantity of material,

FIGURE 23. Rocket, Lorrain, 1630. Substantially as rockets are made today. After the propelling charge has burned completely and the rocket has reached the height of its flight, the fire reaches the charge in the head which bursts and throws out large and small stars, serpents and grasshoppers, or English firecrackers. The container, which is loaded into the head of the rocket, is shown separately with several grasshoppers in the lower right-hand corner of the picture.

and no geometric proportionality applies. Rockets intended to contain an ounce or an ounce and a half should have the following for their compositions.

Take of fine powder (gunpowder) passed through a screen or very fine sieve four ounces, of soft charcoal one ounce, and mix them well together.[28]

Otherwise. Of powder sieved and screened as above one pound, of saltpeter one ounce and a half, of soft charcoal

[28] The charcoal makes the powder burn more slowly, and produces a trail of sparks when the rocket is fired.

one ounce and a half. It does not matter what charcoal it is; that of light wood is best, particularly of wood of the vine.

For rockets weighing two ounces. Take of the above-said powder four ounces and a half, of saltpeter one ounce.

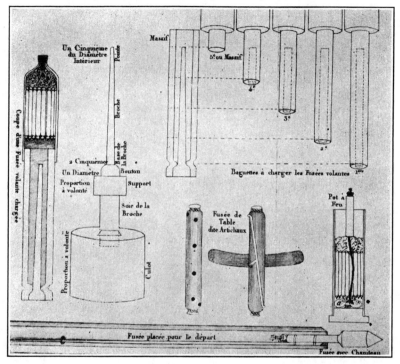

FIGURE 24. Details of Construction of Rocket and of Other Pieces, Audot, 1818. The rocket case, already crimped or constricted, is placed upon the spindle (*broche*); the first portion of the propelling charge is introduced and pounded firmly into place by means of a mallet and the longest of the *drifts* pictured in the upper right-hand corner; another portion of the charge is introduced, a shorter drift is used for tamping it, and so on until the case is charged as shown at the extreme left. A tourbillion (table rocket or artichoke) and a mine charged with serpents of fire are also shown.

Otherwise for the same weight. Take powder two ounces, of soft charcoal half an ounce.

Composition for rockets weighing from 4 to 8 ounces. Take powder as above seventeen ounces, of saltpeter four ounces, of soft charcoal four ounces.

Otherwise and very good. Of saltpeter ten ounces, of sulfur one ounce, of powder three ounces and a half, of charcoal three ounces and a half.

To make them go up more suddenly. Take of powder ten ounces, of saltpeter three ounces and a half, of sulfur one ounce, of charcoal three ounces and a half.

For rockets weighing one pound. Take of powder one pound, of soft charcoal two ounces, and of sulfur one ounce.

Otherwise. Of saltpeter one pound four ounces, of sulfur two ounces, of soft charcoal five ounces and a half.

For rockets weighing three pounds. Of saltpeter 30 ounces, of charcoal 11 ounces, of sulfur 7 ounces and a half.

For rockets weighing four, five, six, and seven pounds. Of soft charcoal ten pounds, of sulfur four pounds and a half, of saltpeter thirty one pounds.[29]

Present practice is illustrated by the specifications tabulated below for 1-ounce, 3-ounce, and 6-pound rockets as now manufactured by an American fireworks company. The diameter of

		Ounce	Ounce	Pound
Size		1	3	6
Composition of charge	Saltpeter	36	35	30
	Sulfur	6	5	5
	No. 3 charcoal	..	5	12
	No. 5 charcoal	12	..	..
	Charcoal dust	7	17	12

	Inch	Inch	Inch
Length of case	3	4 1/4	13
Outside diameter	1/2	11/16	2 3/8
Inside diameter	5/16	7/16	1 1/2
Overall length of spindle	2 3/4	4	12 3/4
Length of taper	2 1/2	3 23/32	12
Choke diameter	5/32	1/4	3/4

the base of the spindle is, of course, the same as the inside diameter of the case. That of the hemispherical tip of the spindle is half the diameter of the choke, that is, half the diameter of the hole in the clay plug at the base of the rocket. The clay rings and plugs, formed into position by high pressure, actually make grooves in the inner walls of the cases, and these grooves hold them in place against the pressures which arise when the rockets are used. The propelling charge is loaded in several successive small portions by successive pressings with hydraulic presses

[29] Lorrain, *op. cit.,* pp. 236-237.

which handle a gross of the 1-ounce or 3-ounce rockets at a time but only three of the 6-pound size. The presses exert a total pressure of 9 tons on the three spindles when the 6-pound rockets are being loaded.

FIGURE 25. Loading Rockets by Means of an Hydraulic Press. (Courtesy National Fireworks Company.)

Rockets of the smaller sizes, for use as toys, are closed at the top with plugs of solid clay and are supplied with conical paper caps. They produce the spectacle only of a trail of sparks streaking skyward. Rockets are generally equipped with sticks to give them balance and direct their flight and are then fired from a trough or frame, but other rockets have recently come on the market which are equipped with vanes and are fired from a level surface while standing in a vertical position.

Large exhibition rockets are equipped with heads which contain stars of various kinds (see below), parachutes, crackers (see grasshoppers), serpents (compare Figure 23), and so on. In these, the clay plug which stands at the top of the rocket case is perforated, and directly below it there is a *heading* of composition which burns more slowly than the propelling charge. In a typical example this is made from a mixture of saltpeter 24 parts, sulfur 6, fine charcoal 4, willow charcoal dust 1½, and dextrin 2; it is loaded while slightly moist, pressed, and allowed to dry before the head of the rocket is loaded. When the rocket reaches the top of its flight, the heading burns through, and its fire, by means of several strands of black match which have been inserted in the perforation in the clay plug, passes into the head. The head is filled with a mixture, say, of gunpowder, Roman candle composition (see below), and stars. When the fire reaches this mixture, the head blows open with a shower of sparks, and the stars, which have become ignited, fall through the air, producing their own specialized effects.

In another example, the head may contain a charge of gunpowder and a silk or paper parachute carrying a flare or a festoon of lights or colored *twinklers*, the arrangement being such that the powder blows the wooden head from the rocket, ejects the parachute, and sets fire to the display material which it carries. In order that the fire may not touch the parachute, the materials which are to receive the fire (by match from the bursting charge) are packed softly in cotton wool and the remaining space is rammed with bran.

The very beautiful liquid fire effect is produced by equipment which is fully assembled only at the moment when it is to be used. The perforation in the clay plug at the top of the rocket is filled with gunpowder, and this is covered with a layer of waterproof cloth well sealed, separating it from the space in the empty head. When the piece is to be fired, the pyrotechnist, having at hand a can containing sticks of yellow phosphorus preserved under water, removes the wooden head from the rocket, empties the water from the can of phosphorus, and dumps the phosphorus, still wet, into the head case, replaces the wooden head, and fires. The explosion of the gunpowder at the top of the rocket's flight tears through the layer of waterproof cloth, ignites the phosphorus, blows off the wooden head, and throws out the liquid fire. A similar effect, with a yellow light, is obtained with metallic sodium.

Roman Candles

Roman candles are repeating guns which shoot projectiles of colored fire and send out showers of glowing sparks between the shots. To the pyrotechnists of the seventeenth century they were known as "star pumps" or "pumps with stars."

FIGURE 26. Ramming Roman Candles. (Courtesy National Fireworks Company.)

For the manufacture of Roman candles, gunpowder and stars and a modified black powder mixture which is known as Roman candle composition, or *candle comp*, are necessary. The candle comp is made from:

	PARTS	
Saltpeter.....................	34	(200 mesh)
Sulfur........................	7	(200 mesh)
No. 4 Charcoal (hardwood).......	15	(about 24 mesh)
No. 3 Charcoal (hardwood).......	3	(about 16 mesh)
No. 2 Charcoal (hardwood).......	3	(about 12 mesh)
Dextrin.......................	1	

The materials are mixed thoroughly, then moistened slightly and rubbed for intimate mixture through a 10-mesh sieve, dried quickly in shallow trays, and sifted through a 10-mesh sieve.

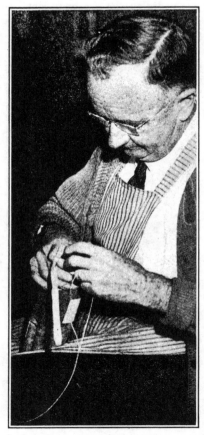

FIGURE 27. Matching a Battery of 10 Ball Roman Candles. (Courtesy National Fireworks Company and the *Boston Globe*.)

Candle comp burns more slowly than black powder and gives luminous sparks. The case is a long, narrow, strong tube of paper plugged at the bottom with clay. Next to the clay is a small quantity of gunpowder (4F); on top of this is a star; and on top of this a layer of candle comp. The star is of such size that it does not fit the tube tightly. It rests upon the gunpowder, and the space between the star and the wall of the tube is partly filled with candle comp. When the three materials have been introduced, they are rammed tightly into place. Then gunpowder, a star, and candle comp again are loaded into the tube and rammed down, and so on until the tube is charged. Damp candle comp, with a piece of black match leading to it and into it, is loaded at the top, pressed tightly into place, and allowed to dry. When a Roman candle is lighted, the candle comp begins to burn and to throw out a fountain of sparks. The fire soon reaches the star, ignites it, and flashes along the side of the star to light the gunpowder which blows the burning star, like a projectile, out of the tube.

Stars

Stars are pellets of combustible material. Those which contain neither aluminum nor magnesium nor Paris green have nothing in their appearance to suggest even remotely the magic which is in them. They are, however, the principal cause of the beauty of aerial pyrotechnic displays.

The components of star composition are mixed intimately and dampened uniformly with some solution which contains a binder, perhaps with gum-arabic water, perhaps with water alone if the composition contains dextrin, perhaps with alcohol if it contains shellac. Several different methods are used for forming the stars.

To make *cut stars*, the damp mixture is spread out in a shallow pan, pressed down evenly, cut into cubes, say $\frac{1}{4}$ to $\frac{3}{4}$ inch on the side, allowed to dry, and broken apart. Because of their corners, cut stars take fire very readily and are well suited for use in rockets and small aerial bombshells. Cylindrical stars are preferred for Roman candles.

For the preparation of a small number of stars, a *star pump* is a convenient instrument. This consists of a brass tube with a plunger which slides within it. The plunger has a handle and, on its side, a peg which works within a slot in the side of the tube— in such manner that it may be fixed in position to leave at the open end of the tube a space equal to the size of the star which it is desired to make. This space is then tightly packed with the damp mixture; the plunger is turned so that the peg may move through the longitudinal slot, and the handle is pushed to eject the star.

For large-scale production, a *star plate* or *star mold* assembly is best. This consists of three flat rectangular plates of hard wood or metal, preferably aluminum. One has a perfectly smooth surface. The second, which rests upon this, has many circular holes of the size of the stars which are desired. The damp mixture is dumped upon this plate, rubbed, pressed, and packed into the holes, and the surface of the plate is then wiped clean. The third

FIGURE 28. A Star Plate or Star Board in Use. (Courtesy National Fireworks Company.)

plate is supplied with pegs, corresponding in number and position to the holes of the second plate, the pegs being slightly narrower than the holes and slightly longer than their depth. The second plate is now placed above a tray into which the stars may fall, and the stars are pushed out by putting the pegged plate upon it. In certain conditions it may be possible to dispense with the pegged plate and to push out the stars by means of a roller of soft crêpe rubber.

Box stars are less likely to crumble from shock, and are accordingly used in large aerial bombshells. They are also used for festoons and for other aerial tableaux effects. Short pieces of

4-ply manila paper tubing, say ¾ inch long and ½ inch in diameter, are taken; pieces of black match long enough to protrude from both ends of the tubes are inserted and held in this position by the fingers while the tubes are pressed full of the damp composition. Box stars require a longer drying than those which are not covered.

White stars, except some of those which contain aluminum, are generally made with potassium nitrate as the oxidizing agent. Various white star compositions are tabulated below. The last three are for white *electric stars*. The last formula, containing perchlorate, was communicated by Allen F. Clark.

Potassium nitrate	70	28	180	20	42	14	28	..
Potassium perchlorate	..	..	..	..	..	..	..	30
Barium nitrate	..	..	..	..	..	..	..	5
Aluminum	..	..	..	..	..	3	5	22
Antimony sulfide	20	..	10	..	..	3	7	..
Antimony metal	..	5	40	..	..	..	..	..
Zinc dust	..	..	..	..	6	..	..	..
Realgar	..	..	..	6	6	..	..	..
Meal powder	..	..	..	..	12	6	3	..
Sulfur	20	8	50	6	23	..	8	..
Charcoal dust	..	..	3	..	..	..	..	..
Dextrin	3	1	6	1	..	1	1	..
Shellac	..	..	..	..	..	..	..	3

Stars which contain aluminum are known as electric stars because of the dazzling brilliancy of their light, which resembles that of an electric arc. Stars which contain chlorate and sulfur or antimony sulfide or arsenic sulfide or picric acid are dangerous to mix, likely to explode if subjected to too sudden shock, and unsafe for use in shells. They are used in rockets and Roman candles. Perchlorate compositions, and chlorate compositions without sulfur, sulfides, and picric acid, will tolerate considerable shock and are used in aerial bombshells.

The following star compositions which contain both chlorate and sulfur are among those recommended by Tessier.[30] Mixtures which contain chlorate and sulfur have a tendency to "sour" with the production of sulfuric acid after they have been wetted,

[30] *Op. cit.*, pp. 338, 343, 344, 345, 347, 349.

and to deteriorate, but the difficulty may be remedied by the addition of an anti-acid, and some of these compositions do in-

FIGURE 29. Aluminum Stars from a Single Rocket.

deed contain carbonates or basic salts which act in that capacity. Tessier recommends that the mixtures be made up while dampened with small quantities of 35 per cent alcohol.

	Red	Lilac	Lilac Mauve	Violet	Blue	Green
Potassium chlorate.......	167	17	17	56	24	48
Strontium carbonate.....	54	9	9	..	..	..
Strontium sulfate........	..	..	..	16	..	..
Barium nitrate..........	..	..	..	..	..	80
Copper oxychloride......	..	2	4	..	..	..
Basic copper sulfate......	..	..	..	8	12	..
Lead chloride..........	..	1	1	3	2	10
Charcoal dust (poplar)...	..	..	..	3	..	..
Sulfur.................	35	7	7	20	8	26
Dextrin...............	7	1	1	3	1	3
Shellac................	16	..	..	..	..	2
Lampblack.............	..	..	..	..	..	2

Weingart[31] reports compositions for cut, pumped, or candle stars which contain chlorate but no sulfur or sulfides, as follows:

	Red		Blue	Green	Yellow	
Potassium chlorate............	12	48	48	12	32	16
Strontium nitrate.............	12	..	..	..	..	..
Strontium carbonate..........	..	8	..	..	..	..
Barium nitrate................	..	..	16	12	12	..
Paris green...................	..	..	18	..	..	..
Calomel......................	..	..	..	1	..	..
Sodium oxalate...............	..	..	..	..	2	7
Fine charcoal................	4	8	..	4	8	1
Dextrin.....................	1	3	3	1	3	1
Shellac......................	2	6	10	2	6	3

For the following perchlorate star formulas the author is indebted to Allen F. Clark.

	Rose	Amber	Green		Violet	
Potassium perchlorate..........	12	10	..	44	12	41
Potassium nitrate..............	6	..	..	..	..	..
Barium perchlorate............	..	..	32	90	..	..
Calcium carbonate.............	..	..	..	..	..	12
Strontium oxalate.............	1	..	..	..	9	3
Copper oxalate................	..	..	..	..	5	..
Sodium oxalate...............	..	4	..	..	..	..
Calomel......................	..	..	..	..	..	3
Sulfur.......................	..	..	..	..	..	14
Lampblack...................	1	..	..	..	..	..
Dextrin.....................	1	..	..	6	..	..
Shellac......................	..	2	3	15	3	..

Illustrative of electric star compositions are the following; those which contain potassium chlorate are reported by Faber,[32] the others, containing perchlorate, were communicated by Allen

[31] *Op. cit.*, p. 114.
[32] *Op. cit.*, Vol. 1, p. 188.

	Red		Gold	Green			Blue	
Potassium perchlorate..........	..	12	..	..	..	6	..	14
Potassium chlorate.............,..	24	..	6	8	..	..	32	..
Barium perchlorate..............	..	..	..	..	12	12	..	..
Barium chlorate.................	..	..	4	16	..	..	..	..
Barium nitrate..................	..	..	..	16	..	..	..	..
Strontium chlorate..............	..	3	..	..	..	..	..	..
Strontium carbonate.............	4	..	..	..	..	..	..	..
Aluminum......................	8	3	4	12	5	8	8	6
Sodium oxalate..................	..	..	2	..	..	..	..	..
Calcium carbonate...............	..	..	..	..	..	..	..	1
Magnesium carbonate............	..	..	1	..	..	..	..	..
Paris green.....................	..	..	..	..	..	..	16	10
Calomel........................	..	..	..	..	..	..	..	2
Fine charcoal..................	1	..	..	3	..	..	..	..
Dextrin.......................	2	1	..	2	..	1	2	1
Red gum.......................	..	..	..	4	..	..	..	..
Shellac.......................	2	1	2	..	1	2	1	2

F. Clark. The last-named authority has also supplied two formulas for magnesium stars. The compositions are mixed while

	Amber	Green
Potassium perchlorate.......	4	..
Barium perchlorate.........	..	12
Magnesium................	1	2
Sodium oxalate............	2	..
Lycopodium powder........	..	1
Shellac.................. ..	1	2

dampened with alcohol which insures that the particles of magnesium are covered with a protective layer of shellac.

Lampblack stars burn with a rather dull soft light. Discharged in large number from a rocket or aerial shell, they produce the beautiful willow-tree effect. They are made, according to Allen F. Clark, by incorporating 3 pounds of lampblack, 4 pounds of meal powder, and ½ pound of finely powdered antimony sulfide with 2 ounces of shellac dissolved in alcohol.

Stars compounded out of what is essentially a modified black powder mixture, given a yellowish or whitish color by the addition of appropriate materials, and used in rockets and shells in the same manner as lampblack stars, produce *gold* and *silver showers*, or, if the stars are larger and fewer in number, *gold* and *silver streamers*. The following formulas are typical.

	Gold	Silver
Potassium nitrate............	16	10
Charcoal.................	1	2
Sulfur....................	4	3
Realgar...................	..	3
Sodium oxalate............	8	..
Red gum.................	1	1

Twinklers are stars which, when they fall through the air, burn brightly and dully by turns. A shower of twinklers produces an extraordinary effect. Weingart in a recent letter has kindly sent the following formula for yellow twinklers:

Meal powder.................	24
Sodium oxalate..............	4
Antimony sulfide.............	3
Powdered aluminum..........	3
Dextrin..................	1

The materials are mixed intimately while dampened with water, and the mixture is pumped into stars about ¾ inch in diameter and ⅞ inch long. The stars are dried promptly. They function only when falling through the air. If lighted on the ground they merely smolder, but when fired from rockets or shells are most effective.

Spreader stars contain nearly two-thirds of their weight of powdered zinc. The remaining one-third consists of material necessary to maintain an active combustion. When they are ignited, these stars burn brightly and throw off masses of burning zinc (greenish white flame) often to a distance of several feet. Weingart[33] gives the two following formulas for spreader stars, the first for "electric spreader stars," the second for "granite stars," so called because of their appearance.

[33] *Op. cit.,* p. 118.

Zinc dust	72	80
Potassium nitrate	..	28
Potassium chlorate	15	..
Potassium dichromate	12	..
Granulated charcoal	12	..
Fine charcoal	..	14
Sulfur	..	5
Dextrin	2	2

The first of these formulas is the more difficult to mix and the more expensive. All its components except the charcoal are first

FIGURE 30. Spreader Stars from a Battery of Rockets.

mixed and dampened; the granulated charcoal, which must be free from dust, is then mixed in, and the stars are formed with a pump. They throw off two kinds of fire when they burn, masses of brightly burning zinc and particles of glowing charcoal. Weingart recommends that the second formula be made into cut stars ⅜ inch on the side. Spreader stars because of the zinc which they contain are much heavier than other stars. Rockets and aerial bombs cannot carry as many of them.

Gerbs

Gerbs produce jets of ornamental and brilliant fire and are used in set pieces. They are rammed or pressed like rockets, on a short nipple instead of a long spindle, and have only a slight depression within the choke, not a long central cavity. They are choked to about one-third the diameter of the tube. The simplest gerbs contain only a modified black powder mixture, say meal powder 4 parts, saltpeter 2, sulfur 1, and charcoal dust 1 or mixed charcoal 2; and are used occasionally for contrast in elaborate set pieces. Similar composition is used for the starting fire of steel gerbs which are more difficult to ignite. If antimony sulfide is used in place of charcoal, as in the mixtures:

Meal powder	2	3
Saltpeter	8	8
Sulfur	3	4
Antimony sulfide	1	2

the gerbs yield compact whitish flames and are used in star and floral designs. Gold gerbs appropriately arranged produce the sunburst effect. Colored gerbs are made by adding small cut stars. In loading the tube, a scoopful of composition is introduced and rammed down, then a few stars, then more composition which is rammed down, and so on. Care must be exercised that no stars containing chlorate are used with compositions which contain sulfur, for an explosion might occur when the charge is rammed. The following compositions are typical. The steel filings

	Steel		Colored Steel	Gold	Colored Gold
Meal powder	6	4	8	40	40
Potassium nitrate	2	..	7	..	..
Sulfur	1	..	..	..	..
Fine charcoal	1	1	2	..	..
Steel filings	1	2	5	..	..
Stars	..	..	5	..	5
Sodium oxalate	..	..	..	6	6
Antimony sulfide	..	..	..	8	9
Aluminum	..	..	..	4	..
Dextrin	..	..	..	..	4

must be protected from rusting by previous treatment with paraffin or linseed oil.

Prismatic fountains, floral bouquets, etc., are essentially colored gerbs. Flower pots are supplied with wooden handles and generally contain a modified black powder composition with lampblack and sometimes with a small amount of granulated black powder. In the charging of fountains and gerbs, a small charge of gunpowder is often introduced first, next to the clay plug which closes the bottom of the tube and before the first scoopful of composition which is rammed or pressed. This makes them finish with a report or *bounce.*

Fountains

Fountains are designed to stand upon the ground, either upon a flat base or upon a pointed wooden stick. They are choked slightly more than gerbs, and have heavier, stronger cases to withstand the greater pressures which eject the fire to greater distances.

The "Giant Steel Fountain" of Allen F. Clark is charged with a mixture of saltpeter 5 parts (200 mesh), cast-iron turnings 1 part (8 to 40 mesh), and red gum 1 part (180 to 200 mesh). For loading, the mixture is dampened with 50 per cent alcohol. The case is a strong paper tube, 20 inches long, 4 inches in external diameter, with walls 1 inch thick, made from Bird's hardware paper. It is rolled on a machine lathe, the paper being passed first through a heavy solution of dextrin and the excess of the gum scraped off. The bottom of the case is closed with a 3-inch plug of clay. The composition will stand tremendous pressures without exploding, and it is loaded very solidly in order that it may stay in place when the piece is burned. The charge is rammed in with a wooden rammer actuated by short blows, as heavy as the case will stand, from a 15-pound sledge. The top is closed with a 3-inch clay plug. A ⅞-inch hole is then bored with an auger in the center of the top, and the hole is continued into the charge to a total depth of 10 inches. The composition is difficult to light, but the ignition is accomplished by a bundle of six strands of black match inserted to the full depth of the cavity and tied into place. This artifice produces a column of scintillating fire, 100 feet or more in height, of the general shape of a red cedar tree. It develops considerable sound, and ends suddenly with a terrifying roar at the moment of its maximum splendor. If loaded at the hydraulic press with a tapered spindle (as is necessary), it finishes its burning with a fountain which grows smaller and smaller and finally fades out entirely.

Wheels

Driving tubes or *drivers*, attached to the periphery of a wheel or to the sides of a square or hexagon of wood which is pivoted

FIGURE 31. Matching Display Wheels. (Courtesy National Fireworks Company and the *Boston Globe*.)

at its center, by shooting out jets of fire, cause the device to rotate and to produce various ornamental effects according to the compositions with which they are loaded. When the fire reaches the bottom of one driver, it is carried by quickmatch to the top of the next. Drivers are loaded in the same manner as gerbs, the compositions being varied slightly according to the size as is done with rockets. A gross of the 1-ounce and 2-ounce sizes in present American practice is loaded at one time by the hydraulic press. Typical wheel turning compositions (Allen F. Clark) for use in 1-ounce and 2-ounce drivers are reported below, the first for a *charcoal spark effect*, the second for an *iron and steel effect*. The speed of the mixtures may be increased by increasing the proportion of gunpowder.

Saltpeter (210 mesh)	10	46
Sulfur (200 mesh)	2	19
Meal powder	6	..
Charcoal dust	..	16
Charcoal (80 mesh)	1	..
6F gunpowder	6	..
7F gunpowder	..	24
Cast-iron turnings (16 mesh)	..	30
Dextrin	..	8

Wheels, gerbs, and colored fires are the parts out of which such display pieces as the "Corona Cluster," "Sparkling Caprice," "Flying Dutchman," "Morning Glory," "Cuban Dragon," "Blazing Sun," and innumerable others are constructed.

Saxons

Saxons are strong paper tubes, plugged with clay at their middles and at both ends, and filled between the plugs with composition similar to that used in drivers. A lateral hole is bored through the middle of the tube and through the central clay plug, and it is around a nail, passed through this hole and driven into a convenient support, that the artifice rotates. Other holes, at right angles to this one, are bored from opposite sides near the ends of the tube, just under the end plugs, through one wall of the tube and into the composition but not through it. A piece of black match in one of these holes ignites the composition. The hot gases, sparks, etc., rushing from the hole cause the device to turn upon its pivot. When the fire reaches the bottom of the charge, it lights a piece of quickmatch, previously connected through a hole at that point and glued to the outside of the case, which carries the fire to the other half of the saxon.

Saxons are generally matched as described, the two halves burning consecutively and rotating it in the same direction. Sometimes they burn simultaneously, and sometimes one half turns it in one direction and the other afterwards "causes the rapid spinning to reverse amid a mad burst of sparks." This effect "is very pleasing and is considered one of the best to be obtained for so small an expenditure."

Pinwheels

To make pinwheels, manila or kraft paper tubes or *pipes*, about 12 inches long and 3/16 inch in diameter, are needed. One end is closed by twisting or folding over. The tubes are filled with composition, the other ends are closed in the same way, and the tubes are wrapped in a moist towel and set aside until they are thoroughly flabby. In this condition they are passed between rollers and flattened to the desired extent. Each tube is then wound in an even spiral around the edge of a cardboard disc which has a hole in its center for the pin, and the whole is placed in a frame which prevents it from uncoiling. Four drops of glue, at the four quarters of the circle, are then brushed on, across the pipes and onto the center disc, and the device is allowed to dry.

Weingart[34] recommends for pinwheels the compositions which are indicated below. The first of these produces both steel and

Meal powder		10	8	2
Gunpowder (fine)	8	5	8	..
Aluminum		..	3	..
Saltpeter	14	4	16	1
Steel filings	6	6	..	..
Sulfur	4	1	3	1
Charcoal	3	1	8	..

charcoal effects, the second steel with much less of the charcoal, the third aluminum and charcoal, and the fourth a circle merely of lilac-colored fire.

Tessier thought highly of pinwheels (*pastilles*). They were, he says,[35]

formerly among the artifices which were called *table fireworks*, the use of which has wholly fallen away since the immense apartments have disappeared which alone provided places where these little pyrotechnic pieces might be burned without too much inconvenience.

The manner of use of these pastilles· calls only for small calibers; also their small dimensions make it possible to turn them out at a low price, and the fireworks makers have always continued to make them the object of current manufacture. But what they have neglected, they still neglect: and that is, to seek to bring them to perfection. Those that

[34] *Op. cit.,* p. 98.
[35] *Op. cit.,* p. 393.

they confine themselves to making serve only for the amusement of children.

However, pastilles may become charming pieces of fireworks, fit to refresh all eyes. They can be made to produce

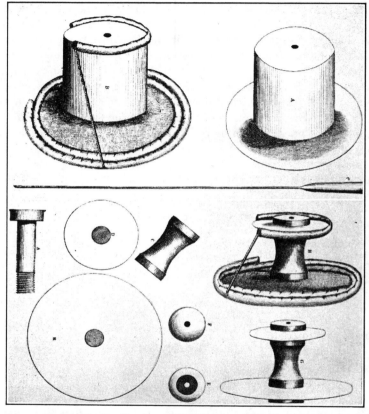

FIGURE 32. Pinwheels, Tessier, 1883. Wheels which show an inner circle of colored fire. Plate 1 (above) pictures pinwheels which are intended to be sold as completely consumable. The instrument represented at the bottom of the plate is the ramrod for tamping the charges in the pipes. Plate 2 (below) represents pinwheels which are intended to be exhibited by the pyrotechnist himself: the wooden parts are to be recovered and used again.

truly marvellous effects, considering the conditions imposed by their size, effects all the more remarkable in as much as, by the very reason of these same conditions, they have no need of a vast theater in which to be fired. The least little garden suffices for them. They burn under the very eye of the spectator, who loses nothing of their splendor, whereas, in general, large pieces of fireworks can be enjoyed only at a distance from the place of firing. Finally, they have over these last the advantage of their low price and the advantage that they can be transported without embarrassment and set in place at the moment of being fired.

Tessier describes ordinary pastilles, diamond pastilles, and pastilles with colored fires. The shorter and more central tubes (Figure 32), wound part way around discs 40 mm. in diameter, hold the colored fire compositions. The longer tubes, forming the larger circles around discs 72 mm. in diameter, are the turning tubes. The latter, it will be seen, are so arranged that they burn for a time before the fire reaches the colored compositions. "The charging of the tubes is commenced, up to a height of about 17 cms., with the four compositions, Nos. 142, 126, 128, and 129, in the order named. The rest of the tube is charged entirely with composition No. 149, or with No. 152, both of which produce scintillating aureoles."[36] The charges are tamped tightly in the tubes by means of a long, thin ramrod and mallet. The compositions in question, designated by Tessier's own numbers, are indicated below.

	142	126	128	129	149	152
Meal powder	16	16	16	32	..	..
Potassium nitrate	1	1	1	4	..	..
Oak charcoal	1	..	..	..	..	..
Litharge	..	2	..	..	..	..
Powdered mica	..	..	2	..	..	..
Antimony sulfide	..	..	..	5	..	..
Plumbic powder No. 1	..	..	..	..	17	17
Cast-iron filings	..	..	..	..	3	..
Steel wool	..	..	..	..	..	3

No. 142 is a composition for ordinary pastilles. Tessier says that it produces "numerous sparks forming a feeble aureole. As this composition is not lively, and as it is not able to make the

[36] *Ibid.*, p. 419.

pastilles turn conveniently, care is taken not to load more of it than a length of 15 mm. in the tube."[37]

Nos. 126, 128, and 129 are also for ordinary pastilles. No. 126 "has not much force; it is incapable alone of making a pastille turn with the necessary rapidity. Hence care is taken in charging it to introduce only a small quantity into the tube." It burns with a white flame "forming a crown, more or less lacy-edged, from which rays and sparks are thrown out."[38]

The two compositions, Nos. 142 and 126, evidently burn while the pastille is turning from the initial twirl given it by the hand. When the fire reaches the next composition, No. 128, the pastille accelerates by its own power. This gives "reddish rays, very straight and very numerous," and No. 129 gives "a white flame around the disc, and numerous and persistent sparks which fall down forming a sort of cascade on each side of the pastille."[39]

Neither No. 128 nor 129 is bright enough to make much of a show if the colored fire is also burning. When they burn to an end, the fire is communicated to the colored composition; at the same time the bright diamond composition, either No. 149 or 152, commences to burn. No. 149 produces "a splendid aureole of silver-white flowers. These flowers are less developed than those produced by steel wool and make a different effect from the latter."[40] No. 152 produces a "splendid effect—no inflamed disc, no reddish sparks—numerous jasmine flowers of all dimensions forming a vast aureole of a striking white."[41]

Plumbic powder No. 1 is made from lead nitrate 12 parts, potassium nitrate 2 parts, and black alder charcoal 3 parts.[42] The materials are powdered and mixed, and then rolled in a wooden ball-mill with balls of hard lead (Pb 5, Sb 1) or brass or bronze.

Tessier[43] gives credit to the earlier French pyrotechnist, Chertier, for the introduction into pyrotechny of lead nitrate (which had been used before his time only for the preparation of slow-

[37] *Ibid.*, p. 408.
[38] *Ibid.*, p. 403.
[39] *Ibid.*, p. 404.
[40] *Ibid.*, p. 411.
[41] *Ibid.*, p. 412.
[42] *Ibid.*, p. 281.
[43] *Op. cit.*, p. 118.

match or fire wick), for the invention[44] of plumbic powder by which a silver shower (*pluie d'argent*) is produced, and for originating[45] the idea of the diamond pastille with colored fires which Chertier called the *dahlia pastille* but for the making of which he did not give precise directions.

Mines

Mines are paper mortars—commonly strong paper tubes each standing vertically on a wooden base into which it is countersunk and glued—arranged to throw into the air a display of stars, serpents, etc. They are often equipped with fountains, Roman candles, etc., which make a display on the ground before the final explosion occurs.

A *serpent mine* (*pot à feu*) is represented in Figure 24. This starts with a steel fountain. When the fire has reached the bottom of the fountain, it is carried by quickmatch to a charge of gunpowder in the paper bag, *a*. Immediately above the paper bag are the serpents. These are small paper tubes, rammed with a mixture of meal powder, gunpowder, saltpeter, sulfur, and mixed charcoal, crimped or plugged with clay at one end, supplied with match (as in the diagram) or merely left open-ended at the other. The lower, matched or open, ends of the serpents take fire from the burning of the gunpowder, which also blows them into the air where they dart and squirm about like little tailless rockets leaving a trail of sparks. In Audot's diagram, directly below the fountain and above the closed ends of the serpents, is a mass of wadding. This tends to offer a slight resistance to the force of the gunpowder, with the result that the serpents receive the fire more surely and are shot farther into the air before they begin to go their several ways.

Saucissons are constructed in the same way as serpents, but are larger, and have, next to the closed end, a small charge of gunpowder which makes them end with a bang. They are used in mines and in rockets.

Mines which discharge serpents, stars, English crackers, etc., are often made by loading these materials into the same paper bags which' contain the blowing charges of granulated gunpowder. About two level teaspoonfuls of blowing powder is used

[44] *Ibid.*, p. 281.
[45] *Ibid.*, p. 414.

per ounce of stars. For making the bags, a board is taken which has had holes bored into it slightly smaller than the internal diameter of the mine case and of a depth suited to the caliber of the mine. A disc of tissue paper is placed over a hole and then punched down into it by a wooden punch or rod with slightly rounded edges which fits rather loosely in the hole. This makes a paper cup into which one end of the fuse is inserted, and around it the stars and blowing charge. The edges of the paper cup are then gathered together and tied with string or wire.

Mines are often made up with a single Roman candle, lacking the plug of clay at the bottom, mounted in the center of the mine case. The fuse leading from the charge in the paper bag is thrust into the bottom of the Roman candle. A mine with a large and short case, carrying a charge of tailed stars, serpents, and English crackers, and having one Roman candle in its center and four others, matched to burn simultaneously, attached to the outside of the case, is known as a *devil among the tailors*.

Comets and Meteors

These are virtually mines which shoot a single large star. A pumped star 1½ inches in diameter is fired, for example, from a tube or mortar 10 inches long and 1¾ inches in internal diameter. A piece of quickmatch (wrapped black match) about 6 inches longer than the mortar is taken; an inch of black match is made bare at one end, bent at right angles, and laid against the base of the star; and the star, with the quickmatch lying along its side, is then enclosed in the middle of a paper cylinder by wrapping a strip, say 4 inches wide, of pasted tissue paper around it. A half teaspoonful of granulated black powder is put into the cup thus formed on the (bottom) side of the star where the black match has been exposed, and the edges of the paper cylinder are brought together over it and tied. The other (upper) end of the paper cylinder is similarly tied around the quickmatch. In using this piece, and in using all others which are lighted by quickmatch, care must be taken that a few inches of the quickmatch have been opened and the black match exposed, before the fire is set to it; otherwise it will be impossible to get away quickly enough. This, of course, is already done in pieces which are offered for public sale.

Comets burn with a charcoal or lampblack effect, meteors with an electric one. The two comet star compositions given below are due to Weingart;[46] that of the green meteor to Allen F. Clark.

	Comets		Green Meteor
Potassium nitrate.........	6	..	..
Barium perchlorate.......	..	..	4
Barium nitrate..........	..	..	2
Meal powder.............	6	3	..
Sulfur..................	1	..	..
Fine charcoal............	3	1	..
Antimony sulfide.........	3	1	..
Lampblack..............	..	2	..
Aluminum...............	..	..	1
Dextrin.................	..	..	1

Some manufacturers apply the name of meteors to artifices which are essentially large Roman candles, mounted on wooden bases and shooting four, six, eight, and ten stars 1½ inches in diameter. They are loaded in the same way as Roman candles except that a special device is used to insure the certain ignition of the stars. Two pieces of black match at right angles to each other are placed under the bottom of the star; the four ends are turned up along the sides of the star and are cut off even with the top of it. The match being held in this position, the star is inserted into the top of the case and pushed down with a rammer onto the propelling charge of gunpowder which has already been introduced. Then coarse candle comp is put in, then gunpowder, then another star in the same manner, and so on. The black match at the side of the star keeps a space open between the star and the walls of the tube, which space is only partly or loosely filled with candle comp. The black match acts as a quickmatch, insuring the early ignition of the propelling charge as well as the sure ignition of the star. Electric stars, spreader stars, and *splitters* are used in meteors. Splitter stars are made from the same composition as snowball sparklers (see below); the composition for stars, however, is moistened with much less water than for sparklers. They split into bright fragments while shooting upward and burst at the top to produce a palm-tree effect.

[46] *Op. cit.,* p. 121

Bombshells

Bombshells are shot from mortars by means of a charge of black powder and burst high in the air with the production of reports, flashes, showers, and other spectacular effects. The smaller ones are shot from paper mortars; the larger, most commonly from mortars of iron. In the past they have often been made in a spherical shape, wood or paper or metal hemispheres pasted heavily over with paper, but now in this country they are made almost exclusively in the form of cylinders. For the same caliber, cylindrical bombshells will hold more stars or other display material than spherical ones, and it is much easier to contrive them in a manner to procure multiple bursts. The materials of construction are paper, paste, and string. The shells are supplied with *Roman fuses* timed to cause them to burst at the top of their flight. The success and safety of bombshells depend upon carefully constructed fuses.

Roman fuses are made by pounding the fuse powder as firmly as is possible into hard, strong, tightly rolled paper tubes. These are commonly made from Bird's hardware paper, pasted all over before it is rolled, and are dried carefully and thoroughly before they are loaded with ramrod and mallet. "When a number of these cases are rolled," says Weingart,[47] "they must be dried in the shade until they are as hard as wood and rattle when struck together." He recommends the first of the following-listed compositions, the Vergnauds the others:[48]

Potassium nitrate	2	4	2
Sulfur	1	2	1
Meal powder	4	6	3
Antimony sulfide	..	1	..

The length of the column of composition determines the duration of the burning. The composition in the fuse must be as hard and as firmly packed as possible; otherwise it will blow through into the shell when in use and will cause a premature explosion. Some manufacturers load the tubes and cut them afterwards with a fine-tooth hack saw. Others prefer to cut them to the desired lengths with a sharp knife while they are prevented from collapsing by a brass rod through them, and afterwards to load

[47] *Ibid.,* p. 130.

[48] A. D. and P. Vergnaud, "Nouveau manuel complet de l'artificier. Pyrotechnie civile." (Ed., G. Petit) (Manuels Roret), Paris, 1906.

the short pieces separately. Different size tubes are often used
for the fuses of different size shells; those for a 4-inch shell
(that is, for a shell to be shot from a 4-inch mortar) are commonly made from tubes 5/16 inch in internal diameter and ⅝

FIGURE 33. Bombshells for 4- and 6-inch Mortars. (Courtesy National
Fireworks Company and the *Boston Globe*.)

inch in external diameter. Fuses are generally attached to the
front end of the bombshell. The forward-pointing end of the
tube, which is outside the shell and receives the fire, is filled flush
with the composition. The other backward-pointing end, inside
the shell, is empty of composition for ¾ inch of its length; a

bundle of stiff 2-inch pieces of black match is inserted into this space and is held in position by a rolled wrapper of paper, glued to the fuse case and tied with a string near the ends of the match, in order that it may not be dislodged by the shock of setback. The match serves to bring the fire more satisfactorily to the bursting charge within the shell.

The preparation of the bombshell is hand work which requires much skill and deserves a fairly full description. We describe the construction of a 4-inch shell to produce a single burst of stars. A strip of bogus or news board paper is cut to the desired length and is rolled tightly on a form without paste. When it is nearly all rolled, a strip of medium-weight kraft paper, 4 inches wider than the other strip, is rolled in and is rolled around the tube several times and is pasted to hold it in position. Three circular discs of pasteboard of the same diameter as the bogus tube (3½ inches) are taken, and a ⅝-inch hole is punched in the center of two of them. The fuse is inserted through the hole in one of them and glued heavily on the inside. When this is thoroughly dry, the disc is glued to one end of the bogus tube, the matched end of the fuse being outside; the outer wrapper of kraft paper is folded over carefully onto the disc, glued, and rubbed down smoothly; and the second perforated disc is placed on top of it.

The shell case is now turned over, there being a hole in the bench to receive the fuse, and it is filled with as many stars (½-inch diameter, ½ inch long) as it will contain. A mixture of 2F gunpowder and candle comp is then added, shaken in, and settled among the stars until the case is absolutely full. A disc of pasteboard is placed over the stars and powder, pressed down against the end of the bogus body and glued, and the outer kraft paper wrapper is folded and glued over the end.

At this point the shell is allowed to dry thoroughly before it is wound with strong jute twine. It is first wound lengthwise; the twine is wrapped as tightly as possible and as firmly against the fuse as may be; each time that it passes the fuse the plane of the winding is advanced by about 10° until 36 turns have been laid on, and then 36 turns are wound around the sides of the cylinder at right angles to the first winding. The shell is now ready to be "pasted in." For this purpose, 50-pound kraft paper is cut into strips of the desired dimensions, the length of the strips being across the grain of the paper. A strip of this paper is folded, rubbed, and twisted in paste until it is thoroughly impregnated. It is then laid out on the bench and the shell is rolled up in it. The cylinder is now stood upright, the fuse end at the top, and the portion of the wet pasted kraft paper wrapper which extends above the body of the shell is torn into strips about ¾ inch wide; these, one by one, are rubbed down carefully and smoothly, one overlapping the other, upon the end of the shell case. They extend up the fuse tube for about ½ inch and are pressed down firmly against it. The shell is now turned over, the fused end resting against a tapered hole in the bench, and a corresponding operation is performed upon the other end. The body of the shell is now about ¼ inch thick on the sides of the cylinder, about ⅜ inch thick at the top end, and about ½ inch at the base end. It is dried outdoors in the sun and breeze, or in a well-ventilated dry-house at 100°F., and, when thoroughly dry, is ready to be supplied with the propelling or *blowing charge.*

A piece of *piped match* (black match in a paper tube) is laid along the side of the bombshell; both are rolled up without paste in 4 thicknesses of 30-pound kraft paper wide enough to extend about 4 inches beyond the ends of the cylinder, and the outer wrapper is tied lightly in place by two strings encircling the cylinder near the ends of the shell case. The cylinder is turned bottom end up. About 3 inches of the paper pipe of the quick-match is removed to expose the black match, a second piece of black match is inserted into the end of the paper pipe, and the pipe is tied with string to hold the match in place. The propelling charge of 2F gunpowder is next introduced; the two inner layers of the outer kraft paper wrapper are folded down upon it and pressed firmly, then the two outer layers are pleated to the center of the cylinder, tied, and trimmed close to the string. The cylinder is then turned to bring the fuse end uppermost. The end of the fuse is scraped clean if it has been touched with paste. Two pieces of black match are crossed over the end of the fuse, bent down along the sides of the fuse tube, and tied in this position with string. The piped match which leads to the blowing charge is now laid down upon the end of the cylinder, up to the bottom of the fuse tube, then bent up along the side of the fuse tube, then bent across its end and down the other side, and then bent back upon itself, and tied in this position. Before it is tied, a small hole is made in the match pipe at the point where it passes the end of the Roman fuse, and a piece of flat black match is inserted. The two inner layers of the kraft paper wrapper are now pleated around the base of the fuse and tied close to the shell. The two outer layers are pleated and tied above the top of the fuse, a 3-foot length of piped match extending from the upper end of the package. A few inches of black match is now bared at the end and an extra piece of black match is inserted and tied in place by a string about 1 inch back from the end of the pipe. The black match, for safety's sake, is then covered with a piece of lance tube, closed at the end, which is to be removed after the shell has been placed in the mortar and is ready for firing.

Maroons

Bombs which explode with a loud report, whether they are intended for use on the ground or in the air, are known as maroons. They are called *marrons* in French, a name which also means large chestnuts in that language—and chestnuts sometimes explode while being roasted.

Maroons are used for military purposes to disconcert the enemy by imitating the sounds of gunfire and shell bursts, and have at times been part of the standard equipment of various armies. A cubical pasteboard box filled with gunpowder is wound in three directions with heavy twine, the successive turns being laid close to one another; an end of miner's fuse is inserted through a hole made by an awl, and the container, already very strong, is made still stronger by dipping it into liquid glue and allowing to dry.

For sharper reports, more closely resembling those of a high-explosive shell, fulminating compositions containing chlorate are used. With these, the necessity for a strong container is not so great; the winding may be done with lighter twine, and the successive turns of twine need not make the closest possible contact. Faber reports two compositions, as follows:

Potassium chlorate . . .	4	1
Sulfur	1	..
Soft wood charcoal. . . .	1	..
Antimony sulfide	..	1

"It is to be noted," he says,[49] "that, while the first formula affords a composition of great strength, the second is still more violent. It is also of such susceptibility that extraordinary care is required in the handling of it, or a premature explosion may result."

Chlorate compositions are not safe for use in maroons. Black powder is not noisy enough. Allen F. Clark has communicated the following perchlorate formulas for reports for maroons. For

Potassium perchlorate	12	6	32
Sulfur	8	2	..
Antimony sulfide	..	3	..
Sawdust	1	..	..
Rosin	..	..	3
Fine charcoal	..	..	3

a *flash report* he uses a mixture of 3 parts of potassium permanganate and 2 of aluminum.

Toy Caps

Toy caps are commonly made from red phosphorus and potassium chlorate, a combination which is the most sensitive, dangerous, and unpredictable of the many with which the pyrotechnist has to deal. Their preparation ought under no conditions to be attempted by an amateur. Powdered potassium chlorate 20 parts is made into a slurry with gum water. It is absolutely essential that the chlorate should be wetted thoroughly before the red phosphorus is mixed with it. Red phosphorus, 8 parts, is mixed with powdered sulfur 1 part and precipitated calcium carbonate 1 part, and the mixture is made into a slurry separately with gum water, and this is stirred into the other until thoroughly mixed. The porridgelike mass is then spotted on paper, and a piece of pasted tissue paper is placed over the spotted surface in a manner to avoid the enclosure of any air bubbles between the two. This is important, for, unless the tissue paper covers the spots snugly, the composition is likely to crumble, to fall out, and to create new dangers. (A strip of caps, for example, may explode between the fingers of a boy who is tearing it.) The moist sheets of caps are piled up between moist blankets in a press, or with a board and weights on top of the pile, and are pressed for an hour or so. They are then cut into strips

[49] *Op. cit.,* Vol. 1, p. 166.

of caps which are dried, packaged, and sold for use in toy repeating pistols. Or they are cut in squares, one cap each, which
are not dried but are used while still moist for making Japanese
torpedoes (see below). The calcium carbonate in this mixture is
an anti-acid, which prevents it from deteriorating under the
influence of moisture during the rather long time which elapses,
especially in the manufacture of torpedoes, before it becomes
fully dry.

Mixtures of potassium chlorate and red phosphorus explode
from shock *and from fire*. They do not burn in an orderly fashion
as do black powder and most other pyrotechnic mixtures. No
scrap or waste ought ever to be allowed to accumulate in the
building where caps are made; it ought to be removed hourly,
whether moist or not, and taken to a distance and thrown upon
a fire which is burning actively.

Silver Torpedoes

These contain silver fulminate, a substance which is as sensitive as the red phosphorus and chlorate mixture mentioned above,
but which, however, is somewhat more predictable. They are
made by the use of a *torpedo board*, that is, a board, say 7/8
inch thick, through which 3/4-inch holes have been bored. A
2-inch square of tissue paper is placed over each hole and
punched into the hole to form a paper cup. A second board of
the same thickness, the *gravel board*, has 1/2-inch holes, bored
not quite through it, in number and position corresponding to
the holes in the torpedo board. Fine gravel, free from dust, is
poured upon it; the holes are filled, and the excess removed.
The torpedo board, filled with paper cups, is inverted and set
down upon the gravel board, the holes matching one another.
Then the two boards, held firmly together, are turned over and
set down upon the bench. The gravel falls down into the paper
cups, and the gravel board is removed. A small amount of silver
fulminate is now put, on top of the gravel, into each of the paper
cups. This is a dangerous operation, for the act of picking up
some of the fulminate with a scoop may cause the whole of it
to explode. The explosion will be accompanied by a loud noise,
by a flash of light, and by a tremendous local disturbance damaging to whatever is in the immediate neighborhood of the ful-

minate but without effect upon objects which are even a few inches away.

In one plant which the present writer has visited, the fulminate destined to be loaded into the torpedoes rests in a small heap in the center of a piece of thin rubber (dentist's dam) stretched over a ring of metal which is attached to a piece of metal weighing about a pound. This is held in the worker's left hand, and a scoop made from a quill, held in the right hand, is used to take up the fulminate which goes into each torpedo. If the fulminate explodes, it destroys the piece of stretched rubber—nothing else. And the rubber, moreover, cushions it so that it is less likely to explode anyway. The pound of metal is something which the worker can hold much more steadily than the light-weight ring with its rubber and fulminate, and it has inertia enough so that it is not jarred from his hand if an explosion occurs. After the fulminate has been introduced into the paper cups, the edges of each cup are gathered together with one hand and twisted with the thumb and forefinger of the other hand which have been moistened with paste. This operation requires care, for the torpedo is likely to explode in the fingers if it is twisted too tightly.

Torpedoes, whether silver, Japanese, or globe, ought to be packed in sawdust for storage and shipment, and they ought not to be stored in the same magazine or shipped in the same package with other fireworks. If a number of them are standing together, the explosion of one of them for any reason is practically certain to explode the others. Unpacked torpedoes ought not to be allowed to accumulate in the building in which they are made.

Japanese Torpedoes

The so-called Japanese torpedoes appear to be an American invention. They contain a paper cap placed between two masses of gravel, and in general require to be thrown somewhat harder than silver torpedoes to make them explode. The same torpedo board is used as in the manufacture of silver torpedoes, but a gravel board which holds only about half as much gravel. After the gravel has been put in the paper cups, a paper cap, still moist, is placed on top of it, more gravel, substantially equal in amount to that already in the cup, is added to each, and the tops are twisted.

Globe Torpedoes[50]

Small cups of manila paper, about ¾ inch in diameter and ⅞ inch deep, are punched out by machine. They are such that two of them may be fitted together to form a box. The requisite amount of powdered potassium chlorate is first introduced into the cups; then, on top of it and without mixing, the requisite

FIGURE 34. Manufacture of Globe Torpedoes. Introducing gravel and closing the paper capsules. (Courtesy National Fireworks Company and the *Boston Globe*.)

amount, already mixed, of the other components of the flash fulminating mixture is added. These other components are antimony sulfide, lampblack, and aluminum. Without disturbing the white and black powders in the bottoms of the cups, workers then fill the cups with clean coarse gravel and put other cups down upon them to form closed ¾ by ¾ inch cylindrical boxes. The little packages are put into a heated barrel, rotating at an angle with the horizontal, and are tumbled together with a solution of water-glass. The solution softens the paper (but later

[50] U. S. Pats. 1,199,775, 1,467,755, 1,783,999.

hardens it), and the packages assume a spherical shape. Small discs of colored paper (punchings) are added a few at a time until the globes are completely covered with them and have lost all tendency to stick together. They are then emptied out of the

FIGURE 35. Manufacture of Globe Torpedoes. Removing the moist spheres from the tumbling barrel. (Courtesy National Fireworks Company and the *Boston Globe*.)

tumbler, dried in steam-heated ovens, and packed in wood shavings for storage and shipment.

Railway Torpedoes

A railway torpedo consists of a flat tin box, of about an ounce capacity, filled with a fulminating composition and having a strip of lead, soldered to it, which may be bent in order to hold it in place upon the railroad track. It explodes when the first wheel of the locomotive strikes it, and produces a signal which is audible to the engineer above the noise of the train. Railway torpedoes were formerly filled with compositions containing chlorate and red phosphorus, similar to those which are used in toy caps; but these mixtures are dangerous and much more sensitive than

FIGURE 36. Packing Globe Torpedoes in Wood Shavings. (Courtesy National Fireworks Company and the *Boston Globe*.)

is necessary. At present, safer perchlorate mixtures without phosphorus are used. The following compositions (Allen F. Clark) can be mixed dry and yield railway torpedoes which will not explode from ordinary shock or an accidental fall.

Potassium perchlorate	6	12
Antimony sulfide	5	9
Sulfur	1	3

English Crackers or Grasshoppers

These devices are old; they were described by John Bate and by Hanzelet Lorrain. English crackers are represented in the lower right-corner of Figure 23, reproduced from Lorrain's book of 1630. They are used in bombshells and, as Lorrain used them, in rockets, where they jump about in the air producing a series of flashes and explosions. Children shoot them on the ground like firecrackers where their movements suggest the behavior of grasshoppers.

English crackers are commonly loaded with granulated gunpowder, tamped into paper pipes like those from which pinwheels are made. The loaded pipes are softened by moisture in the same way, passed between rollers to make them flatter, folded in frames, and, for the best results, tied each time they are folded and then tied over the whole bundle. They are generally supplied with black match for lighting. They produce as many explosions as there are ligatures.

Chinese Firecrackers

Firecrackers have long been used in China for a variety of ceremonial purposes. The houseboat dweller greets the morning by setting off a bunch of firecrackers, for safety's sake in an iron kettle with a cover over it, to keep all devils away from him during the day. For their own use the Chinese insist upon firecrackers made entirely of red paper, which leave nothing but red fragments, for red is a color particularly offensive to the devils. Firecrackers for export, however, are commonly made from tubes of cheap, coarse, brown paper enclosed in colored wrappers. Thirty years ago a considerable variety of Chinese firecrackers was imported into this country. There were "Mandarin crackers," made entirely from red paper and tied at the ends with silk thread; cheaper crackers plugged at the ends with clay (and these never exploded as satisfactorily); "lady crackers," less than an inch long, tied, and no thicker than a match stem; and "cannon crackers," tied with string, 6, 8, and 12 inches long, made of brown paper with brilliant red wrappers. All these were loaded with explosive mixtures of the general nature of black powder, were equipped with fuses of tissue paper twisted around black powder, and were sold, as Chinese firecrackers are now sold, in bunches with their fuses braided together. The composition 4 parts potassium nitrate, 1 of charcoal, and 1 of sulfur has been reported in Chinese firecrackers; more recently mixtures containing both potassium nitrate and a small amount of potassium chlorate have been used; and at present, when the importation of firecrackers over 1¾ inches in length is practically[51] prohibited, flash powders containing aluminum and potassium

FIGURE 37. Chinese Firecrackers. Tying the Tubes into Bundles. (Courtesy Wallace Clark.)

chlorate are commonly used, for they give a sharper explosion than black powder.

The Chinese firecracker industry formerly centered in Canton but, since the Japanese occupation, has moved elsewhere, largely to French Indo-China and Macao in Portuguese territory. Its processes require great skill and manual dexterity, and have long been a secret and a mystery to Europeans. So far as we know, they had not been described in English print until Weingart's book[52] published an account of the manufacture of clay-plugged crackers based upon information received from the manager of a fireworks company at Hong Kong. His account is illustrated with three pen sketches, two of them of workmen

[51] Firecrackers not exceeding 1¾ inches in length and 5/16 inch in diameter carry a duty of 8 cents per pound. For longer crackers the duty is 25 cents per pound, which practically prohibits their importation.

[52] Op. cit., pp. 166–170.

carrying out manual operations and a third which shows some of the tools and instruments. The brief account which follows is

FIGURE 38. Crimping the Back Ends of the Tubes. (Courtesy Wallace Clark.)

based upon conversations with Wallace Clark of Chicago and upon still and moving pictures which he took at a large factory in French Indo-China in January, 1939.

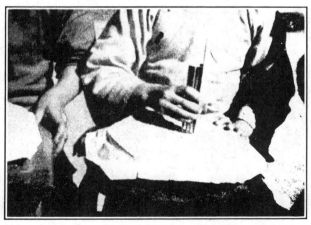

FIGURE 39. Punching Holes for Loading. (Courtesy Wallace Clark.)

The tubes for the firecrackers are rolled and cut to length in outlying villages, and are brought to the factory for loading.

They are tied in hexagonal bundles, Figure 37, each containing 1006 tubes. Since the twine tied tightly around the bundle

Figure 40. Loading. Filling the Tubes with Powder. (Courtesy Wallace Clark.)

crushes the 6 tubes at the corners of the hexagon, and since these are discarded, each bundle contains tubes for 1000 finished crackers.

Figure 41. Fusing and Crimping. (Courtesy Wallace Clark.)

The back ends of the crackers are then crimped; Figure 38. A bamboo stick is placed against the end of the tube and is struck a sharp blow; this forces some of the paper down into the tube and closes it effectively. The operation, like all the other

FIGURE 42. Making the Fuse. (Courtesy Wallace Clark.)

operations in the manufacture of the crackers, is carried out very rapidly.

FIGURE 43. Making the Crackers into Bunches by Braiding Fuses Together. (Courtesy Wallace Clark.)

A sheet of paper is then pasted over the other side of the hexagonal bundle of tubes, closing the ends which are later to carry the fuses. When this is dry, holes corresponding to the tubes are punched in the paper. The operation is carried out by young girls who punch the holes four at a time by means of four bamboo sticks held in one hand while they hold the bundle of tubes steady with the other; Figure 39. The edges of the paper are then bent slightly upward, giving it the form of a shallow saucer with 1000 holes in its bottom. The powder for charging the crackers is then introduced into this saucer, and the whole is

FIGURE 44. Wrapping the Bunches. (Courtesy Wallace Clark.)

shaken gently until all the tubes are full; Figure 40. Then, by a deft movement of the worker's hands and wrists, the excess powder in the saucer, and a portion of the powder in each of the tubes, is emptied out quickly, each of the tubes being left partly full of powder with enough empty space at the top for the fuse and the crimp. This operation, of all those in the manufacture, is considered to be the one which requires the greatest skill. Day after day the average consumption of powder per 1000 or per 100,000 crackers is remarkably constant.

The paper is then torn off from the hexagonal bundle, and the fuses, cut to length, are put in place by one workman while another with a pointed bamboo stick rapidly crimps the paper around them; Figure 41. The fuse is made from narrow strips of tissue paper about 2 feet long. While one end of the strip is clamped to the bench and the other is held in the hand, the strip is shaped by a motion of the worker's other hand into the form of a trough which is then filled with a narrow train of powder, and, by another motion of the hand, the fingers being moistened, is twisted into the finished fuse; Figure 42. This is set aside to dry and is cut into lengths for use in the crackers. The fuses of the finished crackers are braided or pleated together, Figure 43, making the crackers into bunches, and the bunches are wrapped and labeled, Figure 44.

Flash Cracker Composition

Chinese firecrackers and American machine-made salutes are loaded with compositions which contain powdered aluminum and potassium chlorate or perchlorate. They produce a bright flash and an extremely sharp report when they explode. The compositions listed below are typical. The first four in the table have been used in Chinese firecrackers. For the last four the author is indebted to Allen F. Clark.

Potassium perchlorate	..	..	..	..	6	1	7	..
Potassium chlorate	2	3	..	7	..	..	..	..
Potassium nitrate	..	..	5	..	..	..	..	..
Barium nitrate	..	..	..	..	3	1	..	4
Aluminum (fine powder)	1	4	2	1	5	1	5	2
Sulfur	1	3	3	2	..	1	..	1
Antimony sulfide	..	..	..	1	..	..	..	..

The compositions which contain barium nitrate produce a green flash, the others a white one. All of them burn with great rapidity in the open. It is debatable whether the phenomenon of the burning is not really an explosion, or would be one if the material were not allowed to scatter while being burned. With the exception of the third and the last, they are all fulminating explosives when confined. All the mixtures which contain sulfur along with chlorate or perchlorate can be exploded on an iron anvil by a moderately strong blow with an iron hammer.

Sparklers

Snowball sparklers (Allen F. Clark) are made from:

Potassium nitrate	64
Barium nitrate	30
Sulfur	16
Charcoal dust	16
Antimony sulfide	16
Fine aluminum powder	9
Dextrin	16

The ingredients are all powdered to pass a 200-210 mesh sieve. The dry materials are mixed thoroughly and sifted, then moistened little by little with water with thorough mixing until the mixture attains the consistency of heavy molasses. Iron wires (20 gauge) of convenient length are dipped in the mixture and are hung up to dry for 24 hours. These are dipped a second time for size, and allowed to dry for another 24 hours. The sparklers burn with a bright white light and throw out "soft sparks" from the charcoal and occasionally scintillating sparks from the burning of the iron wire.

Other mixtures which produce similar effects are as follows:

Potassium nitrate	64	..
Potassium perchlorate	3	16
Barium nitrate	..	6
Sulfur	18	4
Lampblack	5	..
Red gum	4	4
Fine aluminum powder	6	6
Coarse aluminum powder	..	4

The first of these burns with a lilac-colored flame as contrasted with the flame of the second which appears white. These compositions are applied by adding the intimately mixed dry ingredients to a liquid known as "black wax," procured by melting together 3 pounds of rosin and 1 gallon of liquid roofing-paper tar. The iron wires are dipped two or three times in the resulting slurry, and allowed to dry between dips.

The use of iron and steel filings in the compositions produces a more brilliant display of scintillating sparks. The following formulas are typical. Water is used for applying the compositions. The iron and steel filings which are used in these com-

Barium nitrate	48	48	..
Potassium perchlorate	..	..	6
Fine aluminum powder	7	7	1
Fine iron filings	24	18	..
Fine steel filings	..	9	12
Manganese dioxide	2	1	..
Dextrin	12	12	2
Glucose	..	1	..

positions are coated, before the mixing, with paraffin or linseed oil to protect them from rusting.

Wire Dips and Colored Fire Sticks

These devices are made in the same way as sparklers, by dipping wires or twisted narrow strips of iron or thin sticks of wood, and generally burn with a tranquil flame except for the sparks that come from the burning of the iron wire or strip. Several typical compositions are listed. Alcohol is used for applying the

	Red			Green	White
Potassium chlorate........	..	2	3	..	..
Potassium perchlorate....	..	..	..	..	10
Strontium nitrate........	5	6	16	..	..
Barium chlorate..........	..	..	..	16	..
Fine aluminum powder....	..	..	..	..	7
Coarse aluminum powder .	..	6	..	24	..
Shellac..................	1	1	..	3	..
Red gum.................	..	..	4	..	..
Dextrin.................	..	..	3	..	3

compositions which contain shellac; water, for applying the others which contain dextrin.

Pharaoh's Serpents[53]

Wöhler in 1821 first reported the remarkable property of mercurous thiocyanate that it swells up when it is heated, "winding out from itself at the same time worm-like processes, to many times its former bulk, of a very light material of the color of graphite, with the evolution of carbon disulfide, nitrogen, and mercury." Mercuric thiocyanate, which gives better snakes than the mercurous compound, came early into use for this purpose in pyrotechnic toys. When a heap or pellet of either of these compounds is set on fire, it burns with an inconspicuous blue flame, producing sulfur dioxide and mercury vapor. The resulting pale brown or pale gray snake, if broken, is found to be much darker in the interior, and evidently consists of paracyanogen and mercuric sulfide, the mercury having been burned and vaporized from the outer layer.

Mercuric thiocyanate is prepared by adding a solution of

[53] Cf. Davis, article entitled "Pyrotechnic Snakes," in *J. Chem. Education*, **17**, 268-270 (1940).

potassium, sodium, or ammonium thiocyanate to a solution of mercuric nitrate, a ferric chloride or ferric alum indicator being used to indicate by a red color when enough of the former solution has been added. This is necessary since mercuric thiocyanate is soluble in an excess of either of the reagents by the interaction of which it is produced. The precipitate is collected, washed, dried, powdered, moistened with gum-arabic water in which a little potassium nitrate is dissolved, and made into small pellets by means of a device like a star board or by a pelleting machine. The small pellets are known as *Pharaoh's serpent's eggs*.

Snakes-in-the-grass, volcano snakes, etc., depend upon the use of ammonium dichromate. If this material in the form of a powder is made into a conical heap, and a flame applied to the top of it, a visible but not violent "combustion" proceeds through the mass, which "boils up" to form a large volume of green material resembling tea leaves.

$$(NH_4)_2Cr_2O_7 \longrightarrow N_2 + 4H_2O + Cr_2O_3$$

In practice, more flame is desired than ammonium dichromate alone will give. Weingart[54] recommends a mixture of 2 parts of ammonium dichromate with 1 of potassium nitrate and 1 of dextrin. Tinfoil cones are made from circles of tinfoil shaped on a former, and are introduced by means of the former into conical cavities in a block of wood; they are then about half filled with the powdered mixture, a Pharaoh's serpent's egg is pressed in, and the edges of the tinfoil are turned down upon it to form the base of the cone.

While the fumes from burning mercuric thiocyanate are offensive because of their sulfur dioxide, the small amount of mercury vapor which they contain probably presents no serious danger. The possibility, however, that children may swallow the pellets, with fatal consequences, is a real hazard. For this reason, the sale of mercury snakes has been forbidden by law in many states, and black non-mercury snakes, which are essentially non-poisonous, have come into general use.

Black Non-Mercury Snakes[55]

These are used in the form of *barrel snakes, hat snakes* (black

---

[54] *Op. cit.*, p. 152.
[55] Davis, *loc. cit.*

pellets affixed to black discs of pasteboard to form what look like miniature broad-brimmed black hats), *colored fire snakes,* etc. The best which we have seen are prepared from naphthol pitch by a process described by Weingart.[56] The naphthol pitch is a by-product in the manufacture of β-naphthol. The method of "nitration by kneading" is so unusual that it appears worth while to describe the process in detail.

Preparation of Black Non-Mercury Snakes. Thirty grams of powdered naphthol pitch is mixed intimately with 6 grams of linseed oil, and the material is chilled in a 200-cc. Pyrex beaker surrounded by cracked ice. Twenty-one cubic centimeters of fuming nitric acid (*d.* 1.50) is added in small portions, one drop at a time at first, and the material is stirred over, kneaded, and kept thoroughly mixed at all times by means of a porcelain spatula. The addition of each drop of acid, especially at the beginning of the process, causes an abundance of red fumes, considerable heating, and some spattering. It is recommended that goggles and rubber gloves be worn, and that the operation be carried out in an efficient hood. The heat of the reaction causes the material to assume a plastic condition, and the rate of addition of the acid ought to be so regulated that this condition is maintained. After all the acid has been added, the dark brown, doughlike mass becomes friable on cooling. It is broken up under water with the spatula, washed thoroughly, and allowed to dry in the air. The product is ground up in a porcelain mortar with 10.5 grams of picric acid, made into a moist meal with gum-arabic water, pelleted, and dried. A pellet ½ inch long and ⅜ inch in diameter gives a snake about 4 feet long, smooth-skinned and glossy, with a luster like that of coke, elastic, and of spongy texture within.

The oxidized linseed oil produced during the nitration appears to play an important part in the formation of the snakes. If naphthol pitch alone is nitrated, ground up with picric acid, and made into pellets after moistening with linseed oil, the pellets when fresh do not yield snakes, but do give snakes after they have been kept for several months, during which time the linseed oil oxidizes and hardens. Weingart states in a letter that he has obtained satisfactory results by using, instead of naphthol pitch, the material procured by melting together 60 parts of Syrian asphalt and 40 of roofing pitch. Worked up in the regular way this "yielded fairly good snakes which were improved by rubbing

[56] *Op. cit.,* p. 153.

the finished product up with a little stearine before forming into pellets." The present writer has found that the substitution of β-naphthol for naphthol pitch yields fairly good snakes which, however, are not so long and not so shiny, and are blacker and covered with wartlike protuberances.

Smokes

Smoke shells and rockets are used to produce *smoke clouds* for military signaling and, in daylight fireworks, for ornamental effects. The shell case or rocket head is filled with a fine powder of the desired color, which powdered material need not necessarily be one which will tolerate heat, and this is dispersed in the form of a colored cloud by the explosion of a small bag of gunpowder placed as near to its center as may be. Artificial vermilion (red), ultramarine (blue), Paris green, chrome yellow, chalk, and ivory black are among the materials which have been used, but almost any material which has a bright color when powdered and which does not cake together may be employed.

Colored smokes strictly so called are produced by the burning, in *smoke pots* or smoke cases, of pyrotechnic compositions which contain colored substances capable of being sublimed without an undue amount of decomposition. The substances are volatilized by the heat of the burning compositions to form colored vapors which quickly condense to form clouds of finely divided colored dust. Colored smokes are used for military signaling, and recently have found use in colored moving pictures. Red smokes, for example, were used in the "Wizard of Oz." Colored smoke compositions are commonly rammed lightly, not packed firmly, in cases, say 1 inch in internal diameter and 4 inches long, both ends of which are closed with plugs of clay or wood. Holes, ¼ inch in diameter, are bored through the case at intervals on a spiral line around it; the topmost hole penetrates well into the composition and is filled with starting fire material into which a piece of black match, held in place by meal powder paste, is inserted. According to Faber,[57] the following-listed compositions were used in American airplane smoke-signal grenades during the first World War.

[57] *Op. cit.*, Vol. 1, p. 219.

	Red	Yellow	Green	Blue
Potassium chlorate ...	1	33	33	7
Lactose............	1	24	26	5
Paranitraniline red....	3	..	..	..
Auramine...........	..	34	15	..
Chrysoidine..........	..	9	..	..
Indigo.............	..	..	26	8

The following (Allen F. Clark) are illustrative of the perchlorate colored smoke compositions which have come into use more recently.

	Red	Green	Blue
Potassium perchlorate.........	5	6	5
Antimony sulfide.............	4	5	4
Rhodamine red.............	10	..	..
Malachite green..............	..	10	..
Methylene blue..............	..	..	10
Gum arabic.................	1	1	1

Many other dyestuffs may be used. Paranitraniline Yellow gives a canary yellow smoke, and Flaming Red B gives a crimson-colored smoke by comparison with which the smoke from Paranitraniline Red appears to be scarlet. None of the colored smoke compositions are adapted to indoor use. All the smokes are unpleasant and unwholesome.

White smoke is produced by burning a mixture of potassium chlorate 3 parts, lactose 1, and finely powdered ammonium chloride 1. The smoke, which consists of finely divided ammonium chloride, is not poisonous, and has found some use in connection with the study of problems in ventilation.

For use in trench warfare, for the purpose of obscuring the situation from the sight of the enemy, a very satisfactory dense *white* or *gray smoke* is procured by burning a mixture of zinc dust and hexachloroethane. The mixture requires a strong starting fire. The smoke consists largely of finely divided zinc chloride. For a grayer smoke naphthalene or anthracene is added to the mixture. Torches for *black smoke* have also been used, charged with a mixture of potassium nitrate and sulfur with rosin or pitch and generally with such additional ingredients as sand, powdered chalk, or glue to modify the rate of their burning.

When shells are loaded with certain high explosives which produce no smoke (such as amatol), *smoke boxes* are generally inserted in the charges in order that the artilleryman, by seeing the smoke, may be able to judge the position and success of his fire. These are cylindrical pasteboard boxes containing a mixture of arsenious oxide and red phosphorus, usually with a small amount of stearine or paraffin.

CHAPTER IV

AROMATIC NITRO COMPOUNDS

Aromatic nitro compounds are generally stable but are frequently reactive, especially if they contain groups other than nitro groups in the *meta* position with respect to one another. As a class they constitute the most important of the military high explosives. They are also used as components of smokeless powder, in compound detonators, and in primer compositions. Liquid nitro compounds, and the mixtures which are produced as by-products from the manufacture of pure nitro compounds for military purposes, are used in non-freezing dynamite and other commercial explosives. The polynitro compounds are solvents for nitrocellulose.

The nitro compounds are poisonous. Nitrobenzene, known also as "oil of mirbane," is absorbed through the skin and by breathing its vapors, and has been reported to cause death by the careless wearing of clothing upon which it had been spilled. The less volatile polynitro compounds, like trinitrotoluene, are absorbed through the skin when handled, and may cause injury by the inhalation of their dust or of their vapors when they are melted. Minor TNT sickness may manifest itself by cyanosis, dermatitis, nose bleeding, constipation, and giddiness; the severer form, by toxic jaundice and aplastic anemia.[1] One of the nitro groups is reduced in the body, and dinitrohydroxylaminotoluene may be detected in the urine. Trinitrobenzene is more poisonous than trinitrotoluene, which, in turn, is more poisonous than trinitroxylene, alkyl groups in this series having the same effect as in the

---

[1] J. W. Schereschewsky, "Trinitrotoluol, Practical Points in Its Safe Handling," U. S. Pub. Health Service, Reprint 434 from *Pub. Health Repts.*, Nov. 16, 1917, pp. 1919-1926. C. Voegtlin, C. W. Hooper, and J. M. Johnson, "Trinitrotoluene Poisoning—Its Nature, Diagnosis and Prevention," U. S. Pub. Health Service, *Hyg. Lab. Bull. 126*, 1920. A. Hamilton, "Trinitrotoluene as an Industrial Poison," *J. Ind. Hyg.*, 3, 102-119 (1921). A. L. Leigh Silver, "Treatment and Prevention of Industrial Diseases in Filling Factories," *J. Roy. Army Med. Corps*, July, 1938, pp. 87-96.

phenols, cresol, xylenol, etc., where they reduce the toxicity of the substances but increase their antiseptic strength.

In the manufacture of explosives the nitro groups are always introduced by the direct action of nitric acid on the aromatic substances. The simple reaction involves the production of water and is promoted by the presence of sulfuric acid which thus functions as a dehydrating agent. We shall later see cases in which sulfuric acid is used as a means of hindering the introduction of nitro groups. In consequence of the reaction, the nitrogen

atom of the nitro group becomes attached to the carbon atom of the aromatic nucleus. Nitro groups attached to the nucleus, unless *ortho* and *para* to other nitro groups, are not affected by sulfuric acid as are nitro groups attached to oxygen (in nitric esters) and to nitrogen (in nitroamines), or, ordinarily, by hydrolytic agents as are nitro groups attached to oxygen. Nitric acid is a nitrating agent both at low and at elevated temperatures; its vigor in this respect depends upon the concentration. But it is an oxidizing agent even in fairly dilute solution, and becomes more vigorous if the temperature is raised. Further, it decomposes when heated to produce nitrous acid, which is also a powerful oxidizing agent and may reduce the yield of the desired product. Nitrous acid present in the nitrating acid may also result in the formation of nitrophenols from aromatic amines. Aromatic nitro compounds, such as TNT and picric acid, on refluxing for some hours with nitric acid (*d.* 1.42) and then distilling the mixture, yield appreciable quantities of tetranitromethane, formed by the rupture of the ring and the nitration of the individual carbon atoms. The nitro group "strengthens" the ring against attack by acid oxidizing agents, but makes it more accessible to attack by alkaline ones. The polynitro compounds are destroyed rapidly by warm alkaline permanganate yielding oxalic acid. They combine with aniline, naphthylamine, etc., to form brightly colored molecular compounds. All aromatic nitro compounds give colors, yellow, orange, red, even purple, with alkaline reagents.

The position which the nitro group takes on entering the aromatic nucleus and the ease with which the substitution is accomplished depend upon the group or groups already present on the nucleus. We are accustomed to speak of the orienting or directing effect of the groups already present and of their influence in promoting or inhibiting further substitution. The two simple rules which summarize these effects have important implications and wide applications in the chemistry of aromatic substances.

Effect of Groups on Further Substitution

1. ORIENTING EFFECT. *The Modified Rule of Crum Brown and Gibson.* If the atom attached to the aromatic nucleus is attached to some other atom by an unsaturated linkage (i.e., by any bond which we commonly write as double or triple), then the next entering group takes the *meta* position; otherwise it takes the *ortho* and *para* positions.

The rule relieves us of the necessity for remembering which groups orient *meta* and which *ortho-para*; we may write them down on demand, thus: the —NO_2, —NO, —CO—, —COOH, —CHO, —SO_2—OH, —CN groups orient *meta*; and the —NH_2, —NHR, —NR_2, —OH, —OR, —CH_3, —CH_2–CH_3, —Cl, —Br groups orient *ortho-para*. It is necessary, however, to take note of three or four exceptions, only one of which is important in the chemistry of explosives, namely, that the azo group, —N=N—, orients *ortho-para*; the trichloromethyl group, —CCl_3, *meta*; that such conjugate systems as occur in cinnamic acid, —CH=CH—CO—, orient *ortho-para*; and further that a large excess of strong sulfuric acid reverses to a greater or less extent the normal orienting effects of the methoxy and ethoxy groups, of the amino group wholly and of the monosubstituted and disubstituted amino groups in part.

In all discussions of the application of the rule we make reference to the principal products of the reaction; substitution occurs for the most part in accordance with the rule, or with the exceptions, and small amounts of other materials are usually formed as by-products. In the mononitration of toluene, for example, about 96 per cent of the product is a mixture of *o*- and *p*-nitrotoluene, and about 4 per cent is the *m*-compound. Under the influence of *ortho-para* orienting groups, substitution occurs in the two positions without much preference for either one, but it appears to be the case that, when nitro groups are introduced, low temperatures favor the formation of *p*-compounds. The effect of temperature on sulfonations appears to be exactly the opposite.

2. EASE OF SUBSTITUTION. *Ortho-para* orienting groups promote substitution; *meta* orienting groups hinder it and make it more difficult. The rule may be stated otherwise: that substitution under the influence of *ortho-para* orienting groups occurs under less vigorous conditions of temperature, concentration of reagents, etc., than it does with the unsubstituted aromatic hydrocarbon itself; under the influence of *meta* orienting groups more vigorous conditions than with the unsubstituted hydrocarbon are necessary for its successful accomplishment. The rule may also be stated that *ortho-para* substitution is easier than *meta*. In this last form it fails to make comparison with substitution in the simple hydrocarbon, but does point clearly to the implication, or corollary, that the orienting effect of an *ortho-para* orienting group dominates over that of one which orients *meta*. To the rule in any of these forms, we must add that, when more than one group is already present on the nucleus, the effect of the groups is additive.

Toluene nitrates more easily than benzene; aniline and phenol more easily still. Higher temperature and stronger acid are needed for the introduction of a second nitro group into benzene than for the introduction of the first, for the second is introduced under the influence of the *meta*-orienting first nitro group which tends to make further substitution more difficult. The inhibitory effect of two nitro groups is so great that the nitration of dinitrobenzene to the trinitro compound is extremely difficult. It is more difficult to nitrate benzoic acid than to nitrate nitrobenzene. The common experience of organic chemists indicates that the order of the groups in promoting substitution is about as follows:

$$-OH > -NH_2 > -CH_3 > -Cl > -H > -NO_2 > -SO_2(OH) > -COOH$$

Any one of these groups makes substitution easier than the groups which are printed to the right of it.

Xylene nitrates more easily than toluene. Two methyl groups promote substitution more than one methyl group does, and this appears to be true whether or not the methyl groups agree among themselves in respect to the positions which they activate. Although a nitro group may be said to "activate" a particular position, inasmuch as it points to that position as the one in which substitution will next occur, it nevertheless makes substitution more difficult in that position, as well as in all other positions on the nucleus. The nitroanilines are more difficult to nitrate than aniline because of their inhibiting nitro group, and more easy to nitrate than nitrobenzene because of their promoting amino group. In *o*- and *p*-nitroaniline the amino and nitro groups agree in activating the same positions, and both substances yield 2,4,6-trinitroaniline when they are nitrated. In *m*-nitroaniline, the nitro group "activates" the 5-position, while the amino group activates the 2-, 4-, and 6-positions. Nitration takes place under the influence of the *ortho-para*-orienting amino group, and 2,3,4,6-tetranitroaniline results.

Utilization of Coal Tar

The principal source of aromatic compounds is coal tar, produced as a by-product in the manufacture of coke. Gas tar, of which much smaller quantities are produced, also contains these same materials. Aromatic hydrocarbons occur in nature in Borneo and other petroleums, and they may be prepared artificially by stripping hydrogen atoms from the cycloparaffins which occur in Caucasus petroleum and elsewhere. They are also produced from paraffin hydrocarbons by certain processes of cracking, and it is to be expected that in the future aromatic compounds will be produced in increasing quantity from petroleum which does not contain them in its natural state.

Coal yields about 6 per cent of its weight of tar. One ton of tar on distillation gives:

Light Oil—yielding about 32 lb. of benzene, 5 lb. of toluene, and 0.6 lb. of xylene.

Middle Oil—yielding about 40 lb. of phenol and cresols, and 80-120 lb. of naphthalene.

Heavy Oil—yielding impure cresols and other phenols.

Green Oil—yielding 10-40 lb. of anthracene.

Pitch—1000-1200 lb.

Naphthalene is the most abundant pure hydrocarbon obtained from coal tar. It takes on three nitro groups readily, and four under vigorous conditions, but ordinarily yields no product which is suitable by itself for use as an explosive. Nitrated naphthalenes, however, have been used in smokeless powder and, when mixed with ammonium nitrate and other materials, in high explosives for shells and for blasting.

The phenol-cresol fraction of coal tar yields phenol on distillation, which is convertible to picric acid, and the cresols, of which *m*-cresol is the only one which yields a trinitro derivative directly. Moreover, synthetic phenol from benzene, through chlorobenzene by the Dow process, is purer and probably cheaper in times of stress.

Of the hydrocarbons toluene is the only one which nitrates sufficiently easily and yields a product which has the proper physical and explosive properties. Trinitrotoluene is the most widely used of the pure aromatic nitro compounds. It melts at such temperature that it can be loaded by pouring. It is easily and surely detonated, and is insensitive to shock, though not insensitive enough to penetrate armor-plate without exploding until afterwards. It is powerful and brisant, but less so than trinitrobenzene which would offer certain advantages if it could be procured in sufficient quantity.

Of the xylenes, the *meta* compound yields a trinitro derivative more readily than toluene does, but trinitro-*m*-xylene (TNX) melts somewhat higher than is desirable and is not quite powerful enough when used alone. It has been used in shells in mixtures with TNT and with ammonium nitrate. The other xylenes yield only dinitro derivatives by direct nitration. A mixture of *o*- and *p*-xylene may be converted into an explosive—an oily mixture of a large number of isomers, which has been used in the composition of non-freezing dynamites—by chlorinating at an elevated temperature in the presence of a catalyst (whereby chlorine is substituted both in the side chain and on the nucleus), then nitrating, then hydrolyzing (whereby both chlorines are replaced by hydroxyl groups, the nuclear chlorine being activated by the nitro groups), and finally nitrating once more.

In each step several isomers are formed—only one of the possibilities in each case is indicated above—and the *ortho* and *para* compounds both go through similar series of reactions. The product is too sensitive and in the wrong physical state (liquid) for use as a military explosive. In short, for the manufacture of

FIGURE 45. Marius Marqueyrol, Inspecteur-Général des Poudres, France, 1919. Author of many researches on aromatic nitro compounds, nitrocellulose, smokeless powder, stabilizers and stability, chlorate explosives, etc.—published for the most part in the *Mémorial des poudres* and in the *Bulletin de la société chimique de France*.

military explosives toluene is the most valuable of the materials which occur in coal tar.

In time of war the industries of a country strive to produce as much toluene as possible. The effort results in the production also of increased quantities of other aromatic hydrocarbons, particularly of benzene, and these become cheaper and more abundant. Every effort is made to utilize them profitably for military purposes. As far as benzene is concerned, the problem has been solved through chlorobenzene, which yields aniline and phenol by the Dow process, and hence picric acid, and which gives dinitrochlorobenzene on nitration which is readily convertible, as will be described later, into picric acid and tetryl and several other explosives that are quite as necessary as TNT for military purposes.

Effects of Substituents on Explosive Strength

Bomb experiments show that trinitrobenzene is the most powerful explosive among the nitrated aromatic hydrocarbons. One methyl group, as in TNT, reduces its strength; two, as in TNX, reduce it further; and three, as in trinitromesitylene, still further yet. The amino and the hydroxyl groups have less effect than the methyl group; indeed, two hydroxyl groups have less effect than one methyl—and trinitroresorcinol is a stronger explosive than TNT, though weaker than TNB. TNT is stronger than trinitrocresol, which differs from it in having an hydroxyl group. The figures given below were determined by exploding the materials, loaded at the density indicated, in a small bomb, and measuring the pressure by means of a piston and obturator. *Density of loading* is grams of explosive per cubic centimeter of bomb capacity.

	Pressure: Kilograms per square centimeter		
Density of loading:	0.20	0.25	0.30
Trinitrobenzene...............	2205	3050	4105
Trinitrotoluene...............	1840	2625	3675
Trinitro-*m*-xylene	1635	2340	2980
Trinitromesitylene............	1470	2200	2780
Trinitrophenol (picric acid)......	2150	3055	3865
Trinitroresorcinol (styphinic acid)	2080	2840	....
Trinitroaniline (picramide)......	2080	2885	3940
Trinitro-*m*-cresol........... ...	1760	2480	3360
Trinitronaphthalene...........	....	2045	2670

Similar inferences may be made from the results of lead block tests. Fifteen grams of the explosives produced the expansions indicated below, each figure representing the average from twenty or more experiments:

Trinitrobenzene......... 480 cc.
Trinitrotoluene.......... 452 cc.
Picric acid............. 470 cc.
Trinitrocresol........... 384 cc.
Trinitronaphthalene..... 166 cc.

Mono- and Di-Nitrobenzene

Nitrobenzene is a pale yellow liquid, b.p. 208.0°, which is poisonous and has an almondlike odor closely resembling that of benzaldehyde (which is not poisonous). It is used as a component of certain Sprengel explosives and as a raw material for the preparation of aniline and of intermediates for the manufacture of dyestuffs and medicinals. Its preparation, familiar to every student of organic chemistry, is described here in order that the conditions for the substitution of one nitro group in benzene may serve us more conveniently as a standard for judging the relative ease and difficulty of the nitration of other substances.

Preparation of Nitrobenzene. One hundred and fifty grams of concentrated sulfuric acid (*d.* 1.84) and 100 grams of nitric acid (*d.* 1.42) are mixed in a 500-cc. flask and cooled to room temperature, and 51 grams of benzene is added in small portions at a time with frequent shaking. Shaking at this point is especially necessary lest the reaction suddenly become violent. If the temperature of the mixture rises above 50-60°, the addition of the benzene is interrupted and the mixture is cooled at the tap. After all the benzene has been added, an air condenser is attached to the flask and the material is heated in the water bath for an hour at 60° (thermometer in the water). After cooling, the nitrobenzene (upper layer) is separated from the spent acid, washed once with water (the nitrobenzene is now the lower layer), then several times with dilute sodium carbonate solution until it is free from acid, then once more with water, dried with calcium chloride, and distilled (not quite to dryness). The portion boiling at 206-208° is taken as nitrobenzene.

m-Dinitrobenzene, in accordance with the rule of Crum Brown and Gibson, is the only product which results ordinarily from the nitration of nitrobenzene. Small amounts of the *ortho* and *para* compounds have been procured, along with the *meta*, from the nitration of benzene in the presence of mercuric nitrate.[2] Dinitrobenzene has been used in high explosives for shells in mixtures with more powerful explosives or with ammonium nitrate. Its use

[2] Davis, *J. Am. Chem. Soc.,* **43,** 598 (1921).

as a raw material for the manufacture of tetranitroaniline is now no longer important.

Preparation of Dinitrobenzene. A mixture of 25 grams of concentrated sulfuric acid (*d*. 1.84) and 15 grams of nitric acid (*d*. 1.52) is heated in an open flask in the boiling water bath in the hood, and 10 grams of nitrobenzene is added gradually during the course of half an hour. The mixture is cooled somewhat, and drowned in cold water. The dinitrobenzene separates as a solid. It is crushed with water, washed with water, and recrystallized from alcohol or from nitric acid. Dinitrobenzene crystallizes from nitric acid in beautiful needles which are practically colorless, m.p. 90°.

Trinitrobenzene

1,3,5-Trinitrobenzene (*sym*-trinitrobenzene, TNB) may be prepared only with the greatest difficulty by the nitration of *m*-dinitrobenzene. Hepp[3] first prepared it by this method, and Hepp and Lobry de Bruyn[4] improved the process, treating 60 grams of *m*-dinitrobenzene with a mixture of 1 kilo of fuming sulfuric acid and 500 grams of nitric acid (*d*. 1.52) for 1 day at 100° and for 4 days at 110°. Claus and Becker[5] obtained trinitrobenzene by the action of concentrated nitric acid on trinitrotoluene. Trinitrobenzoic acid is formed first, and this substance in the hot liquid loses carbon dioxide from its carboxyl group.

For commercial production the Griesheim Chem. Fabrik[6] is reported to have used a process in which 1 part of TNT is heated at 150-200° with a mixture of 5 parts of fuming nitric acid and 10 parts of concentrated sulfuric acid. In a process devised by J. Meyer,[7] picryl chloride (2,4,6-trinitrochlorobenzene) is reduced by means of copper powder in hot aqueous alcohol. The reported details are 25 kilos of picryl chloride, 8 kilos of copper

[3] *Ber.*, **9**, 402 (1876); *Ann.*, **215**, 345 (1882).
[4] *Rec. trav. chim.*, **13**, 149 (1894).
[5] *Ber.*, **16**, 1597 (1883).
[6] Ger. Pat. 77,353, 77,559 (1893); 127,325 (1901).
[7] Ger. Pat. 234,726 (1909).

powder, 250 liters of 95 per cent alcohol, and 25 liters of water, refluxed together for 2 hours and filtered hot; the TNB crystallizes out in good yield when the liquid is cooled.

The nitration of m-dinitrobenzene is too expensive of acid and of heat for practical application, and the yields are poor. Toluene and chlorobenzene are nitrated more easily and more economically, and their trinitro compounds are feasible materials for the preparation of TNB. Oxidation with nitrosulfuric acid has obvious disadvantages. The quickest, most convenient, and cheapest method is probably that in which TNT is oxidized by means of chromic acid in sulfuric acid solution.

Preparation of Trinitrobenzene. A mixture of 30 grams of purified TNT and 300 cc. of concentrated sulfuric acid is introduced into a tall beaker, which stands in an empty agateware basin, and the mixture is stirred actively by means of an electric stirrer while powdered sodium dichromate ($Na_2Cr_2O_7 \cdot 2H_2O$) is added in small portions at a time, care being taken that no lumps are formed and that none floats on the surface of the liquid. The temperature of the liquid rises. When it has reached 40°, cold water is poured into the basin and the addition of dichromate is continued, with stirring, until 45 grams has been added, the temperature being kept always between 40° and 50°. The mixture is stirred for 2 hours longer at the same temperature, and is then allowed to cool and to stand over night, in order that the trinitrobenzoic acid may assume a coarser crystalline form and may be filtered off more readily. The strongly acid liquid is filtered through an asbestos filter; the solid material is rinsed with cold water and transferred to a beaker in which it is treated with warm water at 50° sufficient to dissolve all soluble material. The warm solution is filtered, and boiled until no more trinitrobenzene precipitates. The crystals of TNB growing in the hot aqueous liquid often attain a length of several millimeters. When filtered from the cooled liquid and rinsed with water, they are practically pure, almost colorless or greenish yellow leaflets, m.p. 121-122°.

Trinitrobenzene is only moderately soluble in hot alcohol, more readily in acetone, ether, and benzene. Like other polynitro aromatic compounds it forms colored molecular compounds with many aromatic hydrocarbons and organic bases.[8] The compound

[8] Compare Hepp, *Ann.,* **215,** 356 (1882); Sudborough, *J. Chem. Soc.,* **75,** 588 (1889); **79,** 522 (1901); **83,** 1334 (1903); **89,** 583 (1906); **97,** 773 (1910); **99,** 209 (1911); **109,** 1339 (1916); Sachs and Steinert, *Ber.,* **37,** 1745 (1904); Nölting and Sommerhoff, *ibid.,* **39,** 76 (1906); Kremann, *ibid.,* **39,** 1022 (1906); Ciusa, *Gazz. chim. ital.,* **43,** II, 91 (1913); Ciusa and

with aniline is bright red; that with naphthalene, yellow. The compounds with amines are beautifully crystalline substances, procurable by warming the components together in alcohol, and are formed generally in the molecular proportions 1 to 1, although diphenylamine and quinoline form compounds in which two molecules of TNB are combined with one of the base.

Trinitrobenzene gives red colors with ammonia and with aqueous alkalies. On standing in the cold with methyl alcoholic sodium methylate, it yields 3,5-dinitroanisol by a metathetical reaction.[9]

On boiling with alcoholic soda solution it undergoes a partial reduction to form 3,3',5,5'-tetranitroazoxybenzene.[10]

The first product, however, of the reaction of methyl alcoholic caustic alkali on TNB is a red crystalline addition product having the empirical composition $TNB \cdot CH_3ONa \cdot \frac{1}{2}H_2O$, isolated by Lobry de Bruyn and van Leent[11] in 1895. The structure of this substance has been discussed by Victor Meyer,[12] by Angeli,[13] by Meisenheimer,[14] and by Schlenck,[15] and is probably best represented by the formula which Meisenheimer suggested. It is thus

---

Vecchiotti, *Atti accad. Lincei,* **20,** II, 377 (1911); **21,** II, 161 (1912); Sastry, *J. Chem. Soc.,* **109,** 270 (1916); Hammick and Sixsmith, *J. Chem. Soc.,* 972 (1939).

[9] Lobry de Bruyn, *Rec. trav. chim.,* **9,** 208 (1890).

[10] Lobry de Bruyn and van Leent, *ibid.,* **13,** 148 (1894).

[11] *Ibid.,* **14,** 150 (1895).

[12] *Ber.,* **29,** 848 (1896).

[13] *Gazz. chim. ital.,* **27,** II, 366 (1897). Compare also Hantzsch and Kissel, *Ber.,* **32,** 3137 (1899).

[14] *Ann.,* **323,** 214, 241 (1902).

[15] *Ber.,* **47,** 473 (1914).

probably the product of the 1,6-addition of sodium methylate to the conjugate system which runs through the ring and terminates in the oxygen of the nitro group. Busch and Kögel[16] have prepared di- and tri-alcoholates of TNB, and Giua[17] has isolated a compound of the empirical composition TNB·NaOH, to which he ascribed a structure similar to that indicated above. All these compounds when dry are dangerous primary explosives. They are soluble in water, and the solutions after acidification contain red, water-soluble acids which yield sparingly soluble salts with copper and other heavy metals, and the salts are primary explosives. The acids, evidently having the compositions TNB·CH$_3$OH, TNB·H$_2$O, etc., have not been isolated in a state of purity, and are reported to decompose spontaneously in small part into TNB, alcohol, water, etc., and in large part into oxalic acid, nitrous fumes, and colored amorphous materials which have not been identified. All the polynitro aromatic hydrocarbons react similarly with alkali, and the use of alkali in any industrial process for their purification is bad practice and extremely hazardous.

Trinitrobenzene reacts with hydroxylamine in cold alcohol solution, picramide being formed by the direct introduction of an amino group.[18]

Two or three nitro groups on the aromatic nucleus, particularly those in the 2,4-, 2,6-, and 2,4,6-positions, have a strong effect in increasing the chemical activity of the group or atom in the

[16] Ber., 43, 1549 (1910).
[17] Gazz. chim. ital., 45, II, 351 (1915).
[18] Meisenheimer and Patzig, Ber., 39, 2534 (1906).

1-position. Thus, the hydroxyl group of trinitrophenol is acidic, and the substance is called picric acid. A chlorine atom in the same position is like the chlorine of an acid chloride (picryl chloride), an amino group like the amino of an acid amide (trinitroaniline is picramide), and a methoxy like the methoxy of an ester (trinitroanisol has the reactions of methyl picrate). In general the picryl group affects the activity of the atom or group to which it is attached in the same way that the acyl or R—CO— group does. If the picryl group is attached to a carboxyl, the carboxyl will be expected to lose CO_2 readily, as pyruvic acid, CH_3—CO—COOH, does when it is heated with dilute sulfuric acid, and this indeed happens with the trinitrobenzoic acid from which TNB is commonly prepared. TNB itself will be expected to exhibit some of the properties of an aldehyde, of which the aldehydic hydrogen atom is readily oxidized to an acidic hydroxyl group, and it is in fact oxidized to picric acid by the action of potassium ferricyanide in mildly alkaline solution.[19] We shall see many examples of the same principle throughout the chemistry of the explosive aromatic nitro compounds.

Trinitrobenzene is less sensitive to impact than TNT, more powerful, and more brisant. The detonation of a shell or bomb, loaded with TNB, in the neighborhood of buildings or other construction which it is desired to destroy, creates a more damaging explosive wave than an explosion of TNT, and is more likely to cause the collapse of walls, etc., which the shell or bomb has failed to hit. Drop tests carried out with a 5-kilogram weight falling upon several decigrams of each of the various explosives contained in a small cup of iron (0.2 mm. thick), covered with a small iron disc of the same thickness, gave the following figures for the distances through which the weight must fall to cause explosion in 50 per cent of the trials.

	CENTIMETERS
Trinitrobenzene	150
Trinitrotoluene	110
Hexanitrodiphenylamine ammonium salt	75
Picric acid	65
Tetryl	50
Hexanitrodiphenylamine	45

[19] Hepp, *Ann.*, **215**, 344 (1882).

According to Dautriche,[20] the density of compressed pellets of TNB is as follows:

PRESSURE: KILOS PER SQUARE CENTIMETER	DENSITY
275	1.343
685	1.523
1375	1.620
2060	1.641
2750	1.654
3435	1.662

The greatest velocity of detonation for TNB which Dautriche found, namely 7347 meters per second, occurred when a column of 10 pellets, 20 mm. in diameter and weighing 8 grams each, density 1.641 or 1.662, was exploded in a paper cartridge by means of an initiator of 0.5 gram of mercury fulminate and 80 grams of dynamite. The greatest which he found for TNT was 7140 meters per second, 10 similar pellets, density 1.60, in a paper cartridge exploded by means of a primer of 0.5 gram of fulminate and 25 grams of dynamite. The maximum value for picric acid was 7800 meters per second; a column of pellets of the same sort, density 1.71, exploded in a copper tube 20-22 mm. in diameter, by means of a primer of 0.5 gram of fulminate and 80 grams of dynamite. The highest velocity with picric acid in paper cartridges was 7645 meters per second with pellets of densities 1.73 and 1.74 and the same charge of initiator.

Velocity of detonation, other things being equal, depends upon the physical state of the explosive and upon the nature of the envelope which contains it. For each explosive there is an optimum density at which it shows its highest velocity of detonation. There is also for each explosive a minimum priming charge necessary to insure its complete detonation, and larger charges do not cause it to explode any faster. Figures for the velocity of detonation are of little interest unless the density is reported or unless the explosive is cast and is accordingly of a density which, though perhaps unknown, is easily reproducible. The cordeau of the following table[21] was loaded with TNT which was subsequently pulverized *in situ* during the drawing down of the lead tube:

[20] *Mém. poudres*, **16**, 28 (1911-1912).
[21] Desvergnes, *Mém. poudres*, **19**, 223 (1922).

	METERS PER SECOND
Cast trinitrobenzene..................	7441
Cast tetryl..........................	7229
Cast trinitrotoluene.................	7028
Cast picric acid.....................	6777
Compressed trinitrotoluene (d. 0.909) ..	4961
Compressed picric acid (d. 0.862)......	4835
Cordeau.............................	6900

Nitration of Chlorobenzene

The nitration of chlorobenzene is easier than the nitration of benzene and more difficult than the nitration of toluene. Trinitrochlorobenzene (picryl chloride) can be prepared on the plant scale by the nitration of dinitrochlorobenzene, but the process is expensive of acid and leads to but few valuable explosives which cannot be procured more cheaply and more simply from dinitrochlorobenzene by other processes. Indeed, there are only two important explosives, namely TNB and hexanitrobiphenyl, for the preparation of which picryl chloride could be used advantageously if it were available in large amounts. In the laboratory, picryl chloride is best prepared by the action of phosphorus pentachloride on picric acid.

During the early days of the first World War in Europe, electrolytic processes for the production of caustic soda were yielding in this country more chlorine than was needed by the chemical industries, and it was necessary to dispose of the excess. The pressure to produce toluene had made benzene cheap and abundant. The chlorine, which would otherwise have become a nuisance and a menace, was used for the chlorination of benzene. Chlorobenzene and dichlorobenzene became available, and dichlorobenzene since that time has been used extensively as an insecticide and moth exterminator. Dinitrodichlorobenzene was tried as an explosive under the name of *parazol*. When mixed with TNT in high-explosive shells, it did not detonate completely, but presented interesting possibilities because the unexploded portion, atomized in the air, was a vigorous itch-producer and lachrymator, and because the exploded portion yielded phosgene. The chlorine atom of chlorobenzene is unreactive, and catalytic processes[22] for replacing it by hydroxyl and amino groups had

---

[22] Steam and silica gel to produce phenol from chlorobenzene, the Dow process with steam and a copper salt catalyst, etc.

not yet been developed. In dinitrochlorobenzene, however, the chlorine is active. The substance yields dinitrophenol readily by hydrolysis, dinitroaniline by reaction with ammonia, dinitromethylaniline more readily yet by reaction with methylamine. These and similar materials may be nitrated to explosives, and the third nitro group may be introduced on the nucleus much more readily, after the chlorine has been replaced by a more strongly *ortho-para* orienting group, than it may be before the chlorine has been so replaced. Dinitrochlorobenzene thus has a definite advantage over picryl chloride. It has the advantage also over phenol, aniline, etc. (from chlorobenzene by catalytic processes), that explosives can be made from it which cannot be made as simply or as economically from these materials. Tetryl and hexanitrodiphenylamine are examples. The possibilities of dinitrochlorobenzene in the explosives industry have not yet been fully exploited.

Preparation of Dinitrochlorobenzene. One hundred grams of chlorobenzene is added drop by drop to a mixture of 160 grams of nitric acid (*d.* 1.50) and 340 grams of sulfuric acid (*d.* 1.84) while the mixture is stirred mechanically. The temperature rises because of the heat of the reaction, but is not allowed to go above 50-55°. After all the chlorobenzene has been added, the temperature is raised slowly to 95° and is kept there for 2 hours longer while the stirring is continued. The upper layer of light yellow liquid solidifies when cold. It is removed, broken up under water, and rinsed. The spent acid, on dilution with water, precipitates an additional quantity of dinitrochlorobenzene. All the product is brought together, washed with cold water, then several times with hot water while it is melted, and finally once more with cold water under which it is crushed. Then it is drained and allowed to dry at ordinary temperature. The product, melting at about 50°, consists largely of 2,4-dinitrochlorobenzene, m.p. 53.4°, along with a small quantity of the 2,6-dinitro compound, m.p. 87-88°. The two substances are equally suitable for the manufacture of explosives. They yield the same trinitro compound, and the same final products by reaction with methylamine, aniline, etc., and subsequent nitration of the materials which are first formed. Dinitrochlorobenzene causes a severe itching of the skin, both by contact with the solid material and by exposure to its vapors.

Trinitrotoluene (TNT, trotyl, tolite, triton, tritol, trilite, etc.)

When toluene is nitrated, about 96 per cent of the material behaves in accordance with the rule of Crum Brown and Gibson.

In industrial practice the nitration is commonly carried out in three stages, the spent acid from the trinitration being used for the next dinitration, the spent acid from this being used for the mononitration, and the spent acid from this either being fortified

FIGURE 46. TNT Manufacturing Building, Showing Barricades and Safety Chutes. (Courtesy E. I. du Pont de Nemours and Company, Inc.)

for use again or going to the acid-recovery treatment. The principal products of the first stage are *o*- (b.p. 222.3°) and *p*-nitrotoluene (m.p. 51.9°) in relative amounts which vary somewhat according to the temperature at which the nitration is carried out. During the dinitration, the *para* compound yields only 2,4-dinitrotoluene (m.p. 70°), while the *ortho* yields the 2,4- and the 2,6- (m.p. 60.5°). Both these in the trinitration yield 2,4,6-trinitrotoluene or *α*-TNT. 2,4-Dinitrotoluene predominates in the product of the dinitration, and crude TNT generally contains a small amount, perhaps 2 per cent, of this material which has escaped further nitration. The substance is stable and less reactive even than α-TNT, and a small amount of it in the purified TNT, if insufficient to lower the melting point materially, is not regarded as an especially undesirable impurity. The principal impurities arise from the *m*-nitrotoluene (b.p. 230-231°) which is formed to the extent of about 4 per cent in the product of the mononitration. We omit discussion of other impurities, such as the nitrated xylenes which might be present in consequence of impurities in the toluene which was used, except to point out that the same considerations apply to trinitro-*m*-xylene (TNX) as apply to 2,4-dinitrotoluene—a little does no real harm—while the nitro derivatives of *o*- and *p*-xylene are likely to form oils and are extremely undesirable. In *m*-nitrotoluene, the nitro group inhibits further substitution, the methyl group promotes it, the two groups disagree in respect to the positions which they activate, but substitution takes place under the orienting influence of the methyl group.

β-TNT or 2,3,4-trinitrotoluene (m.p. 112°) is the principal product of the nitration of *m*-nitrotoluene; γ-TNT or 2,4,5-trinitrotoluene (m.p. 104°) is present in smaller amount; and of ζ-TNT or 2,3,6-trinitrotoluene (m.p. 79.5°), the formation of which is theoretically possible and is indicated above for that reason, there is not more than a trace.[23] During the trinitration a small amount of the α-TNT is oxidized to trinitrobenzoic acid, finally appearing in the finished product in the form of TNB, which, however, does no harm if it is present in small amount. At the same time some of the material is destructively oxidized and nitrated by the strong mixed acid to form tetranitromethane, which is driven off with the steam during the subsequent boiling and causes annoyance by its lachrymatory properties and unpleasant taste. The product of the trinitration is separated from the spent acid while still molten, washed with boiling water until free from acid, and grained—or, after less washing with hot water, subjected to purification by means of sodium sulfite.

In this country the crude TNT, separated from the wash water, is generally grained by running the liquid slowly onto the refrigerated surface of an iron vessel which surface is continually scraped by mechanical means. In France the material is allowed to cool slowly under water in broad and shallow wooden tubs, while it is stirred slowly with mechanically actuated wooden paddles. The cooling is slow, for the only loss of heat is by radiation. The French process yields larger and flatter crystals, flaky, often several millimeters in length. The crystallized crude TNT is of about the color of brown sugar and feels greasy to the touch. It consists of crystals of practically pure α-TNT coated with an oily (low-melting) mixture of β- and γ-TNT, 2,4-dinitrotoluene, and possibly TNB and TNX. It is suitable for many uses as an explosive, but not for high-explosive shells. The oily mixture of impurities segregates in the shell, and sooner or later exudes through the thread by which the fuze is attached. The exudate is disagreeable but not particularly dangerous. The difficulty is that exudation leaves cavities within the mass of the charge, perhaps a central cavity under the booster which may cause the shell to fail to explode. There is also the possibility that the shock of setback across a cavity in the rear of the charge may cause the shell to explode prematurely while it is still within the barrel of the gun.

The impurities may be largely removed from the crude TNT,

---

[23] 3,5-Dinitrotoluene, in which both nitro groups are *meta* to the methyl, is probably not formed during the dinitration, and δ- and ε-TNT, namely 3,4,5- and 2,3,5-trinitrotoluene, are not found among the final products of the nitration of toluene.

with a corresponding improvement in the melting point and appearance of the material, by washing the crystals with a solvent. On a plant scale, alcohol, benzene, solvent naphtha (mixed xylenes), carbon tetrachloride, and concentrated sulfuric acid have all been used. Among these, sulfuric acid removes dinitrotoluene most readily, and organic solvents the β- and γ-TNT,

FIGURE 47. Commercial Sample of Purified TNT (25×).

but all of them dissolve away a portion of the α-TNT with resulting loss. The material dissolved by the sulfuric acid is recovered by diluting with water. The organic solvents are recovered by distillation, and the residues, dark brown liquids known as "TNT oil," are used in the manufacture of non-freezing dynamite. The best process of purification is that in which the crude TNT is agitated with a warm solution of sodium sulfite. A 5 per cent solution is used, as much by weight of the solution as there is of the crude TNT. The sulfite leaves the α-TNT (and any TNB, TNX, and 2,4-dinitrotoluene) unaffected, but reacts rapidly and completely with the β- and γ-TNT to form red-colored materials which are readily soluble in water. After the reaction, the purified material is washed with water until the washings are colorless.

Muraour[24] believes the sulfite process for the purification of TNT to be an American invention. At any rate, the story of its discovery presents an interesting example of the consequences of working rightly with a wrong hypothesis. The nitro group in the *m*-position in β- and γ-TNT is *ortho*, or *ortho* and *para*, to two other nitro groups, and accordingly is active chemically. It is replaced by an amino group by the action of alcoholic ammonia both in the hot[25] and in the cold,[26] and undergoes similar reactions with hydrazine and with phenylhydrazine. It was hoped that it would be reduced more readily than the unactivated nitro groups of α- or symmetrical TNT, and that the reduction products could be washed away with warm water. Sodium polysulfide was tried and did indeed raise the melting point, but the treated material contained finely divided sulfur from which it could not easily be freed, and the polysulfide was judged to be unsuitable. In seeking for another reducing agent, the chemist bethought himself of sodium sulfite, which, however, does not act in this case as a reducing agent, and succeeded perfectly in removing the β- and γ-TNT.

The reaction consists in the replacement of the nitro by a sodium sulfonate group:

$$+ \text{NaNO}_2$$

and

[24] *Bull. soc. chim.*, IV, **35**, 367 (1924); *Army Ordnance*, **5**, 507 (1924).

[25] Hepp, *Ann.*, **215**, 364 (1882).

[26] Giua, *Atti accad. Lincei*, **23**, II, 484 (1914); *Gazz. chim. ital.*, **45**, I, 345 (1915).

The soluble sulfonates in the deep red solution, if they are thrown into the sewer, represent a loss of about 4 per cent of all the toluene—a serious loss in time of war—as well as a loss of many pounds of nitro group nitrogen. The sulfonic acid group in these substances, like the nitro group which it replaced, is *ortho*, or *ortho* and *para*, to two nitro groups, and is active and still capable of undergoing the same reactions as the original nitro group. They may be converted into a useful explosive by reaction with methylamine and the subsequent nitration of the resulting dinitrotolylmethylamines, both of which yield 2,4,6-trinitrotolyl-3-methylnitramine or *m*-methyltetryl.

m-Methyltetryl, pale yellow, almost white, crystals from alcohol, m.p. 102°, was prepared in 1884 by van Romburgh[27] by the nitration of dimethyl-*m*-toluidine, and its structure was demonstrated fully in 1902 by Blanksma,[28] who prepared it by the synthesis indicated on the next page.

β- and γ-TNT lose their active nitro group by the action of aqueous alkali and yield salts of dinitro-*m*-cresol.[29] The mixed dinitro-*m*-cresols which result may be nitrated to trinitro-*m*-cresol, a valuable explosive. Their salts, like the picrates, are primary explosives and sources of danger. β- and γ-TNT react with lead oxide in alcohol to form lead dinitrocresolates, while α-TNT under the same conditions remains unaffected.

In plant-scale manufacture, TNT is generally prepared by a

[27] *Rec. trav. chim.*, **3**, 414 (1884).
[28] *Ibid.*, **21**, 327 (1902).
[29] Will, *Ber.*, **47**, 711 (1914); Copisarow, *Chem. News*, **112**, 283 (1915).

three-stage process, but processes involving one and two nitrations have also been used.

Preparation of Trinitrotoluene (Three Stages). A mixture of 294 grams of concentrated sulfuric acid (*d*. 1.84) and 147 grams of nitric acid (*d*. 1.42) is added slowly from a dropping funnel to 100 grams of toluene in a tall 600-cc. beaker, while the liquid is stirred vigorously with an electric stirrer and its temperature is maintained at 30° to 40° by running cold water in the vessel in which the beaker is standing. The addition of acid will require from an hour to an hour and a half. The stirring is then continued for half an hour longer without cooling; the mixture is allowed to stand over night in a separatory funnel; the lower layer of spent acid is drawn off; and the crude mononitrotoluene is weighed. One-half of it, corresponding to 50 grams of toluene, is taken for the dinitration.

The mononitrotoluene (MNT) is dissolved in 109 grams of concentrated sulfuric acid (*d*. 1.84) while the mixture is cooled in running water. The solution in a tall beaker is warmed to 50°, and a mixed acid, composed of 54.5 grams each of nitric acid (*d.* 1.50) and sulfuric acid (*d*. 1.84), is added slowly drop by drop from a dropping funnel while the mixture is stirred mechanically. The heat generated by the reaction raises the temperature, and the rate of addition of the acid is regulated so that the temperature of the mixture lies always between 90° and 100°. The addition of the acid will require about 1 hour. After the acid has been added, the mixture is stirred for 2 hours longer at 90-100° to complete the nitration. Two layers separate on standing. The upper layer consists largely of dinitrotoluene (DNT), but probably contains a certain amount of TNT. The trinitration in the laboratory is conveniently carried out without separating the DNT from the spent acid.

While the dinitration mixture is stirred actively at a temperature of about 90°, 145 grams of fuming sulfuric acid (*oleum* containing 15 per cent free SO_3) is added slowly by pouring from a beaker. A mixed acid, composed of 72.5 grams each of nitric acid (d. 1.50) and 15 per cent oleum, is now added drop by drop with good agitation while the heat of the reaction maintains the temperature at 100-115°. After about three-quarters of the acid has been added, it will be found necessary to apply external heat to maintain the temperature. After all the acid has been added (during 1½ to 2 hours), the heating and stirring are continued for 2 hours longer at 100-115°. After the material has stood over night, the upper TNT layer will be found to have solidified to a hard cake, and the lower layer of spent acid to be filled with crystals. The acid is filtered through a Büchner funnel (without filter paper), and the cake is broken up and washed with water on the same filter to remove excess of acid. The spent acid contains considerable TNT in solution; this is precipitated by pouring the acid into a large volume of water, filtered off, rinsed with water, and added to the main batch. All the product is washed three or four times by agitating it vigorously with hot water under which it is melted. After the last washing, the TNT is granulated by allowing it to cool slowly under hot water while the stirring is continued. The product, filtered off and dried at ordinary temperature, is equal to a good commercial sample of crude TNT. It may be purified by dissolving in warm alcohol at 60° and allowing to cool slowly, or it may be purified by digesting with 5 times its weight of 5 per cent sodium hydrogen sulfite solution at 90° for half an hour with vigorous stirring, washing with hot water until the washings are colorless, and finally granulating as before. The product of this last treatment is equal to a good commercial sample of purified TNT. Pure α-TNT, m.p. 80.8°, may be procured by recrystallizing this material once from nitric acid (d. 1.42) and once from alcohol.

Several of the molecular compounds of TNT with organic bases are listed below.[30] TNT and diphenylamine give an orange-brown color when warmed together or when moistened with alcohol, and the formation of a labile molecular compound of the two substances has been demonstrated.[31]

The compound of TNT with potassium methylate is a dark red powder which inflames or explodes when heated to 130-150°, and has been reported to explode spontaneously on standing at ordinary temperature. An aqueous solution of this compound, on the addition of copper tetrammine nitrate, gives a brick-red precipitate which, when dry, detonates violently at 120°. Pure TNT

[30] See references under TNB.
[31] Giua. *Gazz. chim. ital.*, **45**, II, 357 (1915).

MOLECULAR PROPORTIONS	M.P.	DESCRIPTION
TNT: Substance		
1 : 1 Aniline....................	83–84°	Long brilliant red needles.
1 : 1 Dimethylaniline.............	...	Violet needles.
1 : 1 o-Toluidine................	53–55°	Light red needles.
1 : 1 m-Toluidine...............	62–63°	Light red needles.
1 : 1 α-Naphthylamine...........	141.5°	Dark red needles.
1 : 1 β-Naphthylamine...........	113.5°	Bright red prismatic needles.
1 : 1 β-Acetnaphthalide..........	106°	Yellow needles.
1 : 1 Benzyl-β-naphthylamine......	106.5°	Brilliant crimson needles.
1 : 1 Dibenzyl-β-naphthylamine....	108°	Deep brick-red needles.
2 : 1 Benzaldehydephenylhydrazone	84°	Dark red needles.
1 : 1 2-Methylindole.............	110°	Yellow needles.
3 : 2 Carbazole.................	160°	Yellow needles.
1 : 1 Carbazole.................	140–200°	Dark yellow needles.

explodes or inflames when heated to about 230°, but Dupré[32] found that the addition of solid caustic potash to TNT at 160° caused immediate inflammation or explosion. A mixture of powdered solid caustic potash and powdered TNT inflames when heated, either slowly or rapidly, to 80°. A similar mixture with caustic soda inflames at 80° if heated rapidly, but may be heated to 200° without taking fire if the heating is slow. If a small fragment of solid caustic potash is added to melted TNT at 100°, it becomes coated with a layer of reaction product and nothing further happens. If a drop of alcohol, in which both TNT and KOH are soluble, is now added, the material inflames within a few seconds. Mixtures of TNT with potassium and sodium carbonate do not ignite when heated suddenly to 100°.

Since the methyl group of TNT is attached to a picryl group, we should expect it in some respects to resemble the methyl group of a ketone. Although acetone and other methyl ketones brominate with great ease, TNT does not brominate and may even be recrystallized from bromine. The methyl group of TNT, however, behaves like the methyl group of acetone in certain condensation reactions. In the presence of sodium carbonate TNT condenses with p-nitrosodimethylaniline to form the dimethylaminoanilide of trinitrobenzaldehyde,[33] from which trinitrobenzaldehyde and N,N-dimethyl-p-diaminobenzene are produced readily by acid hydrolysis.

[32] "Twenty-eighth Annual Report of H. M. Inspector of Explosives," 1903, p. 26.

[33] Sachs and Kempf, *Ber.*, **35**, 1222 (1902); Sachs and Everding, *ibid.*, **36**, 999 (1903).

If a drop of piperidine is added to a pasty mixture of TNT and benzaldehyde, the heat of the reaction is sufficient to cause the material to take fire. The same substances in alcohol or benzene solution condense smoothly in the presence of piperidine to form trinitrostilbene.[34]

Preparation of Trinitrostilbene. To 10 grams of TNT dissolved in 25 cc. of benzene in a 100-cc. round-bottom flask equipped with a reflux condenser, 6 cc. of benzaldehyde and 0.5 cc. of piperidine are added, and the mixture is refluxed on the water bath for half an hour. The material, while still hot, is poured into a beaker and allowed to cool and crystallize. The crystals, collected on a filter, are rinsed twice with alcohol and recrystallized from a mixture of 2 volumes of alcohol and 1 of benzene. Brilliant yellow glistening needles, m.p. 158°.

Trinitrotoluene, in addition to the usual reactions of a nitrated hydrocarbon with alkali to form dangerous explosive materials, has the property that its methyl group in the presence of alkali condenses with aldehydic substances in reactions which produce heat and which may cause fire. Aldehydic substances from the action of nitrating acid on wood are always present where TNT is being manufactured, and alkali of all kinds ought to be excluded rigorously from the premises.

Giua[35] reports that TNT may be distilled in vacuum without the slightest trace of decomposition. It boils at 210-212° at 10-20 mm. When heated for some time at 180-200°, or when exposed to

[34] Pfeiffer and Monath, *Ber.*, **39**, 1306 (1906); Ullmann and Geschwind, *ibid.*, **41**, 2296 (1908).

[35] Giua, "Chimica delle sostanze esplosive," Milan, 1919, p. 248.

sunlight[36] in open tubes, it undergoes a slow decomposition with a consequent lowering of the melting point. Exposure to sunlight in a vacuum in a sealed tube has much less effect. Verola[37] has found that TNT shows no perceptible decomposition at 150°, but that it evolves gas slowly and regularly at 180°. At ordinary temperatures, and even at the temperatures of the tropics, it is stable in light-proof and air-tight containers—as are in general all the aromatic nitro explosives—and it does not require the same surveillance in storage that nitrocellulose and smokeless powder do.

The solubility[38] of trinitrotoluene in various solvents is tabulated below.

SOLUBILITY OF TRINITROTOLUENE

(Grams per 100 grams of solvent)

Temp.	Water	CCl_4	Ben-zene	Tolu-ene	Ace-tone	95% Alcohol	$CHCl_3$	Ether
0°	0.0100	0.20	13	28	57	0.65	6	1.73
5°	0.0105	0.25	24	32	66	0.75	8.5	2.08
10°	0.0110	0.40	36	38	78	0.85	11	2.45
15°	0.0120	0.50	50	45	92	1.07	15	2.85
20°	0.0130	0.65	67	55	109	1.23	19	3.29
25°	0.0150	0.82	88	67	132	1.48	25	3.80
30°	0.0175	1.01	113	84	156	1.80	32.5	4.56
35°	0.0225	1.32	144	104	187	2.27	45	...
40°	0.0285	1.75	180	130	228	2.92	66	...
45°	0.0360	2.37	225	163	279	3.70	101	...
50°	0.0475	3.23	284	208	346	4.61	150	...
55°	0.0570	4.55	361	272	449	6.08	218	...
60°	0.0675	6.90	478	367	600	8.30	302	...
65°	0.0775	11.40	665	525	843	11.40	442	...
70°	0.0875	17.35	1024	826	1350	15.15	...	...
75°	0.0975	24.35	2028	1685	2678	19.50	...	...
80°	0.1075	...	...	...	...	...	...	...
85°	0.1175	...	...	...	...	...	...	...
90°	0.1275	...	...	...	...	...	...	...
95°	0.1375	...	...	...	...	...	...	...
100°	0.1475	...	...	...	...	...	...	...

[36] Molinari and Quartieri, "Notizie sugli esplodenti in Italia," Milan, 1913, p. 157.

[37] *Mém. poudres*, **16**, 40 (1911-1912).

[38] Taylor and Rinkenbach, *J. Am. Chem. Soc.*, **45**, 44 (1923).

Dautriche found the density of powdered and compressed **TNT** to be as follows:

PRESSURE: KILOS PER SQUARE CENTIMETER	DENSITY
275	1.320
685	1.456
1375	1.558
2060	1.584
2750	1.599
3435	1.602
4125	1.610

Trinitrotoluene was prepared by Wilbrand[39] in 1863 by the nitration of toluene with mixed acid, and in 1870 by Beilstein and Kuhlberg[40] by the nitration of o- and p-nitrotoluene, and by Tiemann[41] by the nitration of 2,4-dinitrotoluene. In 1891 Haussermann[42] with the Griesheim Chem. Fabrik undertook its manufacture on an industrial scale. After 1901 its use as a military explosive soon became general among the great nations. In the first World War all of them were using it.

Trinitroxylene (TNX)

In m-xylene the two methyl groups agree in activating the same positions, and this is the only one of the three isomeric xylenes which can be nitrated satisfactorily to yield a trinitro derivative. Since the three isomers occur in the same fraction of coal tar and cannot readily be separated by distillation, it is necessary to separate them by chemical means. When the mixed xylenes are treated with about their own weight of 93 per cent sulfuric acid for 5 hours at 50°, the o-xylene (b.p. 144°) and the m-xylene (b.p. 138.8°) are converted into water-soluble sulfonic acids, while the p-xylene (b.p. 138.5°) is unaffected. The aqueous phase is removed, diluted with water to about 52 per cent acidity calculated as sulfuric acid, and then heated in an autoclave at 130° for 4 hours. The m-xylene sulfonic acid is converted to m-xylene, which is removed. The o-xylene sulfonic acid, which remains in solution, may be converted into o-xylene by autoclaving at a higher temperature. The nitration of m-xylene is conveniently carried out in three steps. The effect of the two methyl

[39] *Ann.*, **128**, 178 (1863).
[40] *Ber.*, **3**, 202 (1870).
[41] *Ber.*, **3**, 217 (1870).
[42] *Z. angew. Chem.*, 1891, p. 508; *J. Soc. Chem. Ind.*, 1891, p. 1028.

groups is so considerable that the introduction of the third nitro group may be accomplished without the use of fuming sulfuric acid. Pure TNX, large almost colorless needles from benzene, melts at 182.3°.

Trinitroxylene is not powerful enough for use alone as a high explosive, and it does not always communicate an initial detonation throughout its mass. It is used in commercial dynamites, for which purpose it does not require to be purified and may contain an oily mixture of isomers and other nitrated xylenes. Its large excess of carbon suggests that it may be used advantageously in conjunction with an oxidizing agent. A mixture of 23 parts of TNX and 77 parts of ammonium nitrate, ground intimately together in a black powder mill, has been used in high-explosive shells. It was loaded by compression. Mixtures, about half and half, of TNX with TNT and with picric acid are semi-solid when warm and can be loaded by pouring. The eutectic of TNX and TNT contains between 6 and 7 per cent of TNX and freezes at 73.5°. It is substantially as good an explosive as TNT. A mixture of 10 parts TNX, 40 parts TNT, and 50 parts picric acid can be melted readily under water. In explosives such as these the TNX helps by lowering the melting point, but it also attenuates the power of the more powerful high explosives with which it is mixed. On the other hand, these mixtures take advantage of the explosive power of TNX, such as that power is, and are themselves sufficiently powerful and satisfactory for many purposes—while making use of a raw material, namely *m*-xylene, which is not otherwise applicable for use in the manufacture of military explosives.

Nitro Derivatives of Naphthalene

Naphthalene nitrates more readily than benzene, the first nitro group taking the α-position which is *ortho* on one nucleus to the side chain which the other nucleus constitutes. The second nitro group takes one or another of the expected positions, either the position *meta* to the nitro group already present or one of the α-positions of the unsubstituted nucleus. The dinitration of naphthalene in actual practice thus produces a mixture which consists almost entirely of three isomers. Ten different isomeric dinitronaphthalenes are possible, seven of which are derived from α-nitronaphthalene, seven from β-nitronaphthalene, and four from both the α- and the β-compounds. After two nitro groups have been introduced, conflicts of orienting tendencies arise and polynitro compounds are formed, among others, in which nitro groups occur *ortho* and *para* to one another. Only four nitro groups can be introduced into naphthalene by direct nitration.

The mononitration of naphthalene takes place easily with a mixed acid which contains only a slight excess of one equivalent of HNO_3.

For the di-, tri-, and tetranitrations increasingly stronger acids and higher temperatures are necessary. In the tetranitration oleum is commonly used and the reaction is carried out at 130°.

The nitration of α-nitronaphthalene[43] (m.p. 59-60°) yields a mixture of α- or 1,5-dinitronaphthalene (silky needles, m.p. 216°), β- or 1,8-dinitronaphthalene (rhombic leaflets, m.p. 170-172°), and γ- or 1,3-dinitronaphthalene (m.p. 144-145°).

The commercial product of the dinitration melts at about 140°, and consists principally of the α- and β-compounds. The nitration of naphthalene at very low temperatures,[44] −50° to −60°, gives good yields of the γ- compound, and some of this material is undoubtedly present in the ordinary product.

The nitration of α-dinitronaphthalene yields α- or 1,3,5-trinitronaphthalene (monoclinic crystals, m.p. 123°), γ- or 1,4,5-

---

[43] Roussin, *Comp. rend.*, **52**, 796 (1861); Darmstädter and Wickelhaus, *Ann.*, **152**, 301 (1869); Aguiar, *Ber.*, **2**, 220 (1869); **3**, 29 (1870); **5**, 370 (1872); Beilstein and Kuhlberg, *Ann.*, **169**, 86 (1873); Beilstein and Kurbatow, *Ber.*, **13**, 353 (1880); *Ann.*, **202**, 219, 224 (1880); Julius, *Chem. Ztg.*, **18**, 180 (1894); Gassmann, *Ber.*, **29**, 1243, 1521 (1896); Friedländer, *ibid.*, **32**, 3531 (1899).

[44] Pictet, *Comp. rend.*, **116**, 815 (1893).

trinitronaphthalene (glistening plates, m.p. 147°), and δ- or
1,2,5-trinitronaphthalene (m.p. 112-113°). The nitration of β-dinitronaphthalene yields β- or 1,3,8-trinitronaphthalene (monoclinic
crystals, m.p. 218°), and the same substance, along with some
α-trinitronaphthalene, is formed by the nitration of γ-dinitronaphthalene.

All these isomers occur in commercial trinitronaphthalene, known
as *naphtite,* which melts at about 110°.

The nitration of α-, β-, and γ-trinitronaphthalene yields γ- or
1,3,5,8-tetranitronaphthalene (glistening tetrahedrons, m.p. 194-
195°). The nitration of the β-compound also yields β- or 1,3,6,8-
tetranitronaphthalene (m.p. 203°), and that of the δ-trinitro
compound yields δ- or 1,2,5,8-tetranitronaphthalene (glistening
prisms which decompose at 270° without melting), a substance
which may be formed also by the introduction of a fourth nitro
group into γ-trinitronaphthalene. The nitration of 1,5-dinitronaphthalene[45] yields α-tetranitronaphthalene (rhombic crystals,
m.p. 259°) (perhaps 1,3,5,7-tetranitronaphthalene), and this substance is also present in the crude product of the tetranitration,
which, however, consists largely of the β-, γ-, and δ-isomers.

The crude product is impure and irregular in its appearance; it is
commonly purified by recrystallization from glacial acetic acid

[45] Aguiar, *Ber.,* 5, 374 (1872).

The purified material consists of fine needle crystals which melt at about 220° and have the clean appearance of a pure substance but actually consist of a mixture of isomers.

None of the nitrated naphthalenes is very sensitive to shock. α-Nitronaphthalene is not an explosive at all and cannot be detonated. Dinitronaphthalene begins to show a feeble capacity for explosion, and trinitronaphthalene stands between dinitrobenzene and dinitrotoluene in its explosive power. Tetranitronaphthalene is about as powerful as TNT, and distinctly less sensitive to impact than that explosive. Vennin and Chesneau report that the nitrated naphthalenes, charged in a manometric bomb at a density of loading of 0.3, gave on firing the pressures indicated below.[46]

KILOS PER SQUARE CENTIMETER	
Mononitronaphthalene.....	1208
Dinitronaphthalene........	2355
Trinitronaphthalene........	3275
Tetranitronaphthalene.....	3745

The nitrated naphthalenes are used in dynamites and safety explosives, in the Favier powders, *grisounites*, and *naphtalites* of France, in the *cheddites* which contain chlorate, and for military purposes to some extent in mixtures with ammonium nitrate or with other aromatic nitro compounds. Street,[47] who proposed their use in cheddites, also suggested a fused mixture of mononitronaphthalene and picric acid for use as a high explosive. *Schneiderite*, used by France and by Italy and Russia in shells during the first World War, consisted of 1 part dinitronaphthalene and 7 parts ammonium nitrate, intimately incorporated together by grinding in a black powder mill, and loaded by compression. A mixture (MMN) of 3 parts mononitronaphthalene and 7 parts picric acid, fused together under water, was used in drop bombs and was insensitive to the impact of a rifle bullet. A mixture (MDN) of 1 part dinitronaphthalene and 4 parts picric acid melts at about 105-110°; it is more powerful than the preceding and is also less sensitive to shock than picric acid alone. The

---

[46] Vennin and Chesneau, "Les poudres et explosifs et les mesures de sécurité dans les mines de houille," Paris and Liége, 1914, p. 269.

[47] *Mon. Sci.*, 1898, p. 495.

Germans used a mine explosive consisting of 56 per cent potassium perchlorate, 32 per cent dinitrobenzene, and 12 per cent dinitronaphthalene.[48] Their *Tri-Trinal* for small-caliber shells was a compressed mixture of 2 parts of TNT (*Tri*) with 1 of trinitronaphthalene (*Trinal*), and was used with a booster of compressed picric acid.

Trinitronaphthalene appears to be a genuine stabilizer for nitrocellulose, a true inhibitor of its spontaneous decomposition. Marqueyrol found that a nitrocellulose powder containing 10 per cent of trinitronaphthalene is as stable as one which contains 2 per cent of diphenylamine. The trinitronaphthalene has the further effect of reducing both the hygroscopicity and the temperature of combustion of the powder.

Hexanitrobiphenyl

2,2',4,4',6,6'-Hexanitrobiphenyl was first prepared by Ullmann and Bielecki[49] by boiling picryl chloride in nitrobenzene solution with copper powder for a short time. The solvent is necessary in order to moderate the reaction, for picryl chloride and copper powder explode when heated alone to about 127°. Ullmann and Bielecki also secured good yields of hexanitrobiphenyl by working in toluene solution, but found that a small quantity of trinitrobenzene was formed (evidently in consequence of the presence of moisture). Hexanitrobiphenyl crystallizes from toluene in light-yellow thick crystals which contain ½ molecule of toluene of crystallization. It is insoluble in water, and slightly soluble in alcohol, acetone, benzene, and toluene, m.p. 263°. It gives a yellow color with concentrated sulfuric acid, and a red with alcohol to which a drop of ammonia water or aqueous caustic soda has been added. It is neutral, of course, and chemically unreactive toward metals, and is reported to be non-poisonous.

Hexanitrobiphenyl cannot[50] be prepared by the direct nitration

---

[48] Naoum, "Schiess- und Sprengstoffe," Dresden and Leipzig, 1927, p. 62.

[49] *Ber.*, **34**, 2174 (1901).

[50] The effect may be steric, although there is evidence that the dinitrophenyl group has peculiar orienting and resonance effects. Rinkenbach and Aaronson, *J. Am. Chem. Soc.*, **52**, 5040 (1930), report that *sym*-diphenylethane yields only very small amounts of hexanitrodiphenylethane under the most favorable conditions of nitration.

of biphenyl. The most vigorous nitration of that hydrocarbon yields only 2,2',4,4'-tetranitrobiphenyl, yellowish prisms from benzene, m.p. 163°.

Jahn in a patent granted in 1918[51] states that hexanitrobiphenyl is about 10 per cent superior to hexanitrodiphenylamine. Fifty grams in the lead block produced a cavity of 1810 cc., while the same weight of hexanitrodiphenylamine produced one of 1630 cc. Under a pressure of 2500 atmospheres, it compresses to a density of about 1.61.

Picric Acid (melinite, lyddite, pertite, shimose, etc.)

The *ortho-para* orienting hydroxyl group of phenol promotes nitration greatly and has the further effect that it "weakens" the ring and makes it more susceptible to oxidation. Nitric acid attacks phenol violently, oxidizing a portion of it to oxalic acid, and produces resinous by-products in addition to a certain amount of the expected nitro compounds. The carefully controlled action of mixed acid on phenol gives a mixture of *o*-nitrophenol (yellow crystals, m.p. 45°, volatile with steam) and *p*-nitrophenol (white crystals, m.p. 114°, not volatile with steam), but the yields are not very good. When these mononitrophenols are once formed, their nitro groups "activate" the same positions as the hydroxyls do, but the nitro groups also inhibit substitution, and their further nitration may now be carried out more smoothly. *p*-Nitrophenol yields 2,4-dinitrophenol (m.p. 114-115°), and later picric acid. *o*-Nitrophenol yields 2,4- and 2,6-dinitrophenol (m.p. 63-64°), both of which may be nitrated to picric acid, but the nitration of *o*-nitrophenol is invariably accompanied by losses resulting from its volatility. The straightforward nitration of phenol cannot be carried out successfully and with satisfying yields. In practice the phenol is sulfonated first, and the sulfonic acid is then nitrated. The use of sulfuric acid (for the sulfonation) in this process amounts to its use as an inhibitor or moderator of the nitration, for the *meta* orienting sulfonic acid group at first slows down the introduction of nitro groups until it is itself finally replaced by one of them.

[51] U. S. Pat. 1,253,691 (1918).

The sulfonation of phenol at low temperatures produces the
o-sulfonic acid, and at high temperatures the p-sulfonic acid
along with more or less of the di- and even of the trisulfonic acids
according to the conditions of the reaction. All these substances
yield picric acid as the final product of the nitration.[52]

Unless carefully regulated the production of picric acid from
phenol is accompanied by losses, either from oxidation of the
material with the production of red fumes which represent a loss
of fixed nitrogen or from over sulfonation and the loss of uncon-
verted water-soluble nitrated sulfonic acids in the mother liquors.
Olsen and Goldstein[53] have described a process which yields 220
parts of picric acid from 100 parts of phenol. In France, where
dinitrophenol was used during the first World War in mixtures
with picric acid which were loaded by pouring, Marqueyrol and
his associates[54] have worked out the details of a four-stage
process from the third stage of which dinitrophenol may be removed if it is desired. The steps are: (1) sulfonation; (2) nitration
to the water-soluble mononitrosulfonic acid; (3) nitration to
dinitrophenol, which is insoluble in the mixture and separates out,
and to the dinitrosulfonic acid which remains in solution; and
(4) further nitration to convert either the soluble material or
both of the substances to picric acid. The process is economical
of acid and gives practically no red fumes, but the reported
yields are inferior to those reported by Olsen and Goldstein. The

[52] Cf. King, J. Chem. Soc., 119, 2105 (1921).

[53] Olsen and Goldstein, Ind. Eng. Chem., 16, 66 (1924).

[54] Marqueyrol and Loriette, Bull. soc. chim., 25, 376 (1919); Marqueyrol
and Carré, ibid., 27, 195 (1920).

dinitrophenol as removed contains some picric acid, but this is of no disadvantage because the material is to be mixed with picric acid anyway for use as an explosive.

Preparation of Picric Acid (Standard Method). Twenty-five grams of phenol and 25 grams of concentrated sulfuric acid (*d*. 1.84) in a round-

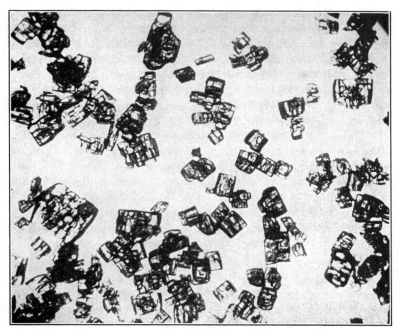

FIGURE 48. Commercial Sample of Picric Acid (25×).

bottom flask equipped with an air condenser are heated together for 6 hours in an oil bath at 120°. After the material has cooled, it is diluted with 75 grams of 72 per cent sulfuric acid (*d*. 1.64). To the resulting solution, in an Erlenmeyer flask in the hood, 175 cc. of 70 per cent nitric acid (*d*. 1.42) is added slowly, a drop at a time, from a dropping funnel. When all the nitric acid has been added and the vigorous reaction has subsided, the mixture is heated for 2 hours on the steam bath to complete the nitration. The next morning the picric acid will be found to have separated in crystals. These are transferred to a porcelain filter, washed with small portions of water until the washings are free from sulfate, and dried in the air. The crude product, which is equal in quality to a good commercial sample, is purified by boiling it with water, in the proportion of 15 grams to the liter, filtering hot, and allowing to cool slowly. The heavy droplets of brown oil which dissolve only slowly during this boiling ought to be discarded. Pure picric acid crystallizes from water in pale yellow flat needles, m.p. 122.5°. It may be obtained in crystals which are almost white by recrystallizing from aqueous hydrochloric acid.

The best process for the production of dinitrophenol is probably the autoclaving of dinitrochlorobenzene with aqueous caustic soda. The product is obtained on acidification and is used as such, or is nitrated to picric acid for the commercial production of that material by the so-called synthetic process.

The "catalytic process" for the production of picric acid directly from benzene in one step by the action of nitric acid in the presence of mercuric nitrate has much theoretical interest and has been applied, though not extensively, in plant-scale manufacture. It yields about as much picric acid as is procurable from the same weight of benzene by the roundabout method of sulfonating the benzene, converting the benzene sulfonic acid into phenol, and nitrating the phenol to picric acid—and the benzene which is not converted to picric acid is for the most part recovered as such or as nitrobenzene. The first mention of the process appears to be in the patent of Wolffenstein and Boeters.[55]

Preparation of Picric Acid (Catalytic Process). Two hundred grams of benzene in a 2-liter round-bottom flask equipped with a sealed-on condenser is refluxed on the sand bath for 7 hours with 600 cc. of nitric acid (d. 1.42) in which 10 grams of mercuric nitrate has been dissolved. The material is then transferred to another flask and distilled with steam. Benzene comes over, then nitrobenzene, then finally and slowly a mixture of dinitrobenzene and dinitrophenol. The distillation is continued until all volatile matter has been removed. The liquid in the flask is filtered hot and allowed to crystallize. If the picric acid is not sufficiently pure, it is recrystallized from hot water.

Mercuric nitrate combines with benzene to form a deep-brown or black addition compound, the probable structure of which is indicated below. This material when warmed with nitric acid is oxidized with the production of red fumes and the formation of

[55] Wolffenstein and Boeters, Ger. Pat. 194,883 (1908); Ger. Pat. 214,045 (1909); Ramy, Brit. Pat. 125,461 (1918); MacDonald and Calvert, Brit. Pats. 126,062, 126,084. 126,675, 126,676 (1918); Brewster, Brit. Pat. 131,403 (1919).

a yellow nitrophenolate of mercuric nitrate. By the continued action of the acid this is nitrated to the trinitrophenolate and decomposed with the formation of picric acid and the regeneration of mercuric nitrate.[56]

The addition of mercuric nitrate is here written as a 1,4-addition, but 1,2-addition would give the same final product, and there is no evidence in the facts concerning benzene which enables us to choose between the alternative hypotheses. Toluene yields trinitro-m-cresol by a similar series of reactions, and it is clear that the nitro group in the addition product of mercuric nitrate and toluene has taken either the 2-, the 4-, or the 6-position, that is, one or the other of the positions activated by the methyl group. In the addition of mercuric nitrate to naphthalene, the nitro group correspondingly may be supposed to go to the active α-position. If the addition is 1,2-, the product on oxidation will yield a derivative of β-naphthol. If it is 1,4-, it will yield a derivative of α-naphthol. The two possibilities are indicated below.

Gentle treatment of naphthalene with nitric acid containing mercuric nitrate yields, 2,4-dinitro-α-naphthol in conformity with the belief that the first addition product is 1,4- as represented by the second of the above formulations.

Picric acid was obtained in 1771 by Woulff, who found that

[56] Davis, Worrall, Drake, Helmkamp, and Young, *J. Am. Chem. Soc.*, **43**, 594 (1921); Davis, *ibid.*, **44**, 1588 (1922). Davis, U. S. Pat. 1,417,368 (1922).

the action of nitric acid on indigo yielded a material which dyed silk yellow. Hausmann[57] isolated the substance in 1778, and reported further studies upon it in 1788, noting particularly its bitter taste. Welter[58] in 1799 obtained picric acid by the action of nitric acid on silk, and the material came to be known generally as "Welter's bitter." Its preparation from indigo, aloes, resin, and other organic substances was studied by many chemists, among them Fourcroy and Vauquelin, Chevreul, Liebig, Wöhler, Robiquet, Piria, Delalande, and Stenhouse. Its preparation from oil of eucalyptus was suggested during the first World War. It was given the name of *acide picrique* by Dumas; *cf.* Greek πικρός = bitter, old English *puckery*. Its relation to phenol was demonstrated in 1841 by Laurent,[59] who prepared it by the nitration of that substance, and its structure was proved fully by Hepp,[60] who procured it by the oxidation of *sym*-trinitrobenzene.

Picric acid is a strong acid; it decomposes carbonates and may be titrated with bases by the use of sodium alizarine sulfonate as an indicator. It is a fast yellow dye for silk and wool. It attacks the common metals, except aluminum and tin, and produces dangerously explosive salts. *Cordeau Lheure,* which was long used extensively in France, was made by filling a tin pipe with fused picric acid and later drawing down to the desired diameter. It had the disadvantage that the metal suffered from the "tin disease," became unduly brittle, and changed to its gray allotropic modification. Picric acid and nitrophenols, when used in ammunition, are not allowed to come in contact with the metal parts. Shells which are to be loaded with these explosives are first plated on the inside with tin or painted with asphaltum varnish or Bakelite.

Dupré[61] in 1901 reported experiments which indicated that the picrates of calcium, lead, and zinc, formed *in situ* from melted picric acid are capable of initiating the explosion of that material. Kast[62] found that the dehydrated picrates are more sensitive than those which contain water of crystallization. The data tabulated

[57] *J. Phys.* **32**, 165 (1788).

[58] *Ann. chim. phys.,* I, **29**, 301 (1799).

[59] *Ann. chim. phys.,* III, **3**, 221 (1841).

[60] Hepp, *loc. cit.*

[61] *Mém. poudres,* **11**, 92 (1901).

[62] *Z. ges. Schiess- u. Sprengstoffw.,* **6, 7**, 31, 67 (1911). See also Will, *ibid.,* **1, 209** (1906); Silberrad and Phillips, *J. Chem. Soc.,* **93**, 474 (1908).

below have been published recently by J. D. Hopper.[63] Explosion temperature was determined as the temperature necessary to cause ignition or explosion in exactly 5 seconds when a thin-walled copper shell containing a few milligrams of the explosive was dipped into a molten metal bath to a constant depth. The minimum drop test was taken as the least distance through which a 2-kilogram weight must fall, in a standard apparatus,[64] to produce detonation or ignition in one or more instances among ten trials.

Substance	Degree of Hydration	Temperature of Drying, °C.	Minimum Drop Test 2-kilo Weight, Inches	Explosion Temperature, °C.
Mercury fulminate	Anhydrous	...	2	210
Tetryl	Anhydrous	...	8	260
TNT	Anhydrous	...	14	470
Picric acid	Anhydrous	...	14	320
Ammonium picrate	Anhydrous	...	17	320
Sodium picrate	1 H_2O	50	17	360
Sodium picrate	Anhydrous	150	15	...
Sodium dinitrophenolate	1 H_2O	100	16	370
Sodium dinitrophenolate	Anhydrous	150	15	...
Copper picrate	3 H_2O	25	19	300
Copper picrate	Anhydrous	150	12	...
Zinc picrate	6 H_2O	25	34	310
Zinc picrate	Anhydrous	150	12	...
Cadmium picrate	8 H_2O	25	35	340
Cadmium picrate	Anhydrous	150	12	...
Nickel picrate	6 H_2O	25	26	390
Nickel picrate	....	100	9	...
Nickel picrate	Anhydrous	150	4	...
Aluminum picrate	10 H_2O	25	36	360
Aluminum picrate	2 H_2O	80	16	...
Aluminum picrate	....	100	16	...
Chromium picrate	13 H_2O	25	36	330
Chromium picrate	....	80	10	...
Chromium picrate	1 H_2O	100	8	...
Ferrous picrate	8 H_2O	25	36	310
Ferrous picrate	....	100	14	...
Ferric picrate	x H_2O	25	36	295
Ferric picrate	....	80	8	...
Ferric picrate	....	100	7	...
Ferric picrate	....	150	6	...

[63] J. Franklin Inst., 225, 219 (1938).
[64] H. S. Deck, Army Ordnance, 7, 34 (1926).

Cast picric acid has a density of about 1.64. The density of pellets of compressed picric acid, according to Dautriche, is as follows.

PRESSURE: KILOS PER SQUARE CENTIMETER	DENSITY
275	1.315
685	1.480
1375	1.614
2060	1.672
2750	1.714
3435	1.731
4125	1.740

The use of picric acid as an explosive appears to have been suggested first in 1867 by Borlinetto,[65] who proposed a mixture of picric acid 35 per cent, sodium nitrate 35 per cent, and potassium chromate 30 per cent for use in mining. Sprengel in 1873 reported that picric acid in conjunction with suitable oxidizing agents is a powerful explosive. In 1885 Turpin[66] patented its use, both compressed and cast, in blasting cartridges and in shells, and shortly thereafter the French government adopted it under the name of *mélinite*. In 1888 Great Britain commenced to use it under the name of *lyddite*. Cast charges require a booster, for which purpose compressed picric acid or tetryl is generally used. The loading of picric acid into shells by pouring is open to two objections, which, however, are not insuperable, namely, the rather high temperature of the melt and the fact that large crystals are formed which may perhaps cause trouble on setback. Both difficulties are met by adding to the picric acid another explosive substance which lowers its melting point. Mixtures are preferred which melt between 70° and 100°, above 70° in order that exudation may be less likely and below 100° in order that the explosive may be melted by hot water. The mixtures are not necessarily eutectics. Two of the favorite French explosives have been DD 60/40, which consists of 60 parts picric acid and 40 parts dinitrophenol; and *crésylite* 60/40, 60 parts trinitro-*m*-cresol and 40 parts picric acid. Others are MDPC, picric acid 55 parts, dinitrophenol 35, and trinitro-*m*-cresol 10; and MTTC, which has the same composition as MDPC except that TNT is used instead of dinitrophenol. All these mixtures melt between

[65] Giua, *op. cit.*, pp. 287, 296.
[66] Fr. Pat. 167,512 (1885).

80° and 90° and are prepared by putting the materials together under water in wooden tanks and blowing in live steam. The water is sometimes acidulated with sulfuric acid to insure the removal of all metallic picrates. An explosive made by mixing 88 parts of picric acid with 12 parts of melted paraffin or stearic acid, and then rolling and graining, gives a compact charge when loaded by compression. It is nearly as powerful and brisant as picric acid, and responds satisfactorily to the impulse of the detonator, but is distinctly less sensitive to mechanical shock.

Ammonium Picrate

Ammonium picrate is less sensitive to shock than picric acid. It is not easily detonated by fulminate, but is commonly used with a

FIGURE 49. Commercial Sample of Ammonium Picrate (25×).

booster of powdered and compressed picric acid or tetryl. The pure substance occurs in two forms, a stable form which is of a bright lemon yellow color and a meta-stable form which is a brilliant red. These differ slightly in their crystal angles but show no detectable difference in their explosive properties. Thallium picrate similarly exists in two forms.

Ammonium picrate is prepared by suspending picric acid in a convenient quantity of hot water, adding strong ammonia water until everything goes into solution and a large excess of ammonia is present, and allowing to cool. The crystals which separate are the red form. A dry sample of this material in a stoppered bottle will remain without apparent change for many years. In contact with its saturated aqueous solution it changes to the yellow form during several months. The yellow form of ammonium picrate is best procured by recrystallizing the red form several times from water.

Pure ammonium picrate melts with decomposition at 265-271°. It is more soluble in warm alcohol than guanidine picrate is, and more soluble in acetone than in alcohol, but it goes into solution very slowly in alcohol and crystallizes out again very slowly when the liquid is allowed to stand.

Solubility of Ammonium Picrate

(Grams per 100 cc. of solution)

Temperature, °C.		Ethyl Acetate	Ethyl Alcohol
0	......	0.290	0.515
10	......	0.300	0.690
20	......	0.338	0.850
30	......	0.380	1.050
40	......	0.420	1.320
50	......	0.450	1.890
60	......	0.500	2.165
70	......	0.540	2.760
80	......	0.560	3.620

Guanidine Picrate

Guanidine picrate is procured as a yellow, finely crystalline precipitate by mixing warm solutions of guanidine nitrate and ammonium picrate. It is even less sensitive to blow and to shock than ammonium picrate; it is not detonated by fulminate and is used with a picric acid booster. The pure material, recrystallized from alcohol or from water, in both of which solvents it is sparingly soluble, melts with decomposition at 318.5-319.5°.

SOLUBILITY OF GUANIDINE PICRATE

(Grams per 100 cc. of solution)

Temperature, °C.	Water	Ethyl Alcohol	Acetone
0	...... 0.005	0.077	0.455
10	...... 0.038	0.093	0.525
20	...... 0.070	0.122	0.605
30	...... 0.100	0.153	0.695
40	...... 0.150	0.200	0.798
50	...... 0.230	0.255	0.920
60	...... 0.350	0.321	1.075
70	...... 0.480	0.413	...
80	...... 0.700	0.548	...
90	...... 1.010	...	...
100	...... 1.380	...	...

Trinitrocresol (cresylite)

This explosive is prepared from m-cresol by a process entirely similar to that by which picric acid is prepared from phenol. The pure material is readily soluble in alcohol, ether, and acetone, soluble in 449 parts of water at 20° and in 123 parts at 100°, yellow needles from water, m.p. 107°. The ammonium sa't, which is sparingly soluble in water, has been used in the composition of certain ammonium nitrate explosives, and it was adopted by the Austrian monarchy under the name of *ecrasite* as an explosive for shells of large caliber.

Trinitroresorcinol (styphnic acid)

Resorcinol nitrates readily to the trinitro compound, yellow prisms from water or alcohol, m.p. 175.5°. Styphnic acid is more expensive and less powerful than picric acid. Liouville[67] found that styphnic acid exploded in a manometric bomb, at a density of loading of 0.2, gave a pressure of 2260 kilos per sq. cm., whereas picric acid under the same conditions gave a pressure of 2350 kilos per sq. cm. It did not agglomerate to satisfactory pellets under a pressure of 3600 kilos per sq. cm. It is a fairly strong dibasic acid, and its salts are notably more violent explosives than the picrates. Lead styphnate has been used to facilitate the ignition of lead azide in detonators.

Trinitroanisol and Trinitrophenetol

2,4,6-Trinitroanisol (2,4,6-trinitrophenyl methyl ether, methyl picrate) has explosive properties comparable with those of picric

[67] *Mém. poudres*, **9**, 139 (1897-1898).

acid and trinitrocresol, but it contains no hydroxyl group and does not attack metals readily with the formation of dangerously explosive salts. In actual use, however, it reacts slowly with moisture and yields some picric acid. It has been colloided with nitrocellulose in the form of a strip powder, flashless and of low hygroscopicity, but the powder in the course of time developed enough picric acid to stain the fingers and to give a yellow solution with water. Its relatively low melting point, 67-68°, gives it an advantage over picric acid for certain purposes. Methyl alcohol is needed for its synthesis, and the present availability of this substance cheaply from high-pressure synthesis further commends it. While anisol is an expensive raw material, and has the further disadvantage that its direct nitration is dangerous, trinitroanisol may be prepared, without it, economically and easily from benzene through the use of dinitrochlorobenzene.

Trinitroanisol was prepared by Cahours[68] in 1849 by the direct nitration of anisol, and the same process has been studied more recently by Broadbent and Sparre.[69] The strongly *ortho-para* orienting methoxy group promotes substitution greatly, the first products of the nitration are explosive, and the temperature of the reaction mixture during the first stages ought never to be allowed to rise above 0°. A small drop of anisol, or of phenetol or other aromatic-aliphatic ether, added to 10 cc. of nitric acid (*d.* 1.42) in a test tube and shaken, causes a remarkable series of color changes; the liquid turns yellow, then green, then blue, and finally reddish purple. A batch of anisol which was being nitrated at ordinary temperature in the author's laboratory detonated without warning and without provocation while showing a bluish-purple color. Small pieces of the 2-liter flask which had contained the mixture were propelled so violently that they punctured the plate-glass windows of the laboratory without, however, breaking or cracking them.

Trinitroanisol may also be prepared by the interaction of methyl iodide and silver picrate, and by the nitration of anisic acid, during which the carboxyl group is lost, but the most convenient method appears to be that of Jackson[70] and his collaborators by which a methoxy group is substituted for chlorine in a nucleus already nitrated. A methyl alcohol solution of picryl

[68] *Ann.,* **69,** 236 (1849).
[69] *Eighth Intern. Congr. Appl. Chem.,* **4,** 15 (1912).
[70] *Am. Chem. J.,* **20,** 448 (1898); **23,** 294 (1901).

chloride, treated with an excess of sodium methylate or of strong caustic soda solution, turns dark red and deposits handsome brilliant red crystals of the empirical composition, trinitroanisol·NaOCH₃. The probable constitution of these crystals is indicated below. On treatment with acid the substance yields trinitroanisol.

The red material is sparingly soluble in alcohol and in water, and is easily decomposed by aqueous acids. It is a primary explosive, stable to moderate heating but decomposing at 165° and exploding violently when introduced into a flame. It is not altered by dry air, but water decomposes it slowly to form first trinitroanisol and later picric acid. On boiling with ethyl alcohol, it yields the sodium ethylate addition product of trinitrophenetol—an interesting reaction analogous to the *ester interchange* in the aliphatic series.

Preparation of Trinitroanisol. Thirty-five grams of picryl chloride is dissolved in 400 cc. of methyl alcohol with warming under reflux, and the solution is allowed to cool to 30-35°. A solution of 23 grams of sodium hydroxide in 35 cc. of water is added slowly through the condenser, while the liquid is cooled, if need be, to prevent it from boiling. The mixture is allowed to stand for an hour or two. The red precipitate is filtered off, washed with alcohol, and stirred up with water while strong hydrochloric acid is added until all red color has disappeared. The slightly yellowish, almost white, precipitate, washed with water for the removal of sodium chloride, dried, and recrystallized from methyl alcohol, yields pale yellow leaflets of trinitroanisol, m.p. 67-68°. From anhydrous solvents the substance separates in crystals which are practically white.

Since the methoxy group exercises a greater effect in promoting substitution than the chlorine atom does, it is to be expected that dinitroanisol would take on a third nitro group more easily than dinitrochlorobenzene (to form picryl chloride), and with less expense for acid and for heat. The reactions indicated below are probably the best for the large-scale commercial production of trinitroanisol.

$$\text{Cl} \xrightarrow{\text{nitration}} \text{Cl}(\text{NO}_2)(\text{NO}_2) \xrightarrow[\text{then acid}]{\text{CH}_3\text{ONa}} \text{OCH}_3(\text{NO}_2)(\text{NO}_2) \xrightarrow{\text{nitration}} \text{OCH}_3(\text{NO}_2)(\text{NO}_2)(\text{NO}_2)$$

During the first World War the Germans used a mixture of trinitroanisol and hexanitrodiphenyl sulfide in bombs.[71]

Trinitrophenetol or ethyl picrate, m.p. 78°, is prepared by the same methods as trinitroanisol. The explosive properties of the two substances have been studied by Desparmets and Calinaud, and by Desvergnes,[72] who has reported the results of the earlier workers together with data of his own and discussions of methods of manufacture and of the explosive properties of mixtures with picric acid, ammonium nitrate, etc. Drop test with a 5-kilogram weight were as follows:

	HEIGHT OF DROP, CENTIMETERS	PER CENT EXPLOSION
Picric acid.........	30	50
Trinitroanisol......	100	20
Trinitroanisol......	110	30
Trinitrophenetol....	100	10
Trinitrophenetol....	110	10

Velocities of detonation (densities not reported) were trinitroanisol 7640 meters per second, trinitrophenetol 6880, and, for comparison, TNT 6880 meters per second. Pellets of the compressed explosives fired in the manometric bomb gave the results tabulated below.

	DENSITY OF LOADING	PRESSURE: KILOS PER SQUARE CENTIMETER
Picric acid...........	0.20	2310
Picric acid...........	0.20	2350
Picric acid...........	0.20	2210
Trinitroanisol........	0.20	2222
Trinitroanisol........	0.20	2250
Trinitroanisol........	0.20	2145
Trinitrophenetol......	0.20	1774
Picric acid...........	0.25	3230
Trinitroanisol........	0.25	2850
Trinitrophenetol......	0.25	2490
Trinitrophenetol......	0.30	3318

[71] Desvergnes, *Mém. poudres*, **19**, 283 (1922).
[72] *Ibid.*

Both trinitroanisol and trinitrophenetol were found to be as satisfactory as compressed TNT for use as a booster charge in 75-mm. shells loaded with *schneiderite*.

Trinitroaniline (picramide)

2,4,6-Trinitroaniline, orange-red crystals from alcohol, m.p. 186°, has but little interest as an explosive for the reason that other more powerful and more valuable explosives may be prepared from the same raw materials. It may be prepared by nitrating aniline in glacial acetic acid solution or by the use of mixed nitric-sulfuric acid in which no large excess of sulfuric acid is present. The presence of nitrous acid must be avoided, as this attacks the amino group, replaces it by hydroxyl, and results in the formation of picric acid. The nitration of aniline in the presence of a large amount of concentrated sulfuric acid yields *m*-nitroaniline[73] and later the nitro compounds which are derived from it.

Tetranitroaniline (TNA)

2,3,4,6-Tetranitroaniline, discovered by Flurscheim,[74] has interesting explosive properties but is such a reactive chemical substance that, when all things are considered, it is unsuitable for use. It was used to some extent during the first World War and was studied very thoroughly at that time.

Flurscheim prepared TNA by a one-stage nitration[75] of *m*-nitroaniline sulfate, that substance being procured by the reduction of *m*-dinitrobenzene with sodium polysulfide. The nitration proceeds smoothly, and the entering groups take the positions indicated by the strongly ortho-para orienting amino group. The yield is about 120 per cent of the weight of the *m*-nitroaniline.

---

[73] van Duin, *Rec. trav. chim.*, **37**, 111 (1917).

[74] *Chem. News*, 1910, 218; Brit. Pats. 3224, 3907 (1910); Ger. Pats. 241,697, 243,079 (1912); U. S. Pat. 1,045,012 (1912); *Z. ges. Schiess- u. Sprengstoffw.*, 1913, 185; *Mon. Sci.*, 1914, 490.

[75] Other studies on the nitration: Stettbacher, *Z. ges. Schiess- u. Sprengstoffw.*, **11**, 114 (1916); van Duin, *loc. cit.* A laboratory method for the preparation of TNA direct from aniline is described in Stettbacher's book, *op. cit.*, p. 201.

Pure TNA, yellowish-brown or greenish-brown crystals from acetone, melts with decomposition at about 210° and deflagrates at about 226°. It is soluble in glacial acetic acid (1 part in 24 at boiling temperature), readily in acetone (1 in 6 at boiling temperature), and sparingly in benzene, ligroin, and chloroform. If a small amount of water is added to an acetone solution of TNA and the liquid is refluxed, the nitro group in the 3-position, having other nitro groups *ortho* and *para* to it, is replaced rapidly by hydroxyl. The resulting trinitroaminophenol, m.p. 176°, is capable of attacking metals to form dangerous explosive salts which are similar to the picrates. If TNA is boiled with aqueous sodium carbonate or bicarbonate both the amino group and the nitro group in the 3-position are hydrolyzed, and trinitroresorcinol is formed.

With alcoholic ammonia TNA yields trinitro-*m*-phenylenediamine, m.p. 288°. Its nitro group in the 3-position reacts with primary and secondary amines, with sodium acid sulfite, etc., in the same way that the *meta* nitro groups of β- and γ-trinitrotoluene do. Marqueyrol found that TNA is attacked rapidly by boiling water, about half of it being converted into trinitroaminophenol, the other half being destroyed with the evolution of gases, largely carbon dioxide and nitrogen along with smaller quantities of carbon monoxide, hydrocyanic acid, and nitric oxide. At 75° the reaction between water and TNA is complete after 4 days; at 60° it is about half complete after 7 days; at 40° it is appreciable after 10 days. Any decomposition of this sort, of course, is too much for an explosive intended for military use.

TNA shows about the same sensitivity as tetryl in the drop test. Lead block experiments have been reported which showed that 10 grams of TNA produced a net expansion of 430 cc., TNT 254 cc., picric acid 297 cc., tetryl 375 cc., guncotton 290 cc., and 75 per cent dynamite 300 cc.[76] Experiments with the manometric bomb gave the results indicated below.

	DENSITY OF LOADING	PRESSURE: KILOS PER SQUARE CENTIMETER
TNA...............	0.20	2356
TNA...............	0.25	3110
Tetryl.............	0.20	2423
Tetryl.............	0.25	3243

Since these data show that tetryl is slightly more powerful than TNA, the superiority of TNA in the lead block test must be interpreted as indicating that TNA has the higher velocity of detonation.

Tetryl (tetralite, pyronite)

Tetryl or 2,4,6-trinitrophenylmethylnitramine was first described by Michler and Meyer[77] in 1879, and was studied soon thereafter by van Romburgh[78] and by Mertens.[79] Van Romburgh proved its structure by synthesizing it from picryl chloride and potassium methylnitramine.

In the early literature of the subject, and to some extent at present, the substance is wrongly designated as tetranitromethylaniline. It results from the nitration of monomethyl- and of

[76] From the pamphlet "Tetra-Nitro-Aniline 'Flurscheim,'" Verona Chemical Company, sole licensed manufacturers for the United States, North Newark, New Jersey, 1917(?), p. 4. Giua, *op. cit.*, p. 317, states that the force of TNA measured in the lead block is 420 compared with picric acid 297.

[77] *Ber.*, **12**, 1792 (1879).

[78] *Rec. trav. chim.*, **2**, 108 (1883); **8**, 215 (1889).

[79] *Ber.*, **19**, 2126 (1886).

dimethylaniline, and is prepared industrially by the nitration of
the latter. The course of the reactions is first the introduction of
two nitro groups in the nucleus, then the removal of one of the

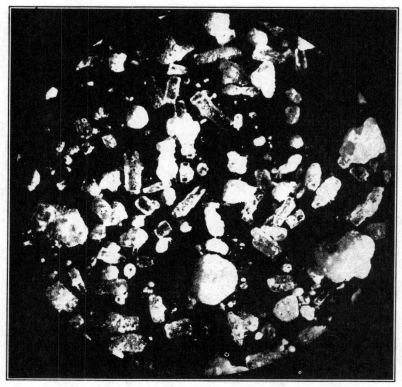

FIGURE 50. Commercial Sample of Tetryl (20×). Material crystallized in this
form pours easily and may be made into pellets by machinery.

methyl groups by oxidation, then the introduction of a third nitro
group in the nucleus, and finally the replacement of the amino
hydrogen by a nitro group.

All the above-indicated intermediates have been isolated from the reaction. The last step is interesting because it is a reversible nitration. If tetryl is dissolved in concentrated (95 per cent) sulfuric acid and allowed to stand, the nitro group on the nitrogen is replaced by hydrogen, and nitric acid and trinitromethylaniline (methylpicramide), m.p. 111.8-112.4°, are formed.[80] Tetryl accordingly gives up this nitro group, and only this one, in the nitrometer. In the industrial preparation of tetryl, the usual method is to dissolve the dimethylaniline in concentrated sulfuric acid and then to carry out all the reactions in one stage. The process has been the subject of many careful studies, among which those of Langenscheidt,[81] van Duin,[82] Knowles,[83] Wride,[84] Desvergnes,[85] and Bain[86] are especially to be noted. The crude tetryl contains impurities which must be removed by boiling the finely comminuted substance in water, and by dissolving the crude material in benzene and filtering for the removal of insoluble materials. For the industrial crystallization of tetryl, either acetone or benzene is commonly used.

Preparation of Tetryl. Twenty grams of dimethylaniline is dissolved in 240 grams of concentrated sulfuric acid (*d.* 1.84), the temperature being kept below 25°, and the solution is allowed to run from a separatory funnel drop by drop into 160 grams of 80 per cent nitric acid (*d.* 1.46), previously warmed to 55° or 60°, while this is stirred continuously and kept at a temperature between 65° and 70°. The addition requires about an hour. After all has been added, the stirring is continued while the temperature of the mixture is maintained at 65° to 70°. The material is allowed to cool; the solid matter is collected on an asbestos filter, washed with water, and boiled for an hour with 240 cc. of water while further water is added from time to time to replace that which boils away. The crude tetryl is filtered off, ground under water to pass a 150-mesh sieve, and boiled twice for 4 hours each time with 12 times its weight of water. The solid is dried and treated with benzene sufficient to dissolve all readily soluble material. The solution is filtered and allowed to evaporate spontaneously, and the residue is recrystal-

[80] Davis and Allen, *J. Am. Chem. Soc.*, **46**, 1063 (1924).
[81] *Z. ges. Schiess- u. Sprengstoffw.*, **7**, 445 (1912).
[82] *Rec. trav. chim.*, **37**, 111 (1917).
[83] *J. Ind. Eng. Chem.*, **12**, 247 (1920).
[84] *Arms and Explosives*, 1920, 6.
[85] *Mém. poudres*, **19**, 217 (1922).
[86] *Army Ordnance*, **6**, 435 (1926).

lized from alcohol. Pure tetryl melts at about 129.4°; a good commercial sample, at about 128.5°.

The nitration of aniline in the presence of a large amount of strong sulfuric acid results wholly in the formation of *m*-nitroaniline, but the similar nitration of dimethylaniline gives principally a mixture of the *ortho*- and *para*-derivatives. Monomethylaniline stands between aniline and dimethylaniline in respect to the orienting effect of its amino group; it yields a considerable amount of the *m*-nitro- compound—and dimethylaniline is preferred for the preparation of tetryl. Commercial dimethylaniline contains a certain amount of monomethylaniline, from which it is extremely difficult to free it, and this in the manufacture of tetryl is converted in part into 2,3,4,6-tetranitrophenylmethylnitramine, or *m*-nitrotetryl, pale yellow, almost white, crystals from benzene, m.p. 146-147.°[87]

No *m*-nitrotetryl is produced if pure dimethylaniline is used in the usual process for the manufacture of tetryl. The amount of this impurity in the usual process depends upon the amount of monomethylaniline which may be present. A large excess of sulfuric acid tends toward the production of *m*-nitro compounds, but a reduction in the amount of sulfuric acid is not feasible for this increases the amount of benzene-insoluble material. *m*-Nitrotetryl reacts with water, as TNA does; the nitro group in the 3-position is replaced by hydroxyl, and *m*-hydroxytetryl or 2,4,6-trinitro-3-methylnitraminophenol, yellow crystals from water, m.p. 183°, is formed. This substance resembles picric acid and forms explosive salts. It is readily soluble in water, and

[87] Van Romburgh, *Rec. trav. chim.*, **8**, 274 (1889). Van Romburgh and Schepers, *Versl. Kon. Akad. Wetenschapen*, **22**, 293 (1913), also prepared this substance by the nitration of dimethylaniline (in 20 times its weight of concentrated sulfuric acid).

m-nitrotetryl is effectively removed from crude tetryl by boiling the finely powdered solid with water.

Crude tetryl commonly contains a small quantity of amorphous-appearing, buff-colored material of high melting point which is insoluble in benzene. The amount of this material is increased by the presence of larger amounts of water in the nitrating acid. Michler and Pattinson[88] found that tetramethylbenzidine is produced when dimethylaniline is heated with concentrated sulfuric acid. The same material is evidently formed during the preparation of tetryl and gives rise to the three substances indicated below, which constitute the benzene-insoluble impurity.

These substances were prepared by Mertens[89] in 1886 by the action of nitric acid on dimethylaniline (I, II, and III) and on monomethylaniline (II and III). Van Romburgh[90] in the same year proved them to be derivatives of benzidine, and at a much later time[91] summarized the work which had been done upon them and synthesized the substances in such manner as to prove the position of the nitro groups.

[88] *Ber.*, **14**, 2161 (1881).
[89] *Loc. cit.*
[90] *Rec. trav. chim.*, **5**, 240 (1886).
[91] *Ibid.*, **41**, 38 (1922).

If the benzene-insoluble material from crude tetryl is dissolved in hot fuming nitric acid and allowed to cool, glistening yellow crystals are procured. These, recrystallized from nitric acid and then from acetone with the addition of two volumes of ligroin, yield cream-colored small crystals of the third of the above-indicated substances, 3,3',5,5'-tetranitro-4,4'-di-(methylnitramino)-biphenyl, or 3,3',5,5'-tetranitrodimethylbenzidinedinitramine. The material decomposes with foaming at 229-230° if its temperature is raised at the rate of 6° per minute. If it is heated more slowly, at 2° per minute, it melts partially and decomposes at 222° with preliminary softening and darkening. Like tetryl and other nitroamines, it gives a blue color with the diphenylamine reagent. Although Willstätter and Kalk[92] have found that monomethylaniline is not convertible into a benzidine derivative by Michler's method, it is nevertheless true that the benzene-insoluble by-products are produced during the preparation of tetryl from monomethylaniline, as indeed Mertens first procured them by the action of nitric acid on that substance.

The usual process for the preparation of tetryl from dimethylaniline has the disadvantage that the by-products, namely, the *m*-nitrotetryl and the benzene-insoluble material, necessitate a rather elaborate purification, and it has the further disadvantage that one of the methyl groups of the dimethylaniline is destroyed by oxidation (expense) with the production of red fumes (nuisance) and the consequent loss of valuable combined nitrogen. All these disadvantages find their origin at points in the reaction earlier than the formation of dinitromonomethylaniline. 2,4-Dinitromonomethylaniline, orange-yellow crystals, m.p. 174°, nitrates smoothly to form tetryl without the production of by-products or red fumes. Synthetic methyl alcohol is now available cheaply and in a quantity which is limited only by the will of the manufacturers to produce it. It reacts with ammonia (from the fixation of nitrogen) at elevated temperatures in the presence of a thorium oxide catalyst to form methylamine,[93] and methylamine reacts with dinitrochlorobenzene to form dinitromonomethylaniline. There seems every reason to believe that tetryl in the future will be manufactured chiefly, or wholly, from dinitrochlorobenzene.

---

[92] *Ber.*, **37**, 3771 (1904).
[93] Davis and Elderfield, *J. Am. Chem. Soc.*, **50**, 1786 (1928).

The solubility of tetryl in various solvents is tabulated below.

SOLUBILITY[94] OF TETRYL

(Grams per 100 grams of solvent)

Temperature, °C.	Water	95% Alcohol	Carbon Tetrachloride	Chloroform	Carbon Disulfide	Ether
0	0.0050	0.320	0.007	0.28	0.0090	0.188
5	0.0058	0.366	0.011	0.33	0.0120	0.273
10	0.0065	0.425	0.015	0.39	0.0146	0.330
15	0.0072	0.496	0.020	0.47	0.0177	0.377
20	0.0075	0.563	0.025	0.57	0.0208	0.418
25	0.0080	0.65	0.031	0.68	0.0244	0.457
30	0.0085	0.76	0.039	0.79	0.0296	0.493
35	0.0094	0.91	0.048	0.97	0.0392	...
40	0.0110	1.12	0.058	1.20	0.0557	...
45	0.0140	1.38	0.073	1.47	0.0940	...
50	0.0195	1.72	0.095	1.78	...	...
55	0.0270	2.13	0.124	2.23	...	...
60	0.0350	2.64	0.154	2.65	...	...
65	0.0440	3.33	0.193	...	...	...
70	0.0535	4.23	0.241	...	...	...
75	0.0663	5.33	0.297	...	...	...
80	0.0810	...	...	...	...	...
85	0.0980	...	...	...	...	...
90	0.1220	...	...	...	...	...
95	0.1518	...	...	...	...	...
100	0.1842	...	...	...	...	...

Tetryl is hydrolyzed rapidly by boiling aqueous sodium carbonate to form sodium picrate, sodium nitrite, and methylamine which escapes. It is not affected by prolonged boiling with dilute sulfuric acid. It reacts with aniline in benzene solution at ordinary temperature; red crystals of 2,4,6-trinitrodiphenylamine, m.p. 179.5-180°, separate after the liquid has stood for a few hours, and extraction of the liquid with water yields an aqueous solution of methylnitramine.

By heating tetryl alone, Farmer[95] and Desvergnes[96] obtained picric acid, and by heating tetryl in high-boiling solvents Mer-

[94] Taylor and Rinkenbach, *J. Am. Chem. Soc.*, **45**, 104 (1923).

[95] *J. Chem. Soc.*, **117**, 1603 (1920).

[96] *Loc. cit.*

tens,[96] van Romburgh,[96] and Davis and Allen[96] obtained methyl-picramide. When refluxed in xylene solution, tetryl gives off nitrous fumes and is converted into a tarlike mass from which picric acid and methylpicramide may be isolated, along with a third, unidentified, buff-colored finely crystalline substance which melts at 240.5°. If pure tetryl is kept at 100°, it gives off nitrous fumes and a small quantity of formaldehyde, and yields after 40 days a mass which remains semi-liquid at ordinary temperature. By heating at 125° it is converted into a viscous liquid after about the same number of hours.

At ordinary temperatures tetryl appears to be perfectly stable. Current methods of purification insure the absence of occluded acid. It is more powerful and more brisant than TNT and picric acid, though distinctly more sensitive to shock, and is probably the best of all the common explosives for use in boosters and reinforced detonators. Koehler[97] reports pressures in the manometric bomb (density of loading = 0.3) and temperatures produced by the explosions, as follows:

	PRESSURE: KILOS PER SQUARE CENTIMETER	TEMPERATURE, °C.
Tetryl..........	4684	2911
Picric acid......	3638	2419
TNT..........	3749	2060
TNB..........	3925	2356

Aranaz[98] reports that the explosion of tetryl produces a temperature of 3339°. Tetryl is slightly more sensitive than picric acid, and considerably more sensitive than TNT, in the drop test. Experimenting with a 5-kilogram weight, Koehler found that a drop of 150 cm. caused the detonation of tetryl 10 times out of 10 trials, a drop of 100 cm. 9 times out of 10, of 50 cm. 5 times out of 10, and of 40 cm. 3 times out of 10. Martin[99] has determined the minimum charges of various primary explosives necessary for the detonation of TNT and tetryl. The explosives were loaded into detonator capsules, and the initiators were compressed upon them at a pressure of 1100 kilos per square centimeter.

[97] Cited by Desvergnes, *loc. cit.*

[98] Aranaz, "Les nuevos explosives," Madrid, 1911, cited by Desvergnes.

[99] Martin, "Ueber Azide und Fulminate," Darmstadt, 1913, cited by Giua, *op. cit.*, p. 320.

	MINIMUM CHARGE FOR DETONATION OF	
	TNT	Tetryl
Mercuric fulminate........	0.36	0.29
Silver fulminate...........	0.095	0.02
Cadmium fulminate.......	0.11	0.008
Mercurous azide...........	0.145	0.045
Silver azide..............	0.07	0.02
Lead azide...............	0.09	0.025
Cadmium azide...........	0.04	0.01

With each of the initiators which was tried, tetryl was more easily detonated than TNT. Taylor and Cope[100] have determined the minimum charges of fulminate-chlorate (90:10) necessary to cause the complete detonation of various mixtures of TNT and tetryl, as follows:

MIXTURE OF TNT-TETRYL		WEIGHT OF INITIATOR, GRAMS
100	0	0.25
90	10	0.22
80	20	0.21
50	50	0.20
0	100	0.19

"Ethyl Tetryl." 2,4,6-Trinitrophenylethylnitramine

The ethyl analogue of tetryl was first prepared by van Romburgh,[101] who procured it both by nitrating monoethylaniline and by nitrating diethylaniline, and reported that it melts at 96°. The present writer has found that the pure material, recrystallized twice from nitric acid (d. 1.42) and once from alcohol, melts at 94°. It is comparable to tetryl in its chemical reactions and in its explosive properties.

"Butyl Tetryl." 2,4,6-Trinitrophenyl-n-butylnitramine

The n-butyl analogue of tetryl[102] has been prepared by two methods: (a) by condensing 2,4-dinitrochlorobenzene with n-butylamine to form 2,4-dinitro-n-butylaniline,[103] and by the nitration of this product; and (b) by the nitration in one step of n-butylaniline. The pure substance crystallizes from alcohol in

---

[100] U. S. Bureau of Mines Technical Paper 145, Washington, 1916.

[101] Rec. trav. chim., **2**, 111 (1883).

[102] Davis, U. S. Pat. 1,607,059 (1926).

[103] Pure 2,4-dinitro-n-butylaniline crystallizes from alcohol in deep yellow or orange needles, m.p. 92.5-93.0°.

lemon-yellow plates which melt at 97.5-98.0°. It is readily soluble in benzene, ethyl acetate, alcohol and acetone, and is insoluble in petroleum ether. It yields sodium picrate when boiled with sodium carbonate solution.

Butyl tetryl is suitable for use in boosters, reinforced detonators, detonating fuse, primer caps, etc. For the detonation of 0.4 gram, it requires 0.19 gram of mercury fulminate. It has a slightly greater shattering effect than TNT in the sand test and shows about the same sensitivity as tetryl in the drop test. It explodes spontaneously at 210°.

Hexanitrodiphenylamine

2,2',4,4',6,6'-Hexanitrodiphenylamine (hexil, hexite, hexamin, etc.) is another explosive which can be prepared most conveniently from dinitrochlorobenzene. Its ammonium salt has been used under the name of *aurantia* as a yellow dye for silk and wool. It has valuable explosive properties but is more poisonous than nitroglycerin and attacks the skin, causing severe blisters which resemble burns. Its dust is injurious to the mucous membranes of the mouth, nose, and lungs. Mertens[104] in 1878 prepared hexanitrodiphenylamine by the nitration of diphenylamine with fuming nitric acid in concentrated sulfuric acid solution. Its behavior as a pseudo-acid has been studied by Alexandrov[105] and by Hantzsch and Opolski.[106] Hausermann[107] in 1891 reported upon its explosive power as compared with trinitrotoluene, and a patent granted in 1909 to Otto Freiherr von Schroetter[108] described an explosive consisting of 80 parts of hexanitrodiphenylamine and 20 parts of trinitrotoluene. The large-scale preparation by the direct nitration of diphenylamine was reported in 1910,[109] and the process from dinitrochlorobenzene, originally described in a patent to the Griesheim Chem. Fabrik,[110] was reported by

---

[104] *Ber.*, 11, 843 (1878). Austen, *ibid.*, 7, 1249 (1874), reported the formation of the substance by the nitration of picryl-*p*-nitroaniline, and Gnehm, *ibid.*, 7, 1399 (1874), by the nitration of methyldiphenylamine.

[105] *J. Russ. Phys. Chem. Soc.*, 39, 1391 (1907).

[106] *Ber.*, 41, 1745 (1908).

[107] *Z. angew. Chem.*, 17, 510 (1891).

[108] U. S. Pat. 934,020 (1909).

[109] *Z. ges. Schiess- u. Sprengstoffw.*, 5, 16 (1910).

[110] Ger. Pat. 86,295 (1895).

Carter[111] in 1913 and studied further by Hoffman and Dame[112] in 1919 and by Marshall[113] in 1920.

Dinitrochlorobenzene reacts with 2 equivalents of aniline, when the materials are warmed together in the absence of solvent or when they are stirred together vigorously with water 80-90°, to form dinitrodiphenylamine in practically quantitative yield, along with 1 equivalent of aniline hydrochloride. The use of the second molecule of aniline to combine with the hydrogen chloride involves unnecessary expense, and the same results may be accomplished by means of some mineral alkali or acid-neutralizing substance like sodium acetate or sodium or calcium carbonate. The product, which is insoluble in water, separates in bright red needles. Pure 2,4-dinitrodiphenylamine, recrystallized from alcohol or from benzene, melts at 156-157°. The crude product is nitrated in one or in two stages to the hexanitro compound.

Preparation of Hexanitrodiphenylamine (Two-Stage Nitration). Seventy grams of aniline and 32 grams of precipitated calcium carbonate are stirred up together with water in such manner as to form a homogeneous suspension, and the mixture is heated to about 60°. Dinitrochlorobenzene, 150 grams, previously melted, is poured in slowly in a fine stream while the stirring is continued and the mixture is heated gradually to about 90°, the rate of heating being regulated by the progress of the reaction. The product is washed with hydrochloric acid to free it from aniline and calcium carbonate, then with water until free from chlorides, and dried in the oven at 100°.

Fifty grams of finely powdered dinitrodiphenylamine is added in small portions at a time to 420 grams of nitric acid (*d.* 1.33), which is stirred vigorously while the temperature is maintained at 50-60°. The progress of the nitration is followed by observing the color change from

[111] *Z. ges. Schiess- u. Sprengstoffw.*, **8**, 205, 251 (1913).

[112] *J. Am. Chem. Soc.*, **41**, 1013 (1919).

[113] *J. Ind. Eng. Chem.*, **12**, 336 (1920).

the red of the dinitro compound to the yellow of the tetranitrodiphenylamine. After all has been added, the temperature is raised to 80-90° and kept there for 2 hours longer while the stirring is continued. After the mixture has cooled, the product is filtered off directly, washed with water until free from acid, and dried in the air or in the oven at 100°.

Fifty grams of the tetranitrodiphenylamine is added slowly, with stirring, during an hour, to a mixture of 250 grams of nitric acid (d. 1.50) and 250 grams of sulfuric acid (d. 1.83). After all has been added, the mixture is allowed to stand for 3 hours at laboratory temperature, and is then drowned in ice water. The hexanitrodiphenylamine is filtered off, washed thoroughly with water, dried in the air, and recrystallized from acetone with the addition of petroleum ether.

Pure hexanitrodiphenylamine, small yellow needles, melts with decomposition at 243.0-244.5°. It is insoluble in chloroform, sparingly soluble in ether and in cold acetic acid, fairly soluble in alcohol, and readily soluble in cold acetone and in warm acetic and nitric acids.

Marshall[114] reports minimum priming charges of fulminate-chlorate (90:10) necessary for the complete detonation of the indicated explosives to be as follows:

	GRAMS
Hexanitrodiphenylamine	0.18
Tetryl	0.20
Tetranitroaniline	0.20
Trinitrotoluene	0.25

He found hexanitrodiphenylamine to be slightly less sensitive in the drop test than tetryl and tetranitroaniline. When 1 pound of the explosive was loaded into a 3.5-inch cubical box of cardboard or tin and fired at with a U.S. Army rifle from a distance of 30 yards, hexanitrodiphenylamine gave no detonations in the cardboard boxes, and 7 detonations and 1 failure in tin; TNT gave no detonation in cardboard, fire and detonation in tin; and tetryl and tetranitroaniline gave detonations in every case with either kind of container. Marshall reported the velocity of detonation of hexanitrodiphenylamine to be 6898 meters per second at density 1.58, and 7150 meters per second at density 1.67. Pellets of the explosive, mixed with 1 per cent of stearic acid, compressed at 5000 pounds per square inch, had a density 1.43, at 10,000

[114] *Loc. cit.*

pounds per square inch, density 1.56; at 15,000 pounds per square inch, density 1.59; and at 20,000 pounds per square inch, density 1.60. The pellets which showed the best homogeneity and the least tendency to crumble were those of density 1.56.

Hexanitrodiphenyl Sulfide

Hexanitrodiphenyl sulfide (picryl sulfide) is formed by the interaction of picryl chloride and sodium thiosulfate in alcohol solution in the presence of magnesium carbonate.[115] It is sparingly soluble in alcohol and ether, more readily in glacial acetic acid and acetone, golden-yellow leaflets from alcohol-acetone, m.p. 234°. It does not stain the fingers yellow and is said to be non-poisonous. Its explosive properties are comparable to those of hexanitrodiphenylamine. Its use in reinforced detonators has been suggested, and the fact that its explosion produces sulfur dioxide has commended it[116] for use in projectiles intended to make closed spaces, such as casemates, holds of ships, etc., untenable. During the first World War the Germans used drop bombs loaded with a mixture of equal parts of TNT and hexanitrodiphenyl sulfide.[117]

Hexanitrodiphenyl Sulfone

The action of nitric acid on hexanitrodiphenyl sulfide yields a substance, faintly yellowish crystals, m.p. 307°, which Stettbacher believes to be the sulfone, not the peroxide as the patent[118] states, for the reason that it is stable at elevated temperatures

[115] Ger. Pat. 275,037 (1912); Brit. Pat. 18,353 (1913).
[116] Brit. Pat. 18,354 (1913).
[117] Alfred Stettbacher, "Die Schiess- und Sprengstoffe," Leipzig, 1919, p. 206.
[118] Ger. Pat. 269,826 (1913).

and is less sensitive to shock than the sulfide. It is a more powerful explosive than hexanitrodiphenyl sulfide.

Hexanitro-oxanilide

This substance, m.p. 295-300°, results from the direct nitration of oxanilide.[119] It is stable and about as powerful as TNT, and is reported to explode with the production of a temperature which is distinctly lower than that produced by many high explosives.

Hexanitrocarbanilide

2,2',4,4',6,6'-Hexanitro-N,N'-diphenylurea   (hexanitrocarbanilide or *sym*-dipicrylurea)[120] may be prepared by the nitration of carbanilide (*sym*-diphenylurea) in one, in two, or in three stages. It is of interest because of its explosive properties and because it supplies one way in which benzene may be converted into an explosive which is valuable both for military and for civil uses. Carbanilide may be prepared by the interaction of aniline and phosgene but is most conveniently and economically procured by heating aniline and urea together at 160-165°.

Preparation of Hexanitrocarbanilide (Two Stages). Forty grams of carbanilide is dissolved in 60 cc. of concentrated sulfuric acid (*d*. 1.84), and the solution is added drop by drop during 4 hours to 96 cc. of nitric acid (*d*. 1.51) while the mixture is stirred vigorously with a mechanical stirrer and its temperature is maintained at 35° to 40°. After all has been added, the stirring is continued and the temperature is raised to 60° during half an hour and maintained at 60° for another hour. The mixture is cooled to room temperature, allowed to stand over night, then treated with cracked ice and water, and filtered. The crude tetra-

---

[119] Fr. Pat. 391,106.
[120] Davis, U. S. Pat. 1,568,502 (1926).

nitrocarbanilide is washed thoroughly with water and allowed to dry in the air.

Ten grams of crude tetranitrocarbanilide is added to a mixture of 16 grams of concentrated sulfuric acid (*d*. 1.84) and 24 grams of nitric acid (*d*. 1.51), and the material is heated on the steam bath for 1 hour with constant stirring. The mixture, after cooling, is treated with cracked ice and water, and filtered. The product, washed with 500 cc. of cold water, then with 500 cc. of hot water, and dried in the air, is hexanitrocarbanilide of satisfactory quality for use as an explosive.

Pure hexanitrocarbanilide crystallizes from acetone-ligroin in pale yellow rosettes which soften and darken at 204° and melt at 208-209° with decomposition. It yields picric acid when warmed with dilute sulfuric acid, and trinitroaniline when boiled with strong ammonia water. A deep ruby-red color is developed when hexanitrocarbanilide is allowed to stand at ordinary temperatures in contact with strong ammonia water. Tetranitrocarbanilide, dinitroaniline, trinitroaniline, picric acid, and dinitrophenol do not give this color.

Hexanitrocarbanilide is a brisant high explosive suitable for use in boosters, reinforced detonators, detonating fuse, primer caps, etc. For the detonation of 0.4 gram, it requires 0.19 gram of mercury fulminate. It is slightly stronger than TNT in the sand test and of about the same sensitivity as tetryl in the drop test. It explodes spontaneously at 345°.

Hexanitroazobenzene

Hexanitroazobenzene may be prepared from dinitrochlorobenzene and hydrazine by the reactions indicated below:

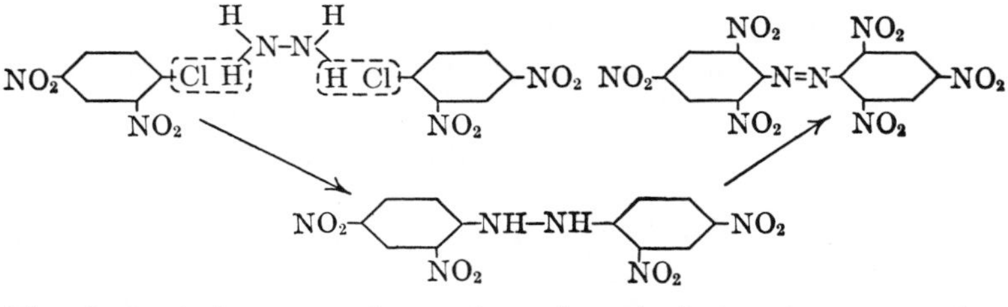

The first of these reactions takes place in hot-water suspension in the presence of sodium or calcium carbonate. The resulting tetranitrohydrazobenzene is both nitrated and oxidized by the mixed acid in the next step. Pure 2,2′,4,4′,6,6′-hexanitroazobenzene crystallizes from acetone in handsome orange-colored needles which melt at 215°. The explosive properties of the substance have not been reported in detail. The azo group makes it more powerful and more brisant than hexanitrodiphenylamine. The accessibility of the raw materials and the simplicity of its preparation commend it for use in boosters and compound detonators.

CHAPTER V

NITRIC ESTERS

Nitric esters or *organic nitrates* contain the nitrate radical, —O—NO$_2$, attached to a carbon atom, or, to express the same idea in a different way, they contain the nitro group, —NO$_2$, attached to an oxygen atom which is attached to a carbon. In *nitro compounds,* strictly so called, the nitro group is attached directly to a carbon; in *nitroamines* or *nitramines* it is attached to an amino nitrogen atom, that is, to a nitrogen which is attached to a carbon. In the nitric esters and in the nitroamines alike, a single atom stands between the nitro group and the carbon atom of the organic molecule. Substances of the two classes are alike in their most characteristic reaction, namely, they are formed by the reversible nitration of alcohols and amines respectively.

During the nitration of glycerin by the action of strong nitric acid or of strong mixed acid upon it, nitro groups are introduced in place of three of the hydrogen atoms of the original molecule. There is therefore a certain propriety in thinking of the product as a nitro compound, and a reasonable warrant for the common practice of calling it by the name of trinitroglycerin or, more commonly, of *nitroglycerin.* The hydrogen atoms which are replaced were attached to oxygen atoms; the product is really a nitric ester, and its proper name is *glyceryl trinitrate.* Similarly, the substances which are commonly called nitroglycol, nitrostarch, nitrosugar, nitrolactose, nitrocotton, etc., are actually nitric esters.

The physical properties of the nitric esters resemble in a general way the physical properties of the alcohols from which they are derived. Thus, methyl and ethyl nitrate, like methyl and ethyl alcohol, are volatile liquids; nitroglycerin is a viscous oil, more viscous and less volatile than glycol dinitrate as glycerin is more viscous and less volatile than glycol. Nitrocellulose from fibrous cellulose yields a tough and plastic colloid, but nitro-starch remains from the evaporation of its solutions as a mass which is brittle and friable.

Methyl Nitrate

Methyl nitrate is a powerful explosive although its physical properties are such that it is not of practical use, and it is of interest only because it is the simplest of the nitric esters. Like ethyl and n-propyl nitrates, it may be prepared by the careful distillation of the alcohol with concentrated nitric acid (d. 1.42) from which, however, the last traces of nitrous acid must first have been removed by the addition of urea. It may also be prepared by adding the alcohol to strong mixed acid at low temperature, stirring, and separating and washing the product without distillation, by a process similar to that which is used for the preparation of nitroglycerin and nitroglycol except that the volatility of the product requires the stirring to be done by mechanical means and not by compressed air. It is a colorless limpid liquid somewhat less viscous than water, boiling point 65–66°, specific gravity 1.2322 at 5°, 1.2167 at 15°, and 1.2032 at 25°. Its vapors have a strongly aromatic odor resembling that of chloroform, and cause headache if they are inhaled. It dissolves collodion nitro-cotton to form a jelly from which the methyl nitrate evaporates readily.

Methyl nitrate has a slightly higher energy content than nitroglycerin and a slightly greater explosive effect. Naoúm [1] reports that 10 grams of methyl nitrate in the Trauzl test with water tamping caused an expansion of 615 cc., while 10 grams of nitroglycerin under the same conditions gave 600 cc. Methyl nitrate is very much more sensitive to initiation than nitroglycerin, a fact which, like its higher velocity of detonation, is probably associated with its lower viscosity. It is less sensitive than nitroglycerin to the mechanical shock of the drop test. In the small lead block test, or lead block compression test, 100 grams of methyl nitrate under slight confinement in a shell of sheet lead 1 mm. thick and tamped with thin cork plates, gave a compres-

---

[1] Phokion Naoúm, "Nitroglycerine and Nitroglycerine Explosives," trans. E. M. Symmes, Baltimore, The Williams and Wilkins Company, 1928, p. 205.

sion of 24.5 mm. while nitroglycol similarly gave 30 mm. and nitroglycerin 18.5 mm.

Methyl nitrate is easily inflammable and burns in an open dish with a large non-luminous flame. Its vapors explode when heated to about 150°.

Berthelot [2] measured the velocity of detonation of methyl nitrate in tubes of such small diameter that the maximum velocity of detonation was not secured, but he was able to make certain interesting inferences both as to the effect of the envelope and as to the effect of the physical state of the explosive. Some of his results are summarized in the table below. The data indicate

TUBE OF	INTERNAL DIAMETER, MILLIMETERS	EXTERNAL DIAMETER, MILLIMETERS	VELOCITY OF DETONATION, METERS PER SECOND
Rubber, canvas covered	5	12	1616
Glass	3	12	2482
Glass	3	7	2191
Glass	5	7	1890
Britannia metal	3	12.6	1230
Steel	3	15	2084
Steel	3	15	2094

that with tubes of the same internal diameter the velocity of detonation is greater in those cases in which the rupture of the tube is more difficult; it is greater in the tubes which have thicker walls and in the tubes which are made of the stronger materials. The extent to which the velocity of detonation builds up depends in some measure upon the pressure which builds up before the container is ruptured. By comparing these results with those from other explosive substances, Berthelot was able to make further inductions.

In fact, nitroglycerin in lead tubes 3 mm. internal diameter gave velocities in the neighborhood of 1300 meters per second, while dynamite in similar metallic tubes attained 2700 meters per second. This sets in evidence the influence of the structure of the explosive substance upon the velocity of propagation of the explosion, pure nitroglycerin, a viscous liquid, transmitting the shock which determines the detonation much more irregularly than the silica impregnated in

[2] *Mém. poudres,* **4,** 13 (1891); *Ann. chim. phys.,* **23, 485** (1901).

a uniform manner with the same liquid. Mica dynamite according to my observations produces effects which are still more considerable, a fact which could be foreseen from the crystalline structure of the mica, a substance which is less deformable than amorphous silica.

This last induction is confirmed by observations on nitromannite, a crystalline solid which appears by reason of this circumstance better suited than liquid methyl nitrate for transmitting detonation. It has in fact given practically constant velocities of 7700 meters per second in lead tubes of 1.9 mm. internal diameter at a density of loading of 1.9. Likewise picric acid, also crystalline, 6500 meters per second. . . .

The influence of the structure of the explosive substance, on the course of the detonation being thus made evident, let us cite new facts which show the effect due to the containing envelope. . . . Compressed guncotton at such densities of loading as 1.0 and 1.27 in lead tubes 3.15 mm. internal diameter gave velocities of 5400 meters per second, while at a density of loading of practically one-half less (0.73) in a lead tube 3.77 mm. internal diameter, a velocity of 3800 meters per second was observed—a difference which is evidently due to the reduced continuity of the material. In supple cordeau, slightly resistant, formed by a single strand or braid, with a density of loading of 0.65, the velocity falls even to 2400 meters per second. But the feeble resistance of the envelope may be compensated by the mass of the explosive which opposes itself, especially in the central portion of the mass, to the instantaneous escape of the gas. Abel, in fact, with cartridges of compressed guncotton, of ten times the diameter of the above-mentioned cordeau, placed end to end, in the open air, has observed velocities of 5300 to 6000 meters per second.[3]

Other Alkyl Nitrates

Ethyl nitrate is a colorless liquid of agreeable odor, boiling point 87°, specific gravity (15°/15°) 1.1159 at 15°, and 1.1044 (25°/25°) at 25°. It has a less favorable oxygen balance than methyl nitrate, and is much less sensitive to initiation than the latter substance. It has only about 48% of the energy content of nitroglycerin, but its lower viscosity tends to give it a higher initial velocity of detonation than nitroglycerin and it performs about 58% as well as nitroglycerin in the sand test.[4] A No. 8

[3] *Mém. poudres*, **4**, 18–19 (1891).
[4] Naoúm, *op. cit.*, p. 207.

blasting cap will not detonate ethyl nitrate unless the explosive is tamped or confined. Mixed with fuller's earth in the proportion 70/30 or 60/40, it yields a brisant explosive which may be detonated without confinement.

n-Propyl nitrate, like ethyl nitrate, can be prepared by mixing the alcohol with nitric acid of density 1.42 or thereabouts, and carefully distilling the mixture. Ethyl alcohol and *n*-propyl alcohol, which contain the methylene group, are easily oxidized; if they are added to nitric acid of greater strength than density 1.42, or if they are added to strong mixed acid, they are likely to react with explosive violence and the abundant production of nitrous fumes, no matter how efficient the cooling. *n*-Propyl nitrate has a pleasant ethereal odor, boiling point 110.5°, specific gravity (15°/15°) 1.0631 at 15°, and (25°/25°) 1.0531 at 25°. It is less sensitive to detonation than ethyl nitrate. Ten grams in a Trauzl block, with water tamping and with a No. 8 blasting cap, detonated only partially and gave an expansion of 45 cc., or 15 cc. more than the cap alone, but 10 grams of it, mixed with 4 grams of fuller's earth to form a moist powder and exploded with a No. 8 cap, gave a sharp explosion and a net expansion of 230 cc.[5]

Isopropyl nitrate, b.p. 101–102°, specific gravity 1.054 at 0°, 1.036 at 19°, is prepared by the interaction of isopropyl iodide and silver nitrate. The hydrogen atom which is attached in isopropyl alcohol to the carbon atom carrying the hydroxyl group is so easily oxidized that it is not feasible to prepare the compound by the action of nitric acid on the alcohol.

Nitroglycerin (Glyceryl trinitrate, NG)

Nitroglycerin was first prepared late in the year 1846 or early in 1847 by the Italian chemist, Ascanio Sobrero (1812–1888), who was at the time professor of applied chemistry at the University of Torino. Sobrero had studied medicine in the same city, and in 1834 had been authorized to practice as a physician. After that he studied with Pelouze in Paris and served as his assistant in his private laboratory from 1840 to 1843. In 1843 he left Paris, studied for several months with Liebig at Giessen, and returned to Torino where he took up the duties of a teacher and in 1845

[5] *Ibid.*, p. 209.

built and equipped a modest laboratory of his own. The earliest printed account of nitroglycerin appears in a letter which Sobrero wrote to Pelouze and which Pelouze caused to be published in *L'Institut* of February 15, 1847.[6] In the same month Sobrero presented to the Academy of Torino a paper, *Sopra alcuni nuovi composti fulminanti ottenuti col mezzo dell'azione dell'acido*

FIGURE 51. Ascanio Sobrero (1812-1888). First prepared nitroglycerin, nitromannite, and nitrolactose, 1846-1847.

nitrico sulle sostanze organiche vegetali,[7] in which he described nitroglycerin, nitromannite, and nitrated lactose. Later in the year he presented another paper, *Sulla Glicerina Fulminante o Piroglycerina*, before the chemistry section of the Ninth Italian Scientific Congress at Venice.[8]

Sobrero found that, if concentrated nitric acid or strong mixed acid is added to glycerin, a violent reaction ensues and red fumes

[6] *L'Institut*, **15**, 53 (1847).
[7] *Mem. Acad. Torino*, [2] **10**, 195 (1847).
[8] *Proc. Ninth Ital. Sci. Congr.*, **3**, 105 (1848).

are evolved, but that, if syrupy glycerin is added to a mixture of two volumes of sulfuric acid (*d.* 1.84) and one volume of nitric acid (*d.* 1.50) with stirring while the mixture is kept below 0°, then the results are entirely different, the glycerin dissolves, and the solution when poured into water gives an oily precipitate of nitroglycerin. He collected the oil, washed it with water until free from acid, dried in a vacuum over sulfuric acid, and procured a transparent liquid of the color and appearance of olive oil. (Pure nitroglycerin is water-white.) Sobrero reported a value for the density which is very close to that which is now generally accepted, observed the ready solubility of nitroglycerin in alcohol and its reprecipitation by water, and reported a number of its chemical reactions—its comportments with acid and with alkali, that

> It detonates when brought into contact with metallic potassium, and evolves oxides of nitrogen in contact with phosphorus at 20° to 30°C., but at higher temperatures it ignites with an explosion. . . . When heated, nitroglycerin decomposes. A drop heated on platinum foil ignites and burns very fiercely. It has, however, the property of detonating under certain circumstances with great violence. On one occasion a small quantity of an ethereal solution of nitroglycerin was allowed to evaporate in a glass dish. The residue of nitroglycerin was certainly not more than 2 or 3 centigrams. On heating the dish over a spirit lamp a most violent explosion resulted, and the dish was broken to atoms. . . . The safest plan for demonstrating the explosive power of nitroglycerin is to place a drop upon a watch glass and detonate it by touching it with a piece of platinum wire heated to low redness. Nitroglycerin has a sharp, sweet, aromatic taste. It is advisable to take great care in testing this property. A trace of nitroglycerin placed upon the tongue, but not swallowed, gives rise to a most violent pulsating headache accompanied by great weakness of the limbs.

For many years Sobrero kept in his laboratory and guarded jealously a sample of the original nitroglycerin which he had prepared in 1847. In 1886 he washed this material with a dilute solution of sodium bicarbonate and took it to the Nobel-Avigliana factory, of which he was a consultant, where he gave verbal testimony of its authenticity and where it has since been stored in one of the magazines. Molinari and Quartieri [9] in a book published in 1913 state that the sample, consisting of about 200 cc. under water in a bottle, was at that time unaltered and that analyses gave values for nitrogen in the neighborhood of 18.35%, close to the theoretical.

Sobrero seems originally to have thought more highly of the solid crystalline nitromannite, which he thought might be used in percussion caps, than of the liquid nitroglycerin, but a spontaneous explosion of 400 grams of the former substance in the laboratory of the arsenal of Torino in 1853 and the extensive damage which resulted caused him to lose interest in the material. After Nobel's invention of dynamite and of the blasting cap had made the use of nitroglycerin safe and practical, Sobrero attempted in 1873 to establish a factory to be operated by Italian capital for the manufacture of an explosive called *melanina*, which was a kind of dynamite formed by absorbing nitroglycerin in a mixture of powdered charcoal and the silicious earth of Santa Fiora in Tuscany.[10] The project did not succeed. Shortly afterwards Sobrero accepted a position as consultant to the Nobel-Avigliana factory, a position which paid a generous salary during his life and a pension to his widow after his death. The high regard in which he was held by the Nobel company is indicated further by the bust of him which was unveiled in 1879 in the Avigliana factory.

Glycerin (glycerol) is a by-product of soap manufacture. All natural fats, whether of animal or vegetable origin, whether solid like beef suet or liquid like olive oil, are glyceryl esters of long-chain fatty acids containing an even number of carbon atoms. When they are warmed with an aqueous solution of strong alkali, they are saponified; soap, which is the alkali salt of the acids of the fats, is formed, and glycerin is produced which remains dissolved in the liquid. Glycerin is also formed from fats by the action of steam; the fatty acids, insoluble in water and generally of higher melting point than the fats, are formed at the same time.

Glycerin is a viscous liquid, colorless and odorless when pure, and possessing a sweet taste. It is hygroscopic, will absorb more

[9] Molinari and Quartieri, "Notizie sugli Esplodenti in Italia," Milano, 1913, p. 15.
[10] *Ibid.,* p. 33.

than half its own weight of moisture from the air, and does not evaporate. Glycerin will solidify in a freezing mixture, and when once frozen melts again at about 17°. It boils at atmospheric pressure at 290° with slight decomposition, and is best purified by distillation in vacuum. Its specific gravity is 1.265 at 15°. Perfectly pure and colorless glycerin yields a water-white nitroglycerin. Dynamite glycerin is a distilled product of high purity, density 1.262 or higher, and contains at least 99% of glycerin and less than 1% of water. It varies in color from pale yellow to dark brown, generally has a faint odor resembling that of burnt sugar, and yields a nitroglycerin of a pale yellow or pale brown color. The explosives makers consider a test nitration on a laboratory scale to be the surest way of estimating the quality of a sample of dynamite glycerin.

Small amounts of glycerin are produced during an ordinary alcoholic fermentation, but the quantity is greatly increased if a considerable amount of sodium sulfite is present. A commercial process based upon this principle was developed and used in Germany during the first World War, when the supply of glycerin from fats was insufficient to fill the needs of the explosives manufacturers, and similar processes have been used to some extent elsewhere and since that time. At the beginning of the second World War an effort was made to increase the production of whale oil for the manufacture of glycerin. Modern methods—harpoons shot from guns, fast Diesel-propelled steel ships—resulted immediately in a tremendous slaughter of whales, and whale oil again has become difficult to procure. Recent advances in synthetic chemistry make it probable that glycerin in the future will be prepared in large quantity from petroleum.

Cracking gas, which is produced when heavy petroleum is cracked to produce gasoline, consists in large part of olefins, particularly ethylene and propylene, and is being used more and more for the manufacture of such materials as glycol and glycerin, glycol dinitrate and nitroglycerin, mustard gas, ethanolamine and pentryl. The olefins under ordinary conditions combine with two atoms of chlorine, adding them readily to the unsaturated linkage, and thereafter react with chlorine no further. It has been found that chlorine does not add to hot propylene in the gas phase, but substitutes instead, one of the hydrogen atoms of the methyl group being replaced and allyl chloride being formed. This at a lower temperature adds chlorine normally to form 1,2,3-trichloropropane which gives glycerin on hydrolysis.

$$
\begin{array}{ccccc}
CH_3 & CH_2-Cl & CH_2-Cl & CH_2-OH & CH_2-ONO_2 \\
| & | & | & | & | \\
CH & \longrightarrow CH & \longrightarrow CH-Cl & \longrightarrow CH-OH & \longrightarrow CH-ONO_2 \\
\| & \| & | & | & | \\
CH_2 & CH_2 & CH_2-Cl & CH_2-OH & CH_2-ONO_2
\end{array}
$$

Nitroglycerin is formed and remains in solution if glycerin is added to a large excess of strong nitric acid. Heat is evolved, and

FIGURE 52. Nitroglycerin Nitrating House. (Courtesy E. I. du Pont de Nemours and Company, Inc.)

cooling is necessary. The nitroglycerin is thrown out as a heavy oil when the solution is diluted with water. A further quantity of the substance is procured by extracting the dilute acid liquors with chloroform. Naoúm [11] reports that 100 grams of glycerin treated in this manner with 1000 grams of 99% nitric acid yields 207.2 grams of nitroglycerin analyzing 18.16% nitrogen (calc. 18.50% N) and containing a small amount of dinitroglycerin (glyceryl dinitrate). The yield of the trinitrate may be improved by the addition to the nitric acid of dehydrating agents such as phosphorus pentoxide, calcium nitrate, or strong sulfuric acid.

[11] Op. cit., pp. 25, 26.

Thus, if 100 grams of glycerin is added with cooling to a solution of 150 grams of phosphorus pentoxide in the strongest nitric acid, phosphoric acid precipitates as a heavy syrupy layer and the supernatant acid liquid on dilution yields about 200 grams of nitroglycerin. The yield is substantially the same if the glycerin is first dissolved in the nitric acid alone and if the phosphorus pentoxide is added afterwards. One hundred grams of glycerin in 500 grams of the strongest nitric acid, 400 grams of anhydrous

FIGURE 53. Nitroglycerin Nitrator. (Courtesy E. I. du Pont de Nemours and Company, Inc.)

calcium nitrate being added and the mixture allowed to stand for some hours, gives on drowning and purification 220 grams of nitroglycerin which contains about 10% of glyceryl dinitrate.

All these methods are too expensive, for the excess of nitric acid is lost or has to be recovered from dilute solution. A process in which the nitroglycerin comes out as a separate phase without the spent acid being diluted is preferable—and it is indeed true that the addition of strong sulfuric acid to a solution of glycerin in strong nitric acid completes the esterification and causes the nitroglycerin to separate out. Since the strongest nitric acid is expensive to manufacture, and since a mixture of less strong nitric acid with oleum (sulfuric acid containing free sulfur trioxide) may be identical in all respects with a mixture of strong nitric and strong sulfuric acids, glycerin is universally nitrated in commercial practice by means of acid already mixed, and the nitroglycerin is procured by means of gravity separation of the phases. One hundred parts by weight of glycerin yield 225 to 235 parts of nitroglycerin.

One part of glycerin is nitrated with about 6 parts of mixed acid, made up by the use of oleum and containing about 40.0%

FIGURE 54. Interior of Nitroglycerin Storage House. (Courtesy E. I. du Pont de Nemours and Company, Inc.)

of nitric acid, 59.5% of sulfuric acid, and 0.5% of water. The nitration in this country is carried out in cast iron or steel nitrators, in Europe in nitrators of lead. The glycerin is commonly added from a cock, controlled by hand, in a stream about the size of a man's finger. The mixture is stirred by compressed air, and the temperature is controlled carefully by means of brine coils, there being usually two thermometers, one in the liquid, one in the gas phase above it. In Great Britain the temperature of the nitration mixture is not allowed to rise above 22°C., in this country generally not above 25°. If the temperature for any reason gets out of control, or if the workman sees red fumes through the window in the nitrator, then the charge is dumped quickly into a drowning tank and the danger is averted. The safety precautions which are everywhere exercised are such that the explosion of a nitroglycerin plant is a rare occurrence. After all the glycerin has been added to the nitrator, agitation and cooling are continued until the temperature drops to about 15°,

FIGURE 55. Nitroglycerin Buggy. (Courtesy Hercules Powder Company.) For transporting nitroglycerin from the storage house to the house where it is mixed with the other ingredients of dynamite. Note the absence of valves and the use of wooden hose clamps as a safety precaution.

and the mixture is then run off to the *separator* where the nitroglycerin rises to the top. The spent acid contains 9 to 10% of nitric acid, 72 to 74% of sulfuric acid, and 16 to 18% of water.

The nitroglycerin from the separator contains about 10% of its weight of dissolved acid (about 8% nitric and about 2% sulfuric). Most of this is removed by a *drowning wash* or *prewash* carried out, in Europe with water at about 15°, in this country with water at 38° to 43°, while the mixture is agitated with compressed air. The higher temperature reduces the viscosity of the nitroglycerin and increases greatly the efficiency of the washing. The nitroglycerin is heavier than water and sinks rapidly to the bottom. It is washed again with water, then with sodium carbonate solution (2 or 3%), and then with water until the washings give no color with phenolphthalein and the nitroglycerin itself is neutral to litmus paper. In this country the nitroglycerin is sometimes given a final wash with a concentrated solution of common salt. This reduces the moisture which is suspended in it, to about the same extent as the filtration to which it is commonly subjected in European practice. The nitroglycerin then goes to storage tanks in a heated building where there is no danger of freezing. It has a milky appearance at first, but this quickly disappears. After one day of storage it generally contains not more than 0.3 or 0.4% of moisture, and this amount does not interfere with its use for the manufacture of dynamite.

Pure nitroglycerin is odorless at ordinary temperatures, but has a faint and characteristic odor at temperatures above 50°. Its specific gravity is 1.6009 at 15° and 1.5910 at 25°.[12] It contracts on freezing. Its vapor pressure has been reported by Marshall and Peace[13] to be 0.00025 mm. at 20°, 0.00083 mm. at 30°, 0.0024 at 40°, 0.0072 at 50°, 0.0188 at 60°, 0.043 at 70°, 0.098 at 80°, and 0.29 mm. at 93.3°. About 5 cc. of nitroglycerin passes over with one liter of water in a steam distillation. Snelling and Storm[14] heated nitroglycerin at atmospheric pressure in a distillation apparatus behind an adequate barricade. They reported that

Nitroglycerin begins to decompose at temperatures as low as 50° or 60°C. . . . At a temperature of about 135°C. the decomposition of nitroglycerin is so rapid as to cause the liquid to become of a strongly reddish color, owing to the absorption of the nitrous fumes resulting from that which is decomposed; and at a temperature of about 145°C. the evolution of decomposition products is so rapid that, at atmospheric pressures, ebullition begins, and the liquid

12 Perkin, *J. Chem. Soc.*, **55**, 685 (1879).
13 *J. Soc. Chem. Ind.*, **109**, 298 (1916).
14 *U. S. Bur. Mines Tech. Paper* 12, "The Behavior of Nitroglycerine When Heated," Washington, 1912.

"boils" strongly. This "boiling" is due in part to the evolution of decomposition products (mainly oxides of nitrogen and water vapor) and in part to the actual volatilization of nitroglycerin itself.

FIGURE 56. C. G. Storm. Author of numerous articles and government publications on the properties, testing, and analysis of smokeless powder and high explosives. Explosives Chemist at Navy Powder Works, 1901-1909, at U. S. Bureau of Mines, 1909-1915; Directing Chemist, Aetna Explosives Company, 1915-1917; Major and Lieutenant-Colonel, Ordnance Department, 1917-1919; Research Chemist, Trojan Powder Company, 1919; Chief Explosives Chemical Engineer, Office of the Chief of Ordnance, War Department, 1919-1942; since early in 1942, Technical Director, National Fireworks, Inc.

. . . At temperatures between 145° and 215°C. the ebullition of nitroglycerin becomes more and more violent; at higher temperatures the amount of heat produced by the decomposing liquid becomes proportionately greater, and at about 218°C. nitroglycerin explodes.[15]

When nitroglycerin is maintained at a temperature between 145° and 210°C., its decomposition goes on rapidly, accompanied by much volatilization, and under these conditions nitroglycerin may be readily distilled. The distillate consists of nitroglycerin, nitric acid, water, and other decomposition products. The residue that remains after heating nitroglycerin under such conditions for some time probably consists mainly of glycerin, with small amounts of dinitroglycerin, mononitroglycerin, and other decomposition products. These substances are far less explosive than ordinary nitroglycerin, and accordingly by heating nitroglycerin slowly it can be caused to "boil" away until the residue consists of products that are practically non-explosive. In a number of experiments nitroglycerin was thus heated, and a copious residue was obtained. By carefully raising the temperature this residue could be made to char without explosion.

Belyaev and Yuzefovich [16] heated nitroglycerin and other explosives in vacuum; and procured the results summarized in the following table. The fact that ignition temperatures are fairly

	B.P. (2 mm.) EXPERIMENTAL, °C.	B.P. (50 mm.) EXPERIMENTAL, °C.	B.P. (760 mm.) MOST PROBABLE VALUE, °C.	IGNITION TEMPERATURE, °C.
Methyl nitrate	...	5	66	...
Glycol dinitrate	70	125	197 ± 3	195–200
TNT	190	245–250	300 ± 10	295–300
Picric acid	195	255	325 ± 10	300–310
TNB	175	250	315 ± 10	...
PETN	160	180	200 ± 10	215
Nitroglycerin	125	180	245 ± 5	200

close to probable boiling points indicates that high concentrations of vapor exist at the moment when the substances ignite. The authors point out that TNT, PETN, and picric acid neither detonate nor burn in vacuum and suggest that this is probably

[15] Munroe had found the "firing temperature" of nitroglycerin to be 203 to 205°, *J. Am. Chem. Soc.*, **12**, 57 (1890).

[16] *Comp. rend. acad. sci. U.S.S.R.*, **27**, 133 (1940).

because the boiling points in vacuum are considerably below the ignition temperatures.

Nitroglycerin crystallizes in two forms, a stable form, dipyramidal rhombic crystals, which melt or freeze at 13.2–13.5°, and a labile form, glassy-appearing triclinic crystals, m.p. 1.9–2.2°. It does not freeze readily or quickly. When cooled rapidly, it becomes more and more viscous and finally assumes the state of a hard glassy mass, but this is not true freezing, and the glassy mass becomes a liquid again at a temperature distinctly below the melting point of the crystalline substance. Nitroglycerin in dynamite freezes in crystals if the explosive is stored for a considerable length of time at low temperatures, the form in which it solidifies being determined apparently by the nature of the materials with which it is mixed.[17] If liquid nitroglycerin is cooled strongly, say to −20° or −60°, stirred with a glass rod, and seeded with particles of one or the other form, then it crystallizes in the form with which it has been seeded. If the solid is melted by warming, but not warmed more than a few degrees above its melting point, it will on being cooled solidify in the form, whether labile or stable, from which it had been melted. If, however, it is warmed for some time at 50°, it loses all preference for crystallizing in one form rather than in the other, and now shows the usual phenomena of supercooling when it is chilled. Crystals of the labile form may be preserved sensibly unchanged for a week or two, but gradually lose their transparency and change over to the stable form. Crystals of the stable form cannot be changed to the labile form except by melting, warming above the melting point, and seeding with the labile form.

Nitroglycerin is miscible in all proportions at ordinary temperatures with methyl alcohol, acetone, ethyl ether, ethyl acetate, glacial acetic acid, benzene, toluene, nitrobenzene, phenol, chloroform, and ethylene chloride, and with homologous nitric esters such as dinitroglycerin, dinitrochlorohydrin, nitroglycol, and trimethyleneglycol dinitrate. Absolute ethyl, propyl, isopropyl, and amyl alcohols mix with nitroglycerin in all proportions if they are hot, but their solvent power falls off rapidly at lower temperatures. One hundred grams of absolute ethyl alcohol dissolves

[17] Hibbert, *Z. ges. Schiess- u. Sprengstoffw.*, **9**, 83 (1914).

37.5 grams of nitroglycerin at 0°, 54.0 grams at 20°. One hundred grams of nitroglycerin on the other hand dissolves 3.4 grams of ethyl alcohol at 0°, 5.5 grams at 20°.

Nitroglycerin dissolves aromatic nitro compounds, such as dinitrotoluene and trinitrotoluene, in all proportions when warm. When the liquids are cooled, 100 grams of nitroglycerin at 20° still holds in solution 35 grams of DNT or 30 grams of TNT. Both nitroglycerin and the polynitro aromatic compounds are solvents or gelatinizing agents for nitrocellulose.

Nitroglycerin dissolves in concentrated sulfuric acid with the liberation of its nitric acid, and may therefore be analyzed by means of the nitrometer (see below).

Nitroglycerin is destroyed by boiling with alcoholic sodium or potassium hydroxide, but glycerin is not formed; the reaction appears to be in accordance with the following equation.

$$C_3H_5(ONO_2)_3 + 5KOH \longrightarrow$$
$$KNO_3 + 2KNO_2 + H-COOK + CH_3-COOK + 3H_2O$$

This however is not the whole story, for resinous products, oxalic acid, and ammonia are also formed. If the reaction with caustic alkali is carried out in the presence of thiophenol, some glycerin is formed and the thiophenol is oxidized to diphenyl sulfide. Alkali sulfides, K_2S, KHS, and CaS, also yield glycerin.

Nitroglycerin vapors cause severe and persistent headache. A workman who is exposed to them constantly soon acquires an immunity. If he is transferred to another part of the plant, he may retain his immunity by paying a short visit every few days to the area in which the nitroglycerin is being used. Workmen appear to suffer no ill effects from handling the explosive continually with the naked hands. Nitroglycerin relaxes the arteries, and is used in medicine under the name of *glonoin. Spirit of glonoin* is a 1% solution of nitroglycerin in alcohol. The usual dose for angina pectoris is one drop of this spirit taken in water, or one lactose or dextrose pellet, containing $\frac{1}{100}$ grain (0.0006 gram) of nitroglycerin, dissolved under the tongue.

Nitroglycerin is not easily inflammable. If a small quantity is ignited, it burns with a slight crackling and a pale green flame—and may be extinguished readily before all is burned. If a larger amount is burned in such manner that the heat accumulates and the temperature rises greatly, or if local overheating occurs as by burning in an iron pot, then an explosion ensues. The explosion of nitroglycerin by heat is conveniently demonstrated by heating a stout steel plate to dull redness, removing the source of heat, and allowing the nitroglycerin to fall drop by drop slowly onto the plate while it is cooling. At first the drops assume the spheroidal condition when they strike the plate and deflagrate or burn with a flash, but when the plate cools somewhat each drop yields a violent explosion.

Nitroglycerin is very sensitive to shock, and its sensitivity is greater if it is warm. A drop of the liquid on a steel anvil, or a drop absorbed by filter paper and the paper placed upon the anvil, is detonated by the blow of a steel hammer. The shock of iron striking against stone, or of porcelain against porcelain, also explodes nitroglycerin, that of bronze against bronze less readily, and of wood against wood much less so. Stettbacher [18] reports drop tests with a 2-kilogram weight: mercury fulminate 4.5 cm., lead azide 9 cm., nitroglycerin 10–12 cm., blasting gelatin 12–15 cm., and tetryl 30–35 cm. He also reports the observations of Kast and Will and of Will that nitroglycerin at 90° requires only half as much drop to explode it as nitroglycerin at ordinary temperature, while the frozen material requires about three times as much.

Nitroglycerin and nitroglycerin explosives, like all other high explosives, show different velocities of detonation under different conditions of initiation and loading. They are sometimes described as having low and high velocities of detonation. Berthelot found for nitroglycerin a velocity of 1300 meters per second in lead or tin tubes of 3 mm. internal diameter. Abel [19] found 1525 meters per second in lead pipe 30 mm. internal diameter, while Mettegang [20] found 2050 meters per second in iron pipes of the same internal diameter. Comey and Holmes [21] working with pipes of 25–37.5 mm. internal diameter found values varying from 1300–1500 to 8000–8500 meters per second, and, with especially strong detonators, they regularly found velocities between 6700

[18] Stettbacher, "Die Schiess- und Sprengstoffe," Leipzig, 1919, p. 124.
[19] *Phil. Trans.*, **156**, 269 (1866); **157**, 181 (1867).
[20] *Internat. Congr.*, **2**, 322 (1903).
[21] *Z. ges. Schiess- u. Sprengstoffw.*, **8**, 306 (1913).

and 7500 meters per second. Naoúm [22] reports that blasting gelatin (92–93% NG, 7–8% collodion nitrocotton) has a low velocity of 1600–2000 meters per second and a high velocity of about 8000. Blasting gelatin filled with air bubbles always shows the higher velocity, while clear and transparent blasting gelatin almost always shows the lower velocity of detonation. Frozen dynamite is more difficult to initiate, but always detonates at the high velocity.[23]

Certain properties of nitroglycerin and of other explosives, reported by Brunswig,[24] are tabulated below and compared in a manner to show the relative power of the substances. The spe-

	Specific Volume, Liters	Explo- sion Temper- ature, °C.	Heat of Explo- sion, Calories	Charac- teristic Product
Nitroglycerin	712	3470	1580	1,125,000
Nitromannite	723	3430	1520	1,099,000
Blasting gelatin (93% NG, 7% NC)	710	3540	1640	1,164,000
75% Guhr dynamite	628	3160	1290	810,000
Nitrocotton (13% N)	859	2710	1100	945,000
Picric acid	877	2430	810	710,000
Black powder	285	2770	685	195,000
Ammonium nitrate	937	2120	630	590,000
Mercury fulminate	314	3530	410	129,000

cific volume is the volume, at 0° and 760 mm., of the gaseous products of the explosion. This number multiplied by the heat of explosion gives the *characteristic product* which Berthelot considered to be a measure of the mechanical work performed by the explosion. The mechanical work has also been estimated, differently, in kilogram-meters by multiplying the heat of explosion by 425, the mechanical equivalent of heat.

Naoúm [25] reports the results of his own experiments with nitroglycerin and with other explosives in the Trauzl lead block test (sand tamping), 10-gram samples, as shown below. The Trauzl test is essentially a measure of brisance, but for explosives of similar

[22] *Op. cit.*, p. 145.
[23] Herlin, *Z. ges. Schiess- u. Sprengstoffw.*, **9**, 401 (1914).
[24] Brunswig, "Explosivstoffe," 1909, cited by Naoúm, *op. cit.*, p. 152.
[25] *Op. cit.*, p. 156.

velocities of detonation it supplies a basis for the comparison of
their total energies.

	EXPANSION, CUBIC CENTIMETERS
Nitroglycerin	550
Nitromannite	560
Compressed guncotton (13.2% N)	420
Blasting gelatin	580
65% Gelatin dynamite	410
75% Guhr dynamite	325
Tetryl	360
Picric acid	300
Trinitrotoluene	285
Mercury fulminate	150

For several years after the discovery of nitroglycerin, the
possibility of using it as an explosive attracted very little interest. Indeed, it first came into use as a medicine, and the first
serious study on its preparation, after the work of Sobrero, was
made by J. E. de Vrij, professor of chemistry in the Medical
School at Rotterdam, and published in the Dutch journal of
pharmacy, *Tijdschrift voor wetensch. pharm.*, in 1855. The next
significant work was done by Alfred Nobel who in 1864
patented [26] improvements both in the process of manufacturing
nitroglycerin and in the method of exploding it. No liquid explosive had been successful in practical use. Nobel believed that he
had solved the difficulty by taking advantage of the property of
nitroglycerin of exploding from heat or from the shock of an
explosion. A small glass vessel containing black powder was to
be immersed in the nitroglycerin and exploded. Another method
was by the local heat of an electric spark or of a wire electrically
heated under the surface of the nitroglycerin. And another was
the percussion cap. Nobel used black powder first in glass bulbs,
later in hollow wooden cylinders closed with cork stoppers, then
a mixture of black powder and mercury fulminate, and later
fulminate in small lead capsules and finally in the copper detonators which are still in general use. The invention of the blasting
cap depended upon the discovery of the phenomenon of initiation, and signalized the beginning of a new era in the history of

[26] Brit. Pat. 1813 (1864).

explosives. Blasting caps were used first for the safe and certain explosion of the dangerous liquid nitroglycerin, but presently they were found to be exactly what was needed for the explosion of the safer and less sensitive dynamites which Nobel also invented.

The first establishment for the manufacture of nitroglycerin in industrial quantities was a laboratory set up by Alfred Nobel and his father, Immanuel Nobel, probably in the autumn of 1863, near the latter's home at Heleneborg near Stockholm. An explosion which occurred there in September, 1864, cost the life of Alfred's younger brother, Emil, and of four other persons. The manufacture of nitroglycerin was prohibited within the city area, but the explosive was already in practical use for the tunnelling operations of the State Railway, and it was desirable to continue its manufacture. The manufacture was removed to a pontoon moored in Malar Lake and was continued there during the late autumn of 1864 and during the following winter until March, 1865, when it was transferred to a new factory, the first real nitroglycerin factory in the world, at Winterwik near Stockholm. Later in the same year the Nobel company commenced manufacturing nitroglycerin in Germany, at a plant near Hamburg, and within a few years was operating explosives factories in the United States and in all the principal countries of Europe.[27]

The first considerable engineering operation in the United States to be accomplished by means of nitroglycerin was the blasting out of the Hoosac tunnel in Massachusetts. The work had been progressing slowly until George M. Mowbray,[28] an "operative chemist" of North Adams, was engaged to manufacture nitroglycerin at the site of the work and to supervise its use. Twenty-six feet of tunnel was driven during May, 1868, 21 during June, 47 during July when the use of nitroglycerin commenced, 44 during August, and 51 feet during September. Mowbray profited by the observation of W. P. Granger that frozen nitroglycerin could not be detonated, and accordingly transported his

[27] Cf. Schück and Sohlman, "The Life of Alfred Nobel," London, William Heinemann, Ltd., 1929.

[28] His experiences and methods are told in a very interesting manner in his book, "Tri-Nitro-Glycerine, as Applied in the Hoosac Tunnel," third edition, rewritten, New York and North Adams, 1874.

material in safety in the frozen condition.[29] He described an explosion which occurred in December, 1870, in which the life of a foreman was lost, and another in March, 1871, in which a large amount of frozen nitroglycerin failed to explode.

The new magazine had hardly been completed, and stored with nitroglycerine, when, on Sunday morning, at half past six o'clock, March twelfth, 1871, the neighborhood was startled by another explosion of sixteen hundred pounds of nitroglycerine. The cause of this last explosion was continuous overheating of the magazine. . . . The watchman con-

[29] During the severe winter of 1867 and 1868, the Deerfield dam became obstructed with ice, and it was important that it should be cleared out without delay. W. P. Granger, Esq., engineer in charge, determined to attempt its removal by a blast of nitroglycerine. In order to appreciate the following details, it must be borne in mind that the current literature of this explosive distinctly asserted that, when congealed, the slightest touch or jar was sufficient to explode nitroglycerine. Mr. Granger desired me to prepare for him ten cartridges, and, as he had to carry them in his sleigh from the west end of the tunnel to the east end or Deerfield dam, a distance of nine miles over the mountain, he requested them to be packed in such a way that they would not be affected by the inclement weather. I therefore caused the nitroglycerine to be warmed up to ninety degrees, warmed the cartridges, and, after charging them, packed them in a box with sawdust that had been heated to the same temperature; the box was tied to the back of the sleigh, with a buffalo robe thrown over it. In floundering across the divide where banks, road, hedge and water courses were indistinguishable beneath the drifted snow, horse, sleigh and rider were upset, the box of cartridges got loose, and were spread indiscriminately over the snow. After rectifying this mishap, picking up the various contents of the sleigh, and getting ready to start again, it occurred to Mr. Granger to examine his cartridges; his feelings may be imagined when he discovered the nitroglycerine frozen solid. To have left them behind and proceeded to the dam, where miners, engineers and laborers were waiting to see this then much dreaded explosive, would never do; so accepting the situation, he replaced them in the case, and, laying it between his feet, proceeded on his way, thinking a heap but saying nothing. Arrived, he forthwith attached fuse, exploder, powder and some guncotton, and inserted the cartridge in the ice. Lighting the fuse, he retired to a proper distance to watch the explosion. Presently a sharp crack indicated that the fuse had done its work, and, on proceeding to the hole drilled in the ice, it was found that fragments of the copper cap were imbedded in the solid cylinder of congealed nitroglycerine, which was driven through and out of the tin cartridge into the anchor ice beneath, but not exploded. A second attempt was attended with like results. Foiled in attempting to explode the frozen nitroglycerine, Mr. Granger thawed the contents of another cartridge, attached the fuse and exploder as before; this time the explosion was entirely successful. From that day I have never transported nitroglycerine except in a frozen condition, and to that lesson are we indebted for the safe transmission of more than two hundred and fifty thousand pounds of this explosive, over the roughest roads of New Hampshire, Vermont, Massachusetts, New York, and the coal and oil regions of Pennsylvania, in spring wagons with our own teams.

Mowbray, *op. cit.,* pp. 45–46.

fessed he had neglected to examine the thermometer, made his fire under the boiler, and gone to bed. . . . Fortunately, this accident involved no damage to life or limb, whilst a very instructive lesson was taught in the following circumstance: Within twelve feet of the magazine was a shed, sixteen feet by eight, containing twelve fifty-pound cans of congealed nitroglycerine ready for shipment. This shed was utterly destroyed, the floor blasted to splinters, the joists rent to fragments, the cans of congealed nitroglycerine driven into the ground, the tin of which they were composed perforated, contorted, battered, and portions of tin and nitroglycerine sliced off but not exploded. Now, this fact proves one of two things: Either that the tri-nitroglycerine made by the Mowbray process, differs from the German nitroglycerine in its properties, or the statements printed in the foreign journals, as quoted again and again, that nitroglycerine, when congealed, is more dangerous than when in the fluid state, are erroneous.[30]

Mowbray used his nitroglycerin in the liquid state, either loaded in cylindrical tin cannisters or cartridges, or poured directly into the bore hole, and exploded it by means of electric detonators. The electric detonators were operated by means of a static electric machine which caused a spark to pass between points of insulated wire; the spark set fire to a priming mixture made from copper sulfide, copper phosphide, and potassium chlorate; and this fired the detonating charge of 20 grains of mercury fulminate contained in a copper capsule, the whole being waterproofed with asphaltum varnish and insulated electrically with gutta-percha. The devices were so sensitive that they could be exploded by the static electricity which accumulated on the body of a miner operating a compressed air drill, and they required corresponding precautions in their use.

Liquid nitroglycerin is still used as an explosive to a limited extent, particularly in the blasting of oil wells, but its principal use is in the manufacture of dynamite and of the propellants, ballistite and cordite.

Dinitroglycerin (Glyceryl dinitrate)

Dinitroglycerin does not differ greatly from nitroglycerin in its explosive properties. It is appreciably soluble in water, and more expensive and more difficult to manufacture than nitrogly-

30 *Ibid.*, pp. 44–45.

cerin. It mixes with the latter substance in all proportions and lowers its freezing point, and was formerly used in Germany in such mixtures in non-freezing dynamites. It has now been superseded entirely for that purpose by dinitrochlorohydrin which is insoluble in water, and cheaper and more convenient to manufacture.

Dinitroglycerin is never formed alone by the nitration of glycerin but is always accompanied by the trinitrate or the mononitrate or both. If the nitration is carried out in a manner to give the best yields of the dinitrate, then considerable trinitrate is formed: if the process is modified to reduce the yield of trinitrate, then the yield of dinitrate is also reduced and some mononitrate is formed. If 3 or 4 parts by weight of nitric acid is added slowly to 1 part of glycerin, so that the glycerin or its nitrated product is always in excess, then the dinitrate is the principal product. If the order of mixing is reversed, so that the glycerin dissolves first in the strong nitric acid, then the yield of trinitrate is more considerable. Dinitroglycerin is formed if glycerin is added to mixed acid which is low in nitric acid or high in water, or which contains insufficient sulfuric acid for the necessary dehydrating action. It is also one of the products of the hydrolysis of nitroglycerin by cold concentrated (95%) sulfuric acid, the trinitrate by this reagent being in part dissolved and in part converted to the dinitrate, the mononitrate, and to glyceryl sulfate according to the relative amount of sulfuric acid which is used. Dinitroglycerin is separated from its mixture with nitroglycerin and obtained pure by treating the oil with about 15 volumes of water, separating the insoluble trinitrate, extracting the aqueous solution with ether, washing the ether with dilute sodium carbonate solution, and evaporating. The resulting dinitroglycerin gives a poor heat test because of the peroxide which it contains from the ether. Material which gives an excellent heat test may be procured by evaporating the aqueous solution in vacuum.

The dinitroglycerin obtained by the nitration of glycerin is a colorless, odorless oil, more viscous and more volatile than nitroglycerin. It causes the same kind of a headache. It has a specific gravity of 1.51 at 15°, boils at 146–148° at 15 mm. with only slight decomposition, and solidifies at −40° to a glassy solid which melts if the temperature is raised to −30°. It is readily soluble in alcohol, ether, acetone, and chloroform, somewhat less soluble than nitroglycerin in benzene, and insoluble in carbon tetrachloride and ligroin. It consists of a mixture of the two possible structural isomers, the 1,2- or α,β-dinitrate, known also as "dinitroglycerin F," and the 1,3- or α,α'-dinitrate or "dinitroglycerin K." Both are uncrystallizable oils, and both are hygroscopic and take up about 3% of their weight of moisture from the air. They are separated by virtue of the fact that the α,α'-dinitrate forms a hydrate [31] with one-third of a molecule of water, $C_3H_6O_7N_2 + \frac{1}{3} H_2O$, water-clear prisms, m.p. 26°. No hydrate of the α,β-dinitrate has ever been isolated in the state of a crystalline solid. If a test portion of the moist mixture of the isomers is mixed with fuller's earth and chilled strongly, it deposits crystals; and if these are used for seeding the principal quantity of the moist dinitroglycerin, then the hydrate of the α,α'-dinitrate crystallizes out. It may be recrystallized from water, or from alcohol, ether, or benzene without losing its water of crystallization, but it yields the anhydrous α,α'-dinitrate if it is dried over sulfuric acid or warmed in the air at 40°.

The chemical relationships between the mononitroglycerins and dinitroglycerins supply all the evidence which is needed for inferring the identities of the isomers. Of the two mononitrates, the β-compound obviously cannot yield any α,α'-dinitrate by nitration; it can yield only the α,β-. That one of the two isomers which yields only one dinitrate is therefore the β-mononitrate, and the dinitrate which it yields is the α,β-dinitrate. The α-mononitrate on the other hand yields both the α,β- and the α,α'-dinitrates.

Nitroglycide

[31] Will, *Ber.*, **41**, 1113 (1908).

Both of the dinitroglycerins on treatment with 30% sodium hydroxide solution at room temperature yield nitroglycide, and this substance on boiling with water gives α-mononitroglycerin, a series of reactions which demonstrates the identity of the last-named compound.

Dinitroglycerin is a feeble acid and gives a wine-red color with blue litmus, but none of its salts appear to have been isolated and characterized. It does not decompose carbonates, but dissolves in caustic alkali solutions more readily than in water. One hundred parts of water alone dissolves about 8 parts at 15° and about 10 parts at 50°.

Dinitroglycerin gelatinizes collodion nitrocotton rapidly at ordinary temperature. The gel is sticky, less elastic, and more easily deformed than a nitroglycerin gel. Unlike the latter it is hygroscopic, and becomes softer and greasier from the absorption of moisture from the air. Water dissolves out the dinitroglycerin and leaves the nitrocellulose as a tough, stiff mass.

Dinitroglycerin has about the same sensitivity to initiation as nitroglycerin, only slightly less sensitivity to shock, and offers no marked advantages from the point of view of safety. It shows a greater stability in the heat test, and a small amount can be evaporated by heat without explosion or deflagration. It gives off red fumes above 150°, and at 170° decomposes rapidly with volatilization and some deflagration, or in larger quantities shows a tendency to explode.

Naoúm [32] reports that a 10-gram sample of dinitroglycerin in the Trauzl test with water tamping gave a net expansion of about 500 cc., or 83.3% as much as the expansion (600 cc.) produced by 10 grams of nitroglycerin under the same conditions. He points out that the ratio here is almost the same as the ratio between the heats of explosion, and that in this case the Trauzl test has supplied a fairly accurate measure of the relative energy contents of the two explosives. In the small lead block test the effect of the greater brisance and higher velocity of detonation of nitroglycerin becomes apparent; 100 grams of dinitroglycerin gave a compression of 21 mm. while the same amount of nitroglycerin gave one of 30 mm.

[32] Op. cit., p. 170.

Mononitroglycerin (Glyceryl mononitrate)

Mononitroglycerin is a by-product in the preparation of dinitroglycerin and is separated from the latter substance by its greater solubility in water. It is usually obtained as a colorless oil, density 1.417 at 15°, more viscous than dinitroglycerin and less viscous than nitroglycerin. This oil is a mixture of the two isomers which are crystalline when separate but show little tendency to crystallize when they are mixed. α-Mononitroglycerin when pure consists of colorless prisms, m.p. 58–59°, specific gravity 1.53 at 15°; it yields both of the dinitrates on nitration. The β-compound crystallizes in dendrites and leaflets, m.p. 54°, and is more readily soluble in ether than the α-compound; it yields only the α,β-dinitrate on nitration. Both isomers boil at 155–160° at 15 mm.

Mononitroglycerin resembles glycerin in being very hygroscopic and miscible in all proportions with water and alcohol, and in being only slightly soluble in ether, but it differs from glycerin in being freely soluble in nitroglycerin. It does not form a satisfactory gel with collodion cotton. Its aqueous solution reacts neutral. It appears to be perfectly stable on moderate heating, but decomposes to some extent at 170°, gives off gas, and turns yellow.

Mononitroglycerin is insensitive to shock. In the form of oil it is not detonated by a No. 8 blasting cap in the Trauzl test. If the oil is absorbed in fuller's earth, 10 grams gives a net expansion of 75 cc. The crystalline material, however, detonates easily; 10 grams gives an expansion of 245 cc. It is interesting to compare these results, reported by Naoúm,[33] with the results which the same author reports for nitroglycide which is the anhydride of mononitroglycerin. Ten grams of liquid nitroglycide with water tamping and a No. 8 detonator gave a net expansion of 430 cc.; 10 grams absorbed in fuller's earth, with sand tamping, gave 310 cc.; and 10 grams gelatinized with 5% collodion nitrocotton, with sand tamping, gave 395 cc.

Nitroglycide

This substance cannot be prepared by the nitration of glycide, for the action of acids upon that substance opens the ethylene

[33] *Ibid.*, pp. 174, 177.

oxide ring, and mononitroglycerin is formed. Nitroglycide was first prepared by Naoúm [34] in 1907 by shaking dinitroglycerin at room temperature with a 30% aqueous solution of sodium hydroxide. The clear solution presently deposited a colorless oil, and this, washed with water and dried in a desiccator, constituted a practically quantitative yield of nitroglycide.

Nitroglycide is a very mobile liquid with a faint but pleasant aromatic odor, specific gravity 1.332 at 20°. It does not freeze at −20°. It boils at 94° at 20 mm., and with some decomposition at 174–175° at atmospheric pressure. It is not hygroscopic but is distinctly soluble in water, 5 grams in 100 cc. at 20°. Ether will extract nitroglycide from the cool aqueous solution; if the solution is boiled, however, the nitroglycide is hydrated to mononitroglycerin. Nitroglycide is miscible in all proportions with alcohol, ether, acetone, ethyl acetate, and nitroglycerin. It gelatinizes collodion nitrocotton and even guncotton rapidly at ordinary temperature. It explodes in contact with concentrated sulfuric acid. If dissolved in dilute sulfuric acid and then treated with strong sulfuric acid, it gives off nitric acid. It is converted into dinitroglycerin and nitroglycerin by the action of nitric acid. It dissolves in concentrated hydrochloric acid with the evolution of considerable heat, and the solution on dilution with water gives a precipitate of monochlorohydrin mononitrate. Nitroglycide reduces ammoniacal silver nitrate slowly on gentle warming; glycide reduces the same reagent in the cold.

When heated rapidly in a test tube nitroglycide explodes with a sharp report at 195–200°. It is more easily detonated than liquid nitroglycerin. Naoúm believes that its great sensitivity results mainly from the easy propagation of the wave of detonation by a liquid of low viscosity.[35] He points out further that mononitroglycerin has 69.5% of the energy content (i.e., heat of explosion) of nitroglycide, but as a crystal powder in the Trauzl test it gives only about 62% as much net expansion, whence it is to be inferred that nitroglycide has the higher velocity of detonation. Nitroglycide has only 52% of the energy content of nitroglycerin, but produces 72% as much effect in the Trauzl test. It is therefore "relatively more brisant than nitroglycerin."

[34] *Ibid.*, p. 176.
[35] *Ibid.*, p. 178.

Dinitrochlorohydrin (Glycerin chlorohydrin dinitrate)

Among the various substances which may be used in admixture with nitroglycerin for the purpose of lowering its freezing point, dinitrochlorohydrin is preferred in Germany but has not found favor in the United States. Since dinitrochlorohydrin is distinctly safer to prepare than nitroglycerin, it is most commonly prepared by itself, that is, by the nitration of chlorohydrin which is substantially pure and contains not more than a small amount of glycerin. The product is used directly for the preparation of certain explosives, or it is mixed with nitroglycerin for the manufacture of non-freezing dynamites.

Chlorohydrin is prepared by autoclaving glycerin with concentrated hydrochloric acid or by treating it at moderate temperature with sulfur chloride. In the former process, in order to avoid the formation of dichlorohydrin, only enough hydrochloric acid is used to convert about 75% of the glycerin. The product is procured by a vacuum distillation. The monochlorohydrin, which consists almost entirely of the α-compound, comes over between 130° and 150° at 12-15 mm. and the unchanged glycerin between 165° and 180°. It is nitrated with the same mixed acid as is used for the preparation of nitroglycerin; less acid is needed of course, less heat is produced, and the process is safer and more rapid.

$$
\begin{array}{lll}
CH_2-OH & CH_2-Cl & CH_2-Cl \\
| & | & | \\
CH-OH \longrightarrow & CH-OH \longrightarrow & CH-ONO_2 \\
| & | & | \\
CH_2-OH & CH_2-OH & CH_2-ONO_2
\end{array}
$$

If a mixture of chlorohydrin and glycerin is nitrated, the resulting mixture of nitrates contains relatively more nitroglycerin than the original mixture contained of glycerin, for the relative increase of molecular weight during the nitration of glycerin is greater.

Commercial dinitrochlorohydrin is usually yellowish or brownish in color, specific gravity about 1.541 at 15°. It boils at atmospheric pressure with decomposition at about 190°. It may be distilled at 13 mm. at 121.5°, or at 10 mm. at 117.5°, but some decomposition occurs for the distillate is acid to litmus.

Dinitrochlorohydrin is non-hygroscopic, distinctly more volatile than nitroglycerin, and it has similar physiological effects. It can be frozen only with great difficulty, shows a strong tendency to supercool, and can be kept for a long time at −20° without depositing crystals. The solubility of dinitrochlorohydrin and nitroglycerin in each other is so great that only small quantities of nitroglycerin can be frozen out from the mixtures, even after seeding, at winter temperatures. A mixture of 75 parts of nitroglycerin and 25 parts of dinitrochlorohydrin is practically nonfreezing, and yields a dynamite which is not significantly less strong than one made from straight nitroglycerin.

Dinitrochlorohydrin does not take fire readily, and, if ignited, burns rather slowly without detonating and with but little of the sputtering which is characteristic of nitroglycerin mixtures. "Even larger quantities of pure dinitrochlorohydrin in tin cans burn without explosion when in a fire, so that liquid dinitrochlorohydrin is permitted on German railroads in tin cans holding 25 kg., as a safe explosive for limited freight service in the 200 kg. class, while liquid nitroglycerin is absolutely excluded." [36] Dinitrochlorohydrin is more stable toward shock than nitroglycerin. Naoúm, working with a pure sample, was not able to secure a first-rate explosion in the drop test.[37] A 2-kilogram weight dropped from a height of 40 cm. or more gave a very slight partial decomposition and a slight report, from a height of 75 cm. or more, a somewhat more violent partial deflagration but in no case a sharp report, and even a 10-kilogram weight dropped from a height of 10 or 15 cm. gave a very weak partial decomposition. The substance, however, is detonated readily by fulminate. It gives in the Trauzl test a net expansion of 475 cc., or 79% of the 600-cc. expansion given by nitroglycerin, although its heat of explosion is only about 71% of the heat of explosion of nitroglycerin.

Dinitrochlorohydrin produces hydrogen chloride when it explodes. This would tend to make it unsuitable for use in mining explosives were it not for the fact that the incorporation into the explosives of potassium or sodium nitrate sufficient to form chloride with the chlorine of the dinitrochlorohydrin prevents it altogether—and this amount of the nitrate is usually present anyway for other reasons.

[36] Naoúm, *ibid.*, p. 187.
[37] *Ibid.*, p. 188.

Acetyldinitroglycerin (monoacetin dinitrate) and formyldi-nitroglycerin (monoformin dinitrate) have been proposed by Vender [38] for admixture with nitroglycerin in non-freezing dynamite. The former substance may be prepared [39] by nitrating monoacetin or by acetylating dinitroglycerin. The latter substance may be procured already mixed with nitroglycerin by warming glycerin with oxalic acid, whereby monoformin (glyceryl monoformate) is formed, and nitrating the resulting mixture of monoformin and glycerin. Formyldinitroglycerin has apparently not yet been isolated in the pure state. These substances are satisfactory explosives but are more expensive to manufacture than dinitrochlorohydrin over which they possess no distinct advantage, and they have not come into general use.

Tetranitrodiglycerin (Diglycerin tetranitrate)

If glycerin is heated with a small amount of concentrated sulfuric acid, ether formation occurs, water splits out, and diglycerin and polyglycerin are formed. If the heating is carried out in the absence of acids, and in such a way that the water which is formed is allowed to escape while the higher-boiling materials are condensed and returned, especially if a small amount of alkali, say 0.5%, or of sodium sulfite is present as a catalyst, then the principal product is diglycerin and not more than a few per cent of polyglycerin is formed. It is feasible for example to convert glycerin into a mixture which consists of 50–60% diglycerin, 4–6% polyglycerin, and the remainder, 34–46%, unchanged glycerin. The diglycerin is ordinarily not isolated in the pure state. The mixture, either with or without the addition of glycerin, is nitrated directly to form a mixture of tetranitrodiglycerin and nitroglycerin which is used for the manufacture of non-freezing dynamite.

Diglycerin when obtained pure by a vacuum distillation is a water-white liquid, more viscous and more dense than glycerin, sweet-tasting, very hygroscopic, b.p. 245–250° at 8 mm. It is nitrated with the same mixed acid as glycerin, although a smaller

[38] *Z. ges. Schiess- u. Sprengstoffw.*, **2**, 21 (1907). *Fourth Internat. Congr.* **2**, 582 (1906).

[39] Ger. Pat. 209,943 (1906); Brit. Pat. 9791 (1906); French Pat. 372,267 (1906); Swiss Pat. 50,836 (1910); U. S. Pat. 1,029,519 (1912).

amount is necessary. Salt solutions are always used for washing the nitrated product, otherwise the separation of the phases is extremely slow.

$$
\begin{array}{ccc}
\text{CH}_2\!-\!\text{OH} & \text{CH}_2\!-\!\text{OH} & \text{CH}_2\!-\!\text{ONO}_2 \\
| & | & | \\
\text{CH}\!-\!\text{OH} & \text{CH}\!-\!\text{OH} & \text{CH}\!-\!\text{ONO}_2 \\
| & | & | \\
\text{CH}_2\!-\!\text{O}[\text{H}] & \text{CH}_2 & \text{CH}_2 \\
\qquad\quad \searrow \text{O} & \searrow \text{O} & \searrow \text{O} \\
\text{CH}_2\!-\![\text{OH}] & \text{CH}_2 & \text{CH}_2 \\
| & | & | \\
\text{CH}\!-\!\text{OH} & \text{CH}\!-\!\text{OH} & \text{CH}\!-\!\text{ONO}_2 \\
| & | & | \\
\text{CH}_2\!-\!\text{OH} & \text{CH}_2\!-\!\text{OH} & \text{CH}_2\!-\!\text{ONO}_2
\end{array}
$$

Tetranitrodiglycerin is a very viscous oil, non-hygroscopic, insoluble in water, and readily soluble in alcohol and in ether. It has not been obtained in the crystalline state. It is not a good gelatinizing agent for collodion cotton when used alone. Its mixture with nitroglycerin gelatinizes collodion cotton more slowly than nitroglycerin alone but gives a satisfactory gel. It is less sensitive to mechanical shock than nitroglycerin, about the same as dinitroglycerin, but is readily exploded by fulminate. According to Naoúm [40] 75% tetranitrodiglycerin guhr dynamite gave in the Trauzl test a net expansion of 274 cc. or 85.6% of the expansion (320 cc.) produced by 75% nitroglycerin guhr dynamite.

Nitroglycol (Ethylene glycol dinitrate, ethylene dinitrate)

Nitroglycol first found favor in France as an ingredient of non-freezing dynamites. It has many of the advantages of nitroglycerin and is safer to manufacture and handle. Its principal disadvantage is its greater volatility. Formerly the greater cost of procuring glycol, which is not as directly accessible as glycerin but has to be produced by synthesis from ethylene, was an impediment to its use, but new sources of ethylene and new methods of synthesis have reduced its cost and increased its accessibility.

Ethylene was formerly procured from alcohol (itself produced from raw material which was actually or potentially a foodstuff) by warming with sulfuric acid, by passing the vapors over heated coke impregnated with phosphoric acid, or by comparable methods. Ethylene combines with bromine to give ethylene dibromide,

[40] *Op. cit.*, p. 201.

which yields glycol by hydrolysis, but bromine is expensive. Ethylene also combines readily with chlorine, but, even if care is exercised always to have the ethylene present in excess, substitution occurs, and tri- and tetrachloroethane are formed along with the ethylene dichloride, and these do not yield glycol by hydrolysis. Ethylene is now produced in large quantities during the cracking of petroleum. Its comportment with chlorine water has been found to be much more satisfactory for purposes of synthesis than its comportment with chlorine gas. Chlorine water contains an equilibrium mixture of hydrogen chloride and hypochlorous acid.

$$Cl_2 + H_2O \leftrightarrows HCl + HOCl$$

Ethylene adds hypochlorous acid more readily than it adds either moist chlorine or hydrogen chloride. Bubbled into chlorine water, it is converted completely into ethylene chlorohydrin, and by the hydrolysis of this substance glycol is obtained. Ethylene chlorohydrin is important also because of its reaction with ammonia whereby mono-, di-, and triethanolamine are formed, substances which are used in the arts and are not without interest for the explosives chemist. Ethylene may be oxidized catalytically in the gas phase to ethylene oxide which reacts with water to form glycol and with glycol to form diglycol which also is of interest to the dynamite maker.

Glycol is a colorless liquid (bluer than water in thick layers), syrupy, sweet tasting, less viscous than glycerin, specific gravity 1.1270 at 0°, 1.12015 at 10°, and 1.11320 at 20°.[41] It shows a tendency to supercool but freezes at temperatures between −13° and −25°, and melts again at −11.5°. It boils at 197.2° at atmospheric pressure. It is very hygroscopic, miscible in all pro-

[41] C. A. Taylor and W. H. Rinkenbach, *Ind. Eng. Chem.*, **18**, 676 (1926).

portions with water, alcohol, glycerin, acetone, and acetic acid, and not miscible with benzene, chloroform, carbon disulfide, and ether.

Nitroglycol is manufactured with the same mixed acid and with the same apparatus as nitroglycerin. Somewhat more heat, is produced by the nitration reaction, and, as glycol is less viscous than glycerin, it is feasible to conduct the operation at a lower temperature. The washing is done with cold water and with less agitation by compressed air, and smaller amounts of wash water are used than are used with nitroglycerin, for nitroglycol is appreciably more volatile and more soluble in water. The tendency of the partially washed product to undergo an acid-catalyzed decomposition is less in the case of nitroglycol than in the case of nitroglycerin.

Nitroglycol is a colorless liquid, only slightly more viscous than water, specific gravity $(x°/15°)$ 1.5176 at 0°, 1.5033 at 10°, and 1.4890 at 20°.[42] It freezes at about −22.3°. Rinkenbach reports the index of refraction of nitroglycol for white light to be 1.4452 at 22.3°, and that of a commercial sample of nitroglycerin under the same conditions to be 1.4713. The same author reports the vapor pressure of nitroglycol to be 0.007 mm. of mercury at 0° and 0.0565 mm. at 22°, and points out that its vapor pressure at 22° is approximately 150 times as great as the vapor pressure, 0.00037 mm., reported by Peace and Marshall [43] for nitroglycerin at 22°. Nitroglycol produces a headache similar to that produced by nitroglycerin, but, corresponding to its greater volatility, the headache is more violent and does not last so long. Nitroglycol is non-hygroscopic. Its comportment with organic solvents is similar to that of nitroglycerin, but it is distinctly more soluble in water than that substance. Naoúm [44] reports that 1 liter of water at 15° dissolves 6.2 grams of nitroglycol, at 20° 6.8 grams of nitroglycol or 1.8 grams of nitroglycerin, and at 50° 9.2 grams of nitroglycol.

Nitroglycol has a slightly larger energy content than nitroglycerin. In the Trauzl test with 10-gram samples and water tamping, Naoúm [45] found that nitroglycol gave a net expansion

---

[42] Rinkenbach, *Ind. Eng. Chem.,* **18,** 1195 (1926).
[43] *Loc. cit.*
[44] Naoúm, *op. cit.,* p. 224.
[45] *Ibid.,* p. 227.

of 650 cc. and nitroglycerin one of 590 cc. Nitroglycol, like nitroglycerin, burns with sputtering and explodes if local overheating occurs, but nitroglycol and nitroglycol explosives in general burn more quietly and show less tendency to explode from heat than the corresponding nitroglycerin preparations. Nitroglycol explodes with a sharp report if heated rapidly to 215°. It is less sensitive to mechanical shock than nitroglycerin. Naoúm [46] reports the height of drop necessary to cause explosion, with a 2-kilogram weight, as follows.

	HEIGHT OF DROP, CENTIMETERS	
	Nitroglycol	Nitroglycerin
Drop absorbed on filter paper..............	20–25	8–10
Blasting gelatin	25–30	12
Guhr dynamite	15	5

Rinkenbach [47] reports tests with a small drop machine having a weight of 500 grams, nitroglycol 110 cm., nitroglycerin 70 cm., and a commercial mixture of nitroglycerin and nitropolyglycerin 90 cm.

Nitroglycol gelatinizes collodion cotton much faster than nitroglycerin and acts at ordinary temperatures, while nitroglycerin requires to be warmed. The greater volatility of nitroglycol does not affect its usefulness in gelatin dynamite, especially in temperate climates, but renders it unsuitable for use during the warm season of the year in ammonium nitrate explosives which contain only a few per cent of the oily nitric ester. It is too volatile for use in double-base smokeless powder, for its escape by evaporation affects the ballistic properties.

Dinitrodiglycol (Diethylene glycol dinitrate)

A study of the preparation and properties of dinitrodiglycol was reported by Rinkenbach [48] in 1927 and a further study of the nitration of diethylene glycol by Rinkenbach and Aaronson [49] in 1931. Dinitrodiglycol is a viscous, colorless, and odorless liquid, specific gravity $(x°/15°)$ 1.4092 at 0°, 1.3969 at 10°, and 1.3846 at 20°, freezing point −11.5°. It is completely miscible at ordinary temperatures with nitroglycerin, nitroglycol, ether, acetone,

[46] Op. cit., p. 225.
[47] Loc. cit.
[48] Ind. Eng. Chem., 19, 925 (1927).
[49] Ind. Eng. Chem., 23, 160 (1931).

methyl alcohol, chloroform, benzene, and glacial acetic acid. It is immiscible, or only slightly soluble, in ethyl alcohol, carbon tetrachloride, and carbon disulfide. It is slightly hygroscopic and is soluble in water to the extent of about 4.1 grams per liter of water at 24°. It can be ignited only with difficulty, and in small quantity is not readily exploded by heat. It is less sensitive than nitroglycol in the drop test. It is so insensitive to initiation that it will not propagate its own detonation under conditions where nitroglycol and nitroglycerin will do it. In 50/50 mixture however with either of these substances it detonates satisfactorily "and shows an explosive effect but little less than that of either of these compounds." It has a vapor pressure of about 0.007 mm. of mercury at 22.4°, and produces headaches similar to those produced by nitroglycerin.

Trinitrophenoxyethyl Nitrate

Another explosive which is preparable from glycol and which may perhaps be of interest for special purposes in the future is the β-2,4,6-trinitrophenoxyethyl nitrate described by Wasmer [50] in 1938. Glycol is converted into its monosodium derivative, and this is made to react with dinitrochlorobenzene at 130° for the production of β-dinitrophenoxyethyl alcohol which gives the explosive by nitration with mixed acid.

Trinitrophenoxyethyl nitrate is procured as a white powder, m.p. 104.5°, insoluble in water and readily soluble in acetone. It gelatinizes collodion nitrocotton, and is intermediate between picric acid and tetryl in its sensitivity to mechanical shock.

[50] *Mém. poudres*, **28**, 171 (1938).

Nitration of Ethylene

By passing ethylene into a mixture of concentrated nitric and sulfuric acids Kekulé [51] obtained an oil, specific gravity 1.47, which broke down when distilled with steam to give glycollic acid, oxalic acid, nitric oxide, and nitric acid. On reduction with sodium amalgam it yielded glycol and ammonia among other products. Wieland and Sakellarios [52] distilled the Kekulé oil in steam and then in vacuum, and obtained nitroglycol, b.p. 105° at 19 mm., and β-nitroethyl nitrate, b.p. 120–122° at 17 mm. These two substances are evidently formed from ethylene by the reactions indicated below.

$$\begin{array}{l} CH_2 \\ \| \\ CH_2 \end{array} + \begin{array}{l} OH \\ | \\ NO_2 \end{array} \longrightarrow \begin{array}{l} CH_2-OH \\ | \\ CH_2-NO_2 \end{array} + HNO_3 \longrightarrow H_2O + \begin{array}{l} CH_2-ONO_2 \\ | \\ CH_2-NO_2 \end{array}$$

β-Nitroethyl nitrate

$$\begin{array}{l} CH_2 \\ \| \\ CH_2 \end{array} + 3HNO_3 \longrightarrow HONO + H_2O + \begin{array}{l} CH_2-ONO_2 \\ | \\ CH_2-ONO_2 \end{array}$$

Nitroglycol

A considerable amount of nitrous acid is present in the spent acid. β-Nitroethyl nitrate is feebly acidic and dissolves in dilute alkali solutions with a yellow color. It is not sufficiently stable for use in commercial explosives. On digestion with warm water or on slow distillation with steam it undergoes a decomposition or sort of hydrolysis whereby nitrous acid and other materials are produced. Numerous patents have been issued for processes of procuring pure nitroglycol from the Kekulé oil. One hundred parts of the last-named material yield about 40 parts of nitroglycol, and the economic success of the process depends upon the recovery of valuable by-products from the β-nitroethyl nitrate which is destroyed.

Öhman [53] in Sweden has developed an ingenious electrolytic process for the production of nitric esters direct from ethylene. The discharge of the nitrate ion (NO_3^-) at the anode liberates the free nitrate radical (NO_3) which in part combines directly with ethylene to form nitroglycol.

$$\begin{array}{l} CH_2 \\ \| \\ CH_2 \end{array} + 2NO_3 \longrightarrow \begin{array}{l} CH_2-ONO_2 \\ | \\ CH_2-ONO_2 \end{array}$$

[51] Ber., 2, 329 (1869).
[52] Ber., 53, 201 (1920).
[53] Z. Elektrochem., 42, 862 (1936); Svensk Kemisk Tid., 50, 84 (1938).

Another portion of the free nitrate radical apparently reacts with itself and with water as indicated below, and the oxygen which becomes available enters into the reaction with the consequent formation of dinitrodiglycol.

$$\begin{cases} 2NO_3 \longrightarrow N_2O_5 + [O] \\ N_2O_5 + H_2O \longrightarrow 2HNO_3 \end{cases}$$

$$\begin{matrix} CH_2 \\ \parallel \\ CH_2 \end{matrix} + [O] + 2NO_3 \longrightarrow \begin{matrix} CH_2-ONO_2 \\ | \\ CH_2 \\ \diagdown \\ O \\ \diagup \\ CH_2 \\ | \\ CH_2-ONO_2 \end{matrix}$$

A platinum gauze anode is used. It is immersed in an acetone solution of calcium nitrate which is kept continuously saturated with ethylene which is bubbled through in such manner that it sweeps over the surface of the platinum gauze. An aluminum cathode is used, in a catholyte consisting of a nitric acid solution of calcium nitrate, and the cathode compartment is filled to a higher level since the liquid moves into the anode compartment as the electrolysis progresses. After the electrolysis, the cathode liquid is fortified with nitric acid for use again. The anode liquid is neutralized with slaked lime, and distilled in vacuum for the recovery of the acetone, and the residue, after the removal of calcium nitrate, washing, and drying, consists of a mixture of nitroglycol and dinitrodiglycol and is known as *Oxinite*. Dynamites made from Oxinite differ but little from those made from nitroglycerin.

Pentryl

Pentryl, or 2,4,6-trinitrophenylnitraminoethyl nitrate, is another explosive which is derived from ethylene. It is a nitric ester, an aromatic nitro compound, and a nitroamine. The substance was described in 1925 by Moran [54] who prepared it by the action of mixed acid on 2,4-dinitrophenylethanolamine (large orange-yellow crystals from alcohol, m.p. 92°) procured by the interaction of dinitrochlorobenzene with ethanolamine. von Herz later prepared pentryl by the nitration of β-hydroxyethylaniline, a material which is more commonly called phenylethanolamine and is now available commercially in this country, and was granted

[54] U. S. Pat. 1,560,427 (1925).

British and German patents [55] for its use for certain military purposes. The genesis of pentryl from ethylene, through the intermediacy both of ethanolamine and of phenylethanolamine, is indicated below. The preparation and properties of pentryl have

been studied extensively by LeRoy V. Clark [56] at the U. S. Bureau of Mines. By the reaction of dinitrochlorobenzene in the presence of sodium hydroxide with ethanolamine in alcohol solution at 70–80° he procured dinitrophenylethanolamine in 70% yield. The alcohol solution was filtered for the removal of sodium chloride, which was found to be mixed with a certain quantity of the by-product tetranitrodiphenylethanolamine (lemon-yellow fine powder, m.p. 222°); it was then concentrated to about one-third its volume, and deposited crystals of the product on cooling. This material, dissolved in concentrated sulfuric acid and nitrated by adding the solution to nitric acid and heating, gave pentryl in yields of about 90%, minute cream-colored crystals from benzene, m.p. 128°.

Clark reports that pentryl has an absolute density of 1.82 and an apparent density of only 0.45. When compressed in a detonator shell at a pressure of 3400 pounds per square inch, it has an apparent density of 0.74. It is soluble to some extent in most of the common organic solvents, and is very readily soluble in nitroglycerin. In the drop test with a 2-kilogram weight, 0.02 gram of pentryl was exploded by a drop of 30 cm., a similar sample of

[55] Brit. Pat. 367,713 (1930); Ger. Pat. 530,701 (1931).
[56] Ind. Eng. Chem., 25, 1385 (1933).

tetryl by one of 27.5 cm., and one of picric acid by a drop of 42.5 cm., while TNT was not exploded by a drop of 100 cm. It is somewhat more sensitive to friction than tetryl, and much more sensitive than picric acid and TNT. Pentryl explodes in 3 seconds at 235°.

The results of Clark's experiments to determine the minimum amounts of primary explosive necessary to initiate the explosion of pentryl and of other high explosives are tabulated below. For

	MINIMUM INITIATING CHARGE (GRAMS) OF		
	Diazodi- nitrophenol	Mercury Fulminate	Lead Azide
Pentryl	0.095	0.150	0.025
Picric acid	0.115	0.225	0.12
TNT	0.163	0.240	0.16
Tetryl	0.075	0.165	0.03
Trinitroresorcinol	0.110	0.225	0.075
Trinitrobenzaldehyde	0.075	0.165	0.05
Tetranitroaniline	0.085	0.175	0.05
Hexanitrodiphenylamine	0.075	0.165	0.05

the purpose of these experiments a half-gram portion of the high explosive was weighed into a No. 8 detonator shell, a weighed amount of primary explosive was introduced on top of it, and both were compressed under a reenforcing capsule at a pressure of 3400 pounds per square inch. Without the reenforcing capsule diazodinitrophenol did not cause detonation, and pentryl required 0.035 gram of lead azide and more than 0.4 gram of fulminate. The results show that pentryl has about the same sensitivity to initiation as tetryl and hexanitrodiphenylamine.

In the small Trauzl test pentryl caused an expansion of 15.8 cc., while the same weight of tetryl caused one of 13.8 cc., TNT 12.2 cc., and picric acid 12.4 cc. In the small lead block compression test, in which 50 grams of the explosive was exploded by means of a detonator on top of a lead cylinder 64 mm. long, it was found that pentryl produced a shortening of the block of 18.5 mm., tetryl 16.6 mm., picric acid 16.4 mm., TNT 14.8 mm., and diazodinitrophenol 10.5 mm. Determinations of velocity of detonation made with half-meter lengths, the explosives being contained in extra light lead tubing ½ inch internal diameter and weighing 12 ounces to the foot, gave the following figures.

	Density	Velocity of Detonation (Meters per second)
Pentryl	0.80	5000
Tetryl	0.90	5400
Picric acid	0.98	4970
TNT	0.90	4450

These are not however the maximum velocities of detonation of the substances.

Hexanitrodiphenylaminoethyl Nitrate

This substance also has been studied by LeRoy V. Clark [57] who prepared it by the nitration with mixed acid of tetranitrodiphenylethanolamine, a by-product from the preparation of pentryl.

He procured the pure substance in the form of pale yellow glistening plates, m.p. 184° (corr.), precipitated from acetone solution by the addition of alcohol. Its explosive properties are not widely different from those of pentryl. Its response to initiation

[57] *Ibid.*, **26**, 554 (1934).

is about the same; it is slightly less sensitive to impact, about 7% less effective in the sand test, and about 3% more effective in the small Trauzl test. In compound detonators it is somewhat better than TNT and somewhat poorer than pentryl, tetryl, and picric acid, as indicated by the lead plate test. When heated rapidly, it ignites at 390–400°.

Trimethylene Glycol Dinitrate

Trimethylene glycol occurs in the glycerin which is produced by fermentation. There is no harm in leaving it in glycerin which is to be used for the manufacture of explosives. It may however be separated by fractional distillation. When pure it is a colorless, odorless, syrupy liquid, specific gravity ($x°/4°$) 1.0526 at 18°. It mixes with water in all proportions and boils at atmospheric pressure at 211° without decomposition. At temperatures above 15° or so, it is oxidized rapidly by nitric acid or by mixed acid. It is accordingly nitrated at 0–10° under conditions similar to those which are used in the preparation of ethyl nitrate and other simple aliphatic nitric esters (except methyl nitrate).

$$\begin{array}{ccc} CH_2{-}OH & & CH_2{-}ONO_2 \\ | & & | \\ CH_2 & \longrightarrow & CH_2 \\ | & & | \\ CH_2{-}OH & & CH_2{-}ONO_2 \end{array}$$

Trimethylene glycol dinitrate is a water-white liquid, very mobile, and scarcely more viscous than nitroglycol, specific gravity (20°/4°) 1.393 at 20°. It boils at 108° at 10 mm. without decomposition. It is less volatile than nitroglycol and more volatile than nitroglycerin. It has about the same solubility relationships as nitroglycerin, and forms a good gelatin with collodion nitrocotton. It causes headache by contact with the skin. When heated slowly it takes fire with a puff and burns tranquilly or decomposes at about 185° and deflagrates at about 225°. It is much less sensitive to shock than nitroglycerin and is much more stable in storage. Naoúm [58] reports that a 10-gram sample in the Trauzl test with water tamping gave an expansion of 540 cc. or about 90% of the expansion produced by nitroglycerin. The calculated energy content of trimethylene glycol dinitrate is only

[58] *Op. cit.,* p. 235.

about 77% of that of nitroglycerin, but the relatively greater brisance results from the low viscosity of the substance which gives it a higher velocity of detonation. Naoúm also reports that a 93% trimethylene glycol dinitrate gelatin with 7% collodion cotton gave an expansion of 470 cc. or about 80% as much as a similar nitroglycerin gelatin.

Propylene Glycol Dinitrate (Methylglycol dinitrate, methyl-nitroglycol)

Propylene occurs along with ethylene in cracking gas. Its use as a raw material for the synthesis of glycerin has already been mentioned in the section on nitroglycerin. It yields propylene glycol when subjected to the same chemical processes as those which are used for the preparation of glycol from ethylene.[59] Propylene glycol shows the same tendency toward oxidation during nitration that trimethylene glycol does, but to a less extent; noticeable decomposition occurs only above 30°.

$$
\begin{array}{ccccccc}
CH_3 & & CH_3 & & CH_3 & & CH_3 \\
| & \xrightarrow{HOCl} & | & \xrightarrow{HOH} & | & \xrightarrow{nitration} & | \\
CH & & CH-OH & & CH-OH & & CH-ONO_2 \\
\| & & | & & | & & | \\
CH_2 & & CH_2-Cl & & CH_2-OH & & CH_2-ONO_2
\end{array}
$$

Propylene glycol dinitrate is a colorless liquid of characteristic aromatic odor, more volatile and less viscous than trimethylene glycol dinitrate with which it is isomeric. Its specific gravity (20°/4°) is 1.368 at 20°. It boils at 92° at 10 mm., and does not freeze at −20°. Its solubilities, gelatinizing power, and explosive properties are substantially the same as those of its isomer. Indeed, Naoúm [60] reports that it gave exactly the same expansion as trimethylene glycol dinitrate in the Trauzl lead block test, namely, 540 cc.

[59] Symmes, in a footnote, p. 375, in his English translation of Naoúm's book, *op. cit.*, cites U. S. patents 1,307,032, 1,307,033, 1,307,034, and 1,371,215 which describe a method for the manufacture of mixed ethylene and propylene glycols from cracking gas, satisfactory methods for the nitration of the mixture and for the stabilization of the mixed nitric esters, and explosives made from the products "which practical tests in actual use showed could not be frozen even at temperatures prevailing in winter along the Canadian border, or −10° to −30° F."

[60] *Op. cit.*, p. 237.

Butylene Glycol Dinitrate

Of the four isomeric butylene glycols, the 1,3-compound appears to be the only one which has attracted any interest as a raw material for the preparation of explosives. Its dinitrate, either alone or in admixture with nitroglycerin, has been proposed for use in non-freezing dynamites.[61] The preparation of the glycol from acetaldehyde has been suggested,[62] the acetaldehyde being condensed to aldol and the aldol reduced to glycol. Since acetaldehyde is produced commercially by the catalyzed hydration of acetylene, then butylene glycol-1,3 can be procured by synthesis from coke.

$$
\begin{array}{ccccc}
\text{Coke} & \text{CH} & \text{CH}_3 & \text{CH}_3 & \text{CH}_3 \\
\downarrow & \parallel\!\parallel & | & | & | \\
 & \text{CH} & \text{CH—OH} & \text{CH—OH} & \text{CH—ONO}_2 \\
\text{CaC}_2 & \text{CH}_3 & \text{CH}_2 \longrightarrow & \text{CH}_2 \longrightarrow & \text{CH}_2 \\
 & | & | & | & | \\
 & \text{CHO} & \text{CHO} & \text{CH}_2\text{—OH} & \text{CH}_2\text{—ONO}_2
\end{array}
$$

Butylene glycol shows a strong tendency to oxidize during nitration, and ought to be nitrated at a temperature of $-5°$ or lower. Butylene glycol dinitrate is a colorless liquid, intermediate in volatility between nitroglycol and nitroglycerin, possessing a specific gravity of 1.32 at 15°. It does not freeze at temperatures as low as $-20°$. It yields a good gelatin with collodion nitrocotton. It deflagrates feebly if heated suddenly. It is very insensitive to mechanical shock but detonates easily by initiation. Naoúm [63] reports that a mixture of 75% butylene glycol dinitrate and 25% kieselguhr gave about 240 cc. expansion in the Trauzl test, and that a gelatin containing 90% butylene glycol dinitrate and 10% collodion nitrocotton gave about 370 cc.

Nitroerythrite (Erythritol tetranitrate)

$$NO_2\text{—O—CH}_2\text{—(CH—ONO}_2)_2\text{—CH}_2\text{—ONO}_2$$

i-Erythrite occurs in algae and in lichens. It is a white, crystalline, sweet-tasting substance, very readily soluble in water, m.p. 120°, b.p. 330°, specific gravity 1.59. The tetranitrate is prepared

[61] U. S. Pats. 994,841 and 994,842 (1911).

[62] U. S. Pat. 1,008,333 (1911).

[63] Op. cit., p. 239.

by dissolving erythrite in strong nitric acid with cooling, and then precipitating by the addition of concentrated sulfuric acid.[64] It crystallizes from alcohol in colorless plates, m.p. 61°. Its use as an addition to smokeless powder has been suggested,[65] but it is as powerful as nitroglycerin, and has the advantage over it that it is a solid, and it would be suitable, if it were cheaper, for the same uses as nitromannite.

Nitromannite (Mannitol hexanitrate)

$$NO_2—O—CH_2—(CH—ONO_2)_4—CH_2—ONO_2$$

d-Mannitol occurs fairly widely distributed in nature, particularly in the *Fraxinus ornus*, the sap of which is *manna*. It may also be procured by the reduction of d-mannose either electrolytically or by means of sodium amalgam, or along with d-sorbite by the reduction of d-fructose. It may be nitrated satisfactorily with the same mixed acid as is used for the nitration of glycerin, or more conveniently, because the mass of crystals is so voluminous, by dissolving in strong nitric acid and precipitating by the addition of concentrated sulfuric acid.

Preparation of Nitromannite. Fifty grams of nitric acid (specific gravity 1.51) is cooled thoroughly in a 300-cc. Erlenmeyer pyrex flask immersed in a freezing mixture of ice and salt. Ten grams of mannite is then introduced in small portions at a time while the flask is tilted from side to side and the contents is stirred gently with a thermometer, care being taken that the temperature does not rise above 0°. After all is dissolved, 100 grams of sulfuric acid (specific gravity 1.84) is added slowly from a dropping funnel while the liquid is stirred and the temperature is maintained below 0°. The porridge-like mass is filtered on a sinter-glass filter, or on a Büchner funnel with a hardened filter paper, washed with water, then with dilute sodium bicarbonate solution, then finally again with water. The crude product is dissolved in boiling alcohol; the solution is filtered if need be, and on cooling deposits white needle crystals of nitromannite, m.p. 112-113°. A second crop of crystals may be obtained by warming the alcoholic mother liquors to boiling, adding water while still boiling until a turbidity appears, and allowing to cool. Total yield about 23 grams.

Nitromannite is readily soluble in ether and in hot alcohol, only slightly soluble in cold alcohol, and insoluble in water.

[64] Stenhouse, *Ann.*, **70**, 226 (1849) ; **130**, 302 (1864).
[65] Ger. Pat. 110,289 (1898) ; Brit. Pat. 27,397 (1898).

While its stability at ordinary temperatures is such that it can be used commercially, at slightly elevated temperatures it is distinctly less stable than nitroglycerin. Nitroglycerin will tolerate

FIGURE 57. Nitromannite, Crystals from Alcohol (5×). (Courtesy Atlas Powder Company.)

heating in a covered glass vessel for several days at 75° before it begins to give off acid fumes; nitroglycol, methylglycol dinitrate, and trimethylene glycol dinitrate are more stable yet,

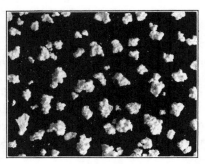

FIGURE 58. Nitromannite, in Grained Form for Charging Detonators (5×). (Courtesy Atlas Powder Company.)

but nitromannite decomposes after a few hours and evolves red fumes. If a small quantity is heated, it decomposes at once at about 150° with copious evolution of nitrous fumes but ordinarily does not deflagrate. With larger samples deflagration occurs at 160–170°.

Kast [66] has reported a velocity of detonation of 8260 meters per second for nitromannite compressed to a density of 1.73 in a column 12.8 mm. in diameter.

Nitromannite is about as sensitive as nitroglycerin to shock and to friction. It detonates under a 4-cm. drop of a 2-kilogram weight, and may be exploded readily on a concrete surface by a blow of a carpenter's hammer. It is not fired by the spit of a fuse, but is made to detonate by the flame of a match which causes local overheating. It is almost, but not quite, a primary explosive. It is used as the high-explosive charge in compound detonators which contain the relatively safe diazodinitrophenol as the primary explosive. A mixture of nitromannite and tetracene is a powerful and brisant primary explosive which detonates from moderate heat.

Nitrodulcite (Dulcitol hexanitrate)

Dulcite is obtained from Madagascar manna by extraction with water and recrystallizing, large monoclinic prisms, m.p. 188°, less soluble than mannite. It may also be procured by the action of sodium amalgam on aqueous solutions of lactose and of d-galactose. Nitrodulcite, isomeric with nitromannite, crystallizes from alcohol in needles which melt at 94–95°.

Nitrosorbite (Sorbitol hexanitrate)

d-Sorbite occurs in the berries of the mountain ash, but is more readily procured by the electrolytic reduction of d-glucose. It crystallizes with one molecule of water in small crystals which lose their water when heated and melt at about 110°. Nitrosorbite, isomeric with nitromannite, exists as a viscous liquid and has never been obtained in the crystalline state. It is used in nonfreezing dynamites.

Nitrated Sugar Mixtures

The sugars are polyhydric alcohols which contain an aldehyde or a ketone group or a cyclic acetal or ketal arrangement within the molecule. They yield nitric esters which are perhaps less stable than the nitric esters of the simple polyhydric alcohols but which probably owe part of their reputation for instability to the

[66] Z. angew. Chem., **36**, 74 (1923).

fact that they are difficult to purify. The nitrosugars resemble the sugars from which they are derived in the respect that often they do not crystallize rapidly and easily. When warmed gently, they frequently soften and become sticky and resinous. In this condition they retain within their masses traces of decomposition products by which further decomposition is provoked; they cannot be washed free from acid, and in the solid or semi-solid state are impossible to stabilize. The stabilization however may be accomplished easily if the nitrosugar is in solution.

A mixture of nitrosucrose and nitroglycerin, prepared by nitrating a solution of 20 parts of cane sugar in 80 parts of glycerin, or of 25 parts in 75, has been used in this country under the name of *nitrohydrene*. It is suitable for use in non-freezing dynamites, and is cheaper than nitroglycerin to the extent that sugar is cheaper than glycerin. The nitrated product is much more viscous than nitroglycerin and forms emulsions readily. It requires repeated washings with soda solution to insure a satisfactory heat test, and then washings with concentrated salt solutions to facilitate the separation of the phases. Nitrohydrene 80/20 (from 80 parts of glycerin and 20 parts of cane sugar) consists of about 86% nitroglycerin and 14% nitrosucrose, and nitrohydrene 75/25 of about 82% nitroglycerin and 18% nitrosucrose. Naoúm[67] reports the following data. The stability of nitro-

| | | LEAD BLOCK EXPANSION, 10-GRAM SAMPLE IN GLASS TUBE | |
	SPECIFIC GRAVITY AT 20°	Sand Tamping, cc.	Water Tamping, cc.
Nitroglycerin	1.596	550	595
Nitrohydrene 80/20	1.605	533	560
Nitrohydrene 75/25	1.612	514	535

hydrene is distinctly poorer than that of nitroglycerin and appears to depend upon the proportion of nitrosucrose which it contains, for nitrohydrene 75/25 gives a poorer heat test than nitrohydrene 80/20 which contains less nitrosucrose. Naoúm[68] points out that the wood meal, etc., which is contained in dynamite made from nitrohydrene apparently acts as a stabilizer and

[67] *Op. cit.*, p. 253.
[68] *Ibid.*, p. 255.

absorbs or reacts chemically with the first decomposition products and destroys them. He says:

> Better still, are very small additions of diphenylamine, which is admirably suited for the stabilization of smokeless powder, since it readily takes up the nitrous acid. Nitrohydrene 80/20 or 75/25, containing only 0.1 to 0.2 per cent of diphenylamine, was stored for seventy-five days at 55°C. without undergoing decomposition. The samples merely showed a coloration and became dark green, a phenomenon which also occurred but to a less extent with a check sample of· nitroglycerin containing the same quantity of diphenylamine. After seventy-five days the nitroglycerin still had a slight odor of diphenylamine, but the nitrohydrene smelled slightly acid, somewhat like sour milk, but not like nitrous or nitric acid.
>
> Similar samples of 100 grams each of the above nitrohydrene containing 0.1 per cent diphenylamine have been stored by the author for more than eight years in diffuse daylight at room temperatures, about 20°C. So far they have remained unchanged, have no acid odor, and show no signs of decomposition. . . . From this it is evident that nitrosugar dissolved in nitroglycerin, although its stability does not reach that of the latter, is sufficiently stable for practical purposes, particularly in the presence of stabilizers.

The individual nitrosugars are stabilized similarly by diphenylamine, and certain ones of them, specifically nitromaltose, nitrolactose, and nitrosucrose, have been able by means of that substance to find a limited industrial application.

Solutions of cane sugar in glycol, and of glucose and lactose in glycerin, have been nitrated to produce mixtures of nitric esters comparable to nitrohydrene.

Nitroarabinose (l-Arabinose tetranitrate), $C_5H_6O(ONO_2)_4$

Nitroarabinose is prepared,[69] as indeed the highly nitrated sugars in general may be prepared, by adding concentrated sulfuric acid drop by drop to a solution of the corresponding sugar in concentrated nitric acid at 0°. It consists of colorless monoclinic crystals which melt at 85° and decompose at 120°. It is readily soluble in alcohol, acetone, and acetic acid, and insoluble in water and ligroin. It reduces Fehling's solution on warming. It is but little stable above 50°, and is easily exploded by shock.

[69] Will and Lenze, *Ber.*, **31**, 68 (1898).

Nitroglucose (d-Glucose pentanitrate), $C_6H_7O(ONO_2)_5$

d-Glucose pentanitrate [69] is a colorless viscous syrup, insoluble in water and in ligroin, readily soluble in alcohol. It becomes hard at 0°. It is unstable above 50°, and if heated slowly to a higher temperature decomposes rapidly at about 135°. It reduces Fehling's solution on warming. *Glucosan trinitrate*, $C_6H_7O_2$ $(ONO_2)_3$, is produced by the nitration of α-glucosan and by the action for several days of mixed acid on d-glucose. It is readily soluble in alcohol and insoluble in water. It has been obtained in the form of aggregates or crusts of crystals which melted not sharply at about 80° and which were probably not entirely free from glucose pentanitrate.

Nitromannose (d-Mannose pentanitrate), $C_6H_7O(ONO_2)_5$

d-Mannose pentanitrate,[69] transparent rhombic needles from alcohol, melts at 81–82° and decomposes at about 124°. It is soluble in alcohol and insoluble in water and reduces Fehling's solution slowly on warming. It undergoes a rapid decomposition if stored at 50°.

Nitromaltose (Maltose octonitrate), $C_{12}H_{14}O_3(ONO_2)_8$

Maltose octonitrate,[69, 70] glistening needles from methyl alcohol, melts with decomposition at 164–165°. If heated quickly, it puffs off at 170–180°. It decomposes slowly at 50°. If fused and allowed to solidify, it has a specific gravity of 1.62. It is readily soluble in methyl alcohol, acetone, and acetic acid, difficultly soluble in ethyl alcohol, and insoluble in water. It reduces warm Fehling's solution more rapidly than nitrosucrose.

Nitrolactose (Lactose octonitrate), $C_{12}H_{14}O_3(ONO_2)_8$

Lactose octonitrate,[69, 71] monoclinic needles from methyl or ethyl alcohol, melts at 145–146° with decomposition. Its specific gravity is 1.684. It is readily soluble in methyl alcohol, hot ethyl alcohol, acetone, and acetic acid, difficultly soluble in cold ethyl alcohol, and insoluble in water. It reduces Fehling's solution on warming.

[70] Pictet and Vogel, *Helv. Chim. Acta,* **10,** 588 (1927).
[71] Gé, *Ber.,* **15,** 2238 (1882).

Lactose hexanitrate, $C_{12}H_{16}O_5(ONO_2)_6$, has been found in the alcoholic mother liquors from the crystallization of the octonitrate, white, amorphous material melting not sharply at about 70°.

Crater [72] in 1934 described explosives containing nitrolactose, one consisting, say, of nitrolactose 25%, ammonium nitrate 65%, sodium nitrate 6%, and vegetable absorbent material 4%, another made by treating wood pulp with an acetone solution of nitrolactose and dinitrotoluene and containing about 78% nitrolactose, about 9% DNT, and about 13% wood pulp. For this use the nitrolactose ought to be stabilized with diphenylamine.

Nitrosucrose (Sucrose octonitrate), $C_{12}H_{14}O_3(ONO_2)_8$

The nitration of cane sugar [69, 73] yields sucrose octonitrate, white glistening needles, which melt at 85.5°. If heated slowly, nitrosucrose decomposes at about 135° and if heated rapidly deflagrates at about 170°. The fused and solidified material has a specific gravity of 1.67. It is readily soluble in methyl alcohol, ether, and nitrobenzene, difficultly soluble in ethyl alcohol and benzene, and insoluble in water and in petroleum ether. It reduces Fehling's solution on warming. It is relatively stable when pure. Monasterski reports that it gives a feeble puff under a 20-cm. drop of a 2-kilogram weight, a puff with one of 25 cm., and a detonation with one of 30 cm. He states that samples of 10 grams in the Trauzl test gave average net expansions of 296 cc.

Other Nitrosugars

The nitration of *d*-xylose [69] yields *d-xylose tetranitrate,* $C_5H_6O(ONO_2)_4$, an oily substance insoluble in water, and a crystalline by-product, m.p. 141°, insoluble in water, which is evidently the *trinitrate,* $C_5H_7O_2(ONO_2)_3$. *Xylosan dinitrate,* $C_5H_6O_2(ONO_2)_2$, has been prepared by the action of mixed acid on *d*-xylose. It consists of little spherical crystal aggregates, soluble in alcohol and melting at 75–80°.

l-Rhamnose tetranitrate,[69] $C_6H_8O(ONO_2)_4$, crystallizes in compact short rhombs which melt with decomposition at 135°. It is

[72] U. S. Pat. 1,945,344 (1934).

[73] Hoffman and Hawse, *J. Am. Chem. Soc.,* 41, 235 (1919). Monasterski, *Z. ges. Schiess- u. Sprengstoffw.,* 28, 349 (1933). Wyler, U. S. Pats. 2,081,161 (1938), 2,105,390 (1938), 2,165,435 (1939).

readily soluble in acetone, acetic acid, and in methyl and ethyl alcohol, and is relatively stable. It reduces Fehling's solution on warming. *l-Rhamnose trinitrate*,[74] $C_6H_9O_2(ONO_2)_3$, results from the action of mixed acid on *l*-rhamnose. It is a white amorphous material, melting below 100°, readily soluble in alcohol and insoluble in water. It explodes feebly under a hammer blow.

α-Methylglucoside tetranitrate,[69] $C_7H_{10}O_2(ONO_2)_4$, crystallizes from alcohol in quadrilateral plates which melt at 49–50° and decompose at 135°. It is more stable than the nitrate of the free sugar. It reduces Fehling's solution slowly on warming.

α-Methylmannoside tetranitrate,[69] $C_7H_{10}O_2(ONO_2)_4$, from the nitration of *d-α*-methylmannoside, crystallizes in fine asbestos-like needles which melt at 36°. It is relatively stable at 50°.

d-Galactose pentanitrate α, $C_6H_7O(ONO_2)_5$, from the nitration of *d*-galactose [69] crystallizes in bundles of transparent needles which melt at 115–116° and decompose at 126°. It is sparingly soluble in alcohol. It decomposes slowly at 50°, and reduces Fehling's solution slowly on warming. The alcoholic mother liquors from the α-form yield *d-galactose pentanitrate β*, transparent monoclinic needles which melt at 72–73° and decompose at 125°. This substance is readily soluble in alcohol, decomposes rapidly at 50°, and reduces hot Fehling's solution. *Galactosan trinitrate*, $C_6H_7O_2(ONO_2)_3$, results from the action during several days of mixed acid on *d*-galactose. It is deposited from alcohol in crusts of small crystals.

Fructosan trinitrate α, $C_6H_7O_2(ONO_2)_3$, is produced by the action of mixed acid at 0–15° on *d*-fructose or on laevulosan,[75] colorless, quickly effluorescing needles from alcohol, which melt at 139–140° and decompose at about 145°. It is readily soluble in methyl and ethyl alcohol, acetic acid, and acetone, and insoluble in water. It is relatively stable at 50°. It reduces hot Fehling's solution. The alcoholic mother liquors from the α-form yield *fructosan trinitrate β*, crusts of white crystals which melt at 48–52° and decompose at 135°. The material decomposes slowly at 50°. It reduces Fehling's solution rapidly on warming.

The action of mixed acid on *d*-sorbose at 15° yields *sorbosan trinitrate*, $C_6H_7O_2(ONO_2)_3$, a crystalline substance which melts not sharply at 40–45°.

[74] Hlasiewetz and Pfaundler, *Ann.*, **127**, 362 (1863).
[75] Pictet and Reilly, *Helv. Chim. Acta*, **4**, 613 (1921).

d-α-Glucoheptose hexanitrate, $C_7H_8O(ONO_2)_6$, from the nitration of *d-α*-glucoheptose,[69] crystallizes from alcohol in transparent needles which melt at 100°. It reduces Fehling's solution on warming.

Trehalose octonitrate, $C_{12}H_{14}O_3(ONO_2)_8$, from the nitration of trehalose,[69] crystallizes from alcohol in birefringent pearly leaflets which melt at ·124° and decompose at 136°. It reduces Fehling's solution on warming.

Raffinose hendecanitrate, $C_{18}H_{21}O_5(ONO_2)_{11}$, from the nitration of raffinose,[69] exists in the form of amorphous aggregates which melt at 55–65° and decompose at 136°. It reduces Fehling's solution on warming. It decomposes rapidly when kept at 50°.

α-Tetraamylose octonitrate, $[C_6H_8O_3(ONO_2)_2]_4$, from α-tetraamylose,[76] crystallizes from acetic acid in fine glistening needles which decompose at 204°. It is readily soluble in ethyl acetate, amyl acetate, pyridine, and nitrobenzene, and sparingly soluble or insoluble in alcohol, ether, benzene, and water. *α-Diamylose hexanitrate,*[76] $[C_6H_7O_2(ONO_2)_3]_2$, prepared from α-diamylose or as the final product of the nitration of tetraamylose, crystallizes from acetone in plates which puff off at 206–207°. It is difficultly soluble in acetic acid, and is reported to be but little stable. The alcohol extract of the crude hexanitrate yields a certain amount of the amorphous *tetranitrate.*[76] *β-Triamylose hexanitrate,*[76] $[C_6H_8O_3(ONO_2)_2]_3$, is procured by dissolving either β-triamylose or β-hexaamylose in strong nitric acid at 0° and adding concentrated sulfuric acid drop by drop, and extracting the crude product with alcohol. It crystallizes from the alcoholic extract in aggregates of microscopic cubes, m.p. 203°. The residue which is insoluble in hot alcohol is recrystallized from acetic acid and yields crystalline crusts of *β-triamylose enneanitrate,*[76] $[C_6H_7O_2(ONO_2)_3]_3$, m.p. 198°.

Early History of Nitrated Carbohydrates

The history of modern explosives commenced with the discoveries of nitroglycerin and of nitrocellulose. At about the time that Sobrero first prepared nitroglycerin, Schönbein at Basel and Böttger at Frankfort-on-the-Main independently of each other nitrated cotton, perceived the possibilities in the product, and

[76] Leibowitz and Silmann, *Ber.,* **58,** 1889 (1925).

soon cooperated with each other to exploit its use in artillery. Pelouze had nitrated paper at an earlier time, and the question may indeed be raised whether he was not the first discoverer of nitrocellulose. Before that, Braconnot, professor of chemistry at Nancy, had prepared a nitric ester from starch. The principal events in the early history of these substances are summarized below.[77]

1833. Braconnot [78] found that starch dissolved in concentrated nitric acid and that the liquid on dilution with water gave a curdy precipitate of material which, after washing, dried out to a white, pulverulent, tasteless, and neutral mass. The product gave a brown color with a solution of iodine. It was not affected by bromine. It did not dissolve in boiling water but softened to a sticky mass. Dilute sulfuric acid did not affect it. Concentrated sulfuric acid dissolved it, and the solution gave no precipitate if it was diluted with water. The material, to which Braconnot gave the name of *xyloïdine*, dissolved in acetic acid very readily on heating, and the solution, if evaporated slowly, gave a transparent film which retained its transparency when placed in water. Applied to paper or cloth it yielded a brilliant, varnish-like coating which was impervious to water. Xyloïdine took fire very readily. It carbonized and liquefied if heated upon a piece of cardboard or heavy paper while the cardboard or paper, though exposed directly to the heat, was not appreciably damaged. Sawdust, cotton, and linen yielded products which Braconnot considered to be identical with the xyloïdine from starch.

1838. Pelouze [79] studied xyloïdine further. He found that if starch was dissolved in concentrated nitric acid and if the solution was diluted immediately with water, xyloïdine precipitated and the acid filtrate on evaporation yielded practically no residue. If the solution of starch in nitric acid was allowed to stand before being precipitated with water, then the amount of xyloïdine was less. If it was allowed to stand for 2 days, or perhaps only for some hours, the xyloïdine was entirely destroyed, a new acid was formed, no precipitate appeared when the solution was diluted,

[77] The papers which are cited in this connection have been published in English in the book by George W. MacDonald, "Historical Papers on Modern Explosives," Whittaker & Co., London and New York, 1912.

[78] *Ann. chim. phys.*, [2] **52**, 290 (1833).

[79] *Compt. rend.*, **7**, 713 (1838).

and the liquid on evaporation gave the new acid in the form of a solid, white, non-crystalline, deliquescent mass of considerably greater weight than the starch which was taken for the experiment. Neither carbon dioxide nor oxalic acid was produced during the reaction, but the new acid on long standing, or on boiling, with nitric acid was converted to oxalic acid without the formation of carbon dioxide. Pelouze considered xyloïdine to be a nitrate of starch. He observed that it was readily combustible,

FIGURE 59. Théophile-Jules Pelouze (1807-1867). (Courtesy E. Berl.) Made many important contributions to organic and inorganic chemistry—ethereal salts, the first nitrile, borneol, glyceryl tributyrate, pyroxylin, improvements in the manufacture of plate glass. He nitrated paper in 1838 and was thus probably the first to prepare nitrocellulose. Reproduced from original in Kekulé's portrait album.

that it ignited at a temperature of 180° and burned with very considerable violence leaving practically no residue. The observation, he says, led him to make certain experiments which, he believed, might have practical application in artillery. Paper, dipped into nitric acid of specific gravity 1.5 and left there long enough for the acid to penetrate into it (generally 2 or 3 minutes), removed, and washed thoroughly, gave a parchment-like material which was impervious to moisture and was extremely combustible. Pelouze had nitrocellulose in his hands, but evidently did not recognize that the material, which had not changed greatly in its physical form, was nevertheless nitrated throughout its mass, for he believed that the products which he obtained from paper and from cotton and linen fabrics owed their new properties to the xyloïdine which covered them.

1846. Schönbein announced his discovery of guncotton at a meeting of the Society of Scientific Research at Basel on May 27, 1846. In an article, probably written in 1847 but published in the *Archives des sciences physiques et naturelles* of 1846, he described some of his experiences with the material and his efforts

FIGURE 60. Christian Friedrich Schönbein (1799-1868). (Courtesy E. Berl.) Discovered guncotton, 1846. Discovered ozone, worked on hydrogen peroxide, auto-oxidation, the passivity of iron, hydrosulfites, catalysts, and prussic acid. Professor of Chemistry at Basel from 1829 until the time of his death. He published more than 300 papers on chemical subjects. Reproduced from original in Kekulé's portrait album.

to put it to practical use and discussed the controversial question of priority of discovery; he described the nitration of cane sugar but deliberately refrained from telling how he had prepared his nitrocellulose. He was led to perform the experiments by certain theoretical speculations relative to ozone which he had discovered a few years before. One volume of nitric acid (1.5) and 2 volumes of sulfuric acid (1.85) were mixed and cooled to 0°, finely powdered sugar was stirred in so as to form a paste, the stirring was continued, and after a few minutes a viscous mass separated from the acid liquid without the disengagement of gas. The pasty mass was washed with boiling water until free from acid, and was dried at a low temperature. The product was brittle at low temperatures, could be molded like jalap resin at slightly elevated ones, was semi-fluid at 100°, and at high temperatures gave off red fumes. When heated more strongly, it deflagrated suddenly and with violence. Schönbein also experimented with other organic substances, and states that in experiments carried out during December, 1845, and the first few months of 1846 he discovered, one after another, all those substances about which so much had lately been said in the French Academy. In March he sent specimens of the new compounds, among them guncotton, to several of his friends, notably, Faraday, Herschel, and Grove.

About the middle of April, 1846, Schönbein went to Württemberg where he carried out experiments with guncotton at the arsenal at Ludwigsburg in the presence of artillery officers and at Stuttgart in the presence of the king. During May, June, and July he experimented at Basel with small arms, mortars, and cannon. On July 28 he fired for the first time a cannon which was loaded with guncotton and with a projectile. Shortly afterward he used guncotton to blast rocks at Istein in the Grand Duchy of Baden and to blow up some old walls in Basel.

In the middle of August Schönbein received news from Professor Böttger of Frankfort-on-the-Main that he too had succeeded in preparing guncotton, and the names of the two men soon became associated in connection with the discovery and utilization of the material. There were, moreover, several other chemists who at about the same time, or within a few months, also worked out methods of preparing it. In a letter [80] to Schönbein dated November 18, 1846, Berzelius congratulated him on the discovery as interesting as it was important, and wrote, "Since Professor Otto of Brunswick made known a method of preparing the guncotton, this discovery has perhaps occupied a greater number of inquisitive persons than any other chemical discovery ever did. I have likewise engaged in experiments upon it."

In August Schönbein went to England where, with the help of the engineer Richard Taylor of Falmouth, he carried out experiments with guncotton in the mines of Cornwall. He also demonstrated his material successfully with small arms and with artillery at Woolwich, at Portsmouth, and before the British

---

[80] MacDonald, *op. cit.*, pp. 47, 48.

Association. He did not apply for an English patent in his own name but communicated his process to John Taylor of Adelphi, Middlesex, who was granted English patent 11,407, dated October 8, 1846, for "Improvements in the Manufacture of Explosive Compounds, communicated to me from a certain foreigner residing abroad." [81] He entered into an agreement for three years

FIGURE 61. Rudolf Böttger (1806-1887). (Courtesy E. Berl.) Professor at Frankfort-on-the-Main. Discovered guncotton independently of Schönbein but somewhat later, in the same year, 1846. He also invented matches, and made important studies on the poisoning of platinum catalysts. Reproduced from original in Kekulé's portrait album.

with Messrs. John Hall & Sons of Faversham that they should have the sole right in England to manufacture guncotton by his process and in return should pay him one-third of the net profit with a minimum of £1000 down and the same each year. The first factory for the manufacture of guncotton was erected at Faversham. On July 14, 1847, within less than a year, the factory was destroyed by an explosion with the loss of twenty-one lives. After this, Messrs. John Hall & Sons refused to continue the manu-

[81] *Ibid.,* pp. 42–44.

facture. About the same time disastrous guncotton explosions occurred at Vincennes and at Le Bouchet, and these produced such an unfavorable effect that no more guncotton was manufactured in England or in France for about sixteen years.

Schönbein offered his process to the Deutscher Bund for 100,000 thalers, and a committee was formed to consider the matter, Liebig representing the state of Hesse and Baron von Lenk, who was secretary, representing Austria. The committee continued to sit until 1852 when it finally decided to take no action. At the suggestion of von Lenk, Austria then acquired the process for 30,000 gulden.

1846. The *Comptes rendus* of 1846 contains several papers on the nitration of cellulose, which papers were presented to the French Academy before the details of Schönbein's process were yet known. Among these, the papers by Dumas and Pelouze are especially interesting. Dumas [82] stated that certain details of the manufacture of guncotton had already been published in Germany. Professor Otto of Brunswick dipped the cotton for half a minute in concentrated fuming nitric acid, pressed between two pieces of glass, washed until free from acid, and afterwards dried.

The explosive property can be considerably increased by several dippings, and I have found that a product of extreme force is obtained after an immersion of 12 hours. A point of extreme importance is the care which ought to be exercised in washing the cotton. The last traces of acid are very difficult to remove, and should any remain it will be found that, on drying, the substance smells strongly of oxides of nitrogen, and when ignited also produces a strong acid smell. The best test of a sample of guncotton is to ignite it upon a porcelain plate. Should it burn slowly, leaving a residue upon the plate, it must be considered as unsatisfactory. A good guncotton burns very violently without leaving any residue. It is also of very great importance that when the guncotton is withdrawn from the acid, it should be washed immediately in a large quantity of water. Should small quantities of water be used it will be found that the guncotton becomes very hot, and that spots of a blue or green color are produced, which are very difficult to remove, and the guncotton is very impure.

Dr. Knopp of the University of Leipzig used a mixture of equal parts of concentrated sulfuric and nitric acids, and immersed the

[82] *Compt. rend.*, 806 (1846); MacDonald, *op. cit.*, pp. 15-17.

cotton in it for several minutes at ordinary temperature. Dumas stated that satisfactory guncotton could be obtained without observing any great exactitude in the proportion of the two acids or in the duration of the immersion. Dr. Bley of Bernberg had discovered that sawdust, treated in the same way as cotton, yielded an explosive which, he believed, might replace gunpowder in firearms and in blasting.

1846. Pelouze [83] made clear distinction between xyloïdine and guncotton. "I shall call *pyroxyline* or *pyroxyle* the product of the action of monohydrated nitric acid on cotton, paper, and ligneous substances, when this action has taken place without having caused the solution of the cellulose." Braconnot in 1833 had prepared xyloïdine from starch; Pelouze had prepared pyroxylin in 1838. He pointed out that xyloïdine dissolves readily in strong nitric acid and, in the course of a day, is destroyed by it and converted to a deliquescent acid. Pyroxylin does not dissolve in concentrated nitric acid. Xyloïdine is very inflammable and explodes when struck, but it leaves a considerable residue of carbon when heated in a retort and may be analyzed like an ordinary organic substance by heating with copper oxide. Pyroxylin explodes when heated to 175° or 180° and cannot be distilled destructively. Pelouze found that 100 parts of starch, dissolved in nitric acid and precipitated immediately, yielded at most 128 to 130 parts of xyloïdine. One hundred parts of cotton or paper, after a few minutes' or after several days' immersion in concentrated nitric acid, yielded 168 to 170 parts of washed and dried pyroxylin. The acid mother liquors, both from the nitration of the starch and from the nitration of the cotton, contained not more than mere traces of organic matter.

1846. Schönbein's process soon became known through the publication of the English patent to John Taylor (cited above). He carried out the nitration by means of a mixture of 1 volume of strong nitric acid (1.45 to 1.5) and 3 volumes of strong sulfuric acid (1.85). The cotton was immersed in this acid at 50–60°F. for 1 hour, and was then washed in a stream of running water until free from acid. It was pressed to remove as much water as possible, dipped in a very dilute solution of potassium carbonate (1 ounce to the gallon), and again pressed as dry as possible.

[83] *Ibid.*, 809, 892 (1846); MacDonald, *op. cit.*, pp. 17–20.

It was then rinsed with a very dilute solution of potassium nitrate (1 ounce to the gallon). The patent states that "the use of this solution appears to add strength to the compound, but the use of this solution and also potassium carbonate are not essential and may be dispensed with." The product is pressed, opened out, and dried at 150°F., and when dried it is fit for use. The patent also covers the possibility of using instead of cotton "other matters of vegetable origin and the possibility of carrying out the nitration with nitric acid alone or with mixed acids of inferior strength."

1846. Teschemacher [84] studied the preparation of guncotton and demonstrated that no sulfuric acid is consumed by the reaction.

1847. Gladstone [85] by exercising special precautions was able to carry out combustion analyses of xyloïdine and of pyroxylin prepared according to the directions of Schönbein. Nitrogen was determined by the differential method. The pyroxylin was found to contain 12.75% nitrogen and was thought to correspond to a pentanitrate while the xyloïdine corresponded more nearly to a trinitrate.

1847. Crum [86] nitrated cotton until he could introduce no further nitrogen into the molecule, and analyzed the product for nitric acid by the method which is used in the nitrometer. His result calculated as nitrogen gives a figure of 13.69%. It is interesting to note that Crum's cotton was "bleached by boiling in caustic soda and put in a solution of bleaching powder; then caustic soda again, and afterwards weak nitric acid. It was well washed and beaten in a bag with water after each operation. . . . The cotton, dried and carded after bleaching, was exposed in parcels of 10 grains each for several hours to the heat of a steam bath, and each parcel was immersed, while hot, into a 1 oz. measure of the following mixture: Sulphuric acid (1.84) 1 measure, and 3 measures of pale lemon-colored nitric acid (1.517). After one hour it was washed in successive portions of water until no trace of acid remained, and was then dried in the open air"—or, for analysis, was dried completely in a vacuum desiccator over sulfuric acid.

[84] *Mem. of the Chem. Soc.*, 253 (1846); MacDonald, *op. cit.*, pp. 28–31.
[85] *Ibid.*, 412 (1847); MacDonald, *op. cit.*, pp. 31–41.
[86] *Proc. Phil. Soc. Glasgow*, 163 (1847); MacDonald, *op. cit.*, pp. 21–27.

1852. The Austrian government acquired the use of Schönbein's process (as mentioned above) and the Emperor of Austria appointed a committee to investigate the use of guncotton for military purposes. This committee, of which von Lenk was the leading spirit, continued to function with some interruptions until 1865. In 1853 a factory was erected at Hirtenberg for the manufacture of guncotton by the method of von Lenk which involved a more elaborate purification than Schönbein's original process. The product was washed for 3 weeks, then boiled with dilute potassium carbonate solution for 15 minutes, washed again for several days, impregnated with water glass, and finally dried. Von Lenk constructed 12-pounder guns which were shot with guncotton cartridges, but they were much damaged by the firing. About 1860 he tried bronze guns, which were less likely to burst than iron ones, and with propelling charges of guncotton fired from them shells which were filled with bursting charges of guncotton. The shells often burst within the barrel, for the acceleration produced by the propelling charge of guncotton was much too sudden and shocking. They could be shot out without exploding when a propelling charge of black gunpowder was used. On July 20, 1863, the magazine at Hirtenberg exploded, and the Austrian government thereupon decided to abandon the use of guncotton as a propellent explosive. Von Lenk was permitted to communicate his process to other nations. In 1862 and 1863, under the name of Révy, he took out English patents to protect his method of purification.[87] In 1863 he visited England and described his process to a committee of the British Association. In the same year Messrs. Prentice and Co. commenced the manufacture of guncotton at Stowmarket by von Lenk's process, but an explosion soon occurred at their establishment. In 1865 a guncotton magazine at Steinfelder Heath, near Vienna, exploded, and on October 11 of that year the manufacture of guncotton in Austria was officially forbidden.

1862. Tonkin's English patent [88] deserves our notice because it mentions the pulping of guncotton—and it was the pulping of guncotton, introduced later by Abel, which remedied in large measure the difficulties of stability which had given guncotton a bad repute and brought it back again to the favorable con-

[87] Brit. Pats. 1090 (1862), 2720 (1863).
[88] Brit. Pat. 320 (1862); MacDonald, *op. cit.*, p. 44.

sideration of the users of explosives. The patent describes the nitration of the cotton with mixed acid, the washing with running water, the pressing, and the dipping in a very dilute solution of potassium carbonate. "The fibre is then taken in the wet state and converted into pulp in the same manner as is practiced by paper-makers, by putting the fibre into a cylinder, having knives revolving rapidly, working close to fixed knives." The patent makes no claim to the pulping of guncotton, but only claims the use of pulped guncotton in an explosive consisting of sodium nitrate 65%, charcoal 16%, sulfur 16%, and guncotton pulp 3%.

1865. Abel's patent [89] for "Improvements in the Preparation and Treatment of Guncotton" claims the pulping and the pressing of it into sheets, discs, cylinders, and other forms and was probably designed to cover the process of getting it into a state where it would burn less violently in the gun. The compressed blocks were an improvement over the yarn of von Lenk, but they were still much too fast; they damaged the guns and were not ballistically uniform in performance. The blocks of compressed guncotton, however, have continued to find use in blasting. And the outstanding advantage of Abel's pulping was that it converted the guncotton into a state where the impurities were more easily washed out of it, and resulted thereby in a great improvement in stability.

1866–1867. Abel's "Researches on Guncotton" [90] demonstrated that guncotton, after proper purification, is far more stable than it had been thought to be. Moisture does not harm it, or exposure to sunlight, and it decomposes only slowly at elevated temperatures; the principal cause of its decomposition is acid, and this is removed by the pulping. Abel wrote:

> In reducing the material to a very fine state of division by means of the ordinary beating and pulping machines, the capillary power of the fibre is nearly destroyed, and the guncotton is, for a considerable period, very violently agitated in a large volume of water. It would be very difficult to devise a more perfect cleansing process than that to which the guncotton is submitted; and the natural result of its application is that the material thus additionally purified acquires considerably increased powers of resisting the de-

[89] Brit. Pat. 1102 (1865); MacDonald, op. cit., pp. 45-46.
[90] Loc. cit.

structive effects of heat. Samples of the pulped guncotton, even in the most porous conditions, have been found to resist change perfectly upon long-continued exposure to temperatures which developed marked symptoms of decomposition in the guncotton purified only as usual. The pulping process applied to guncotton affords, therefore, important additional means of purifying the material, the value of which may be further enhanced by employing a slightly alkaline water in the pulping machine. The slightest change sustained by guncotton is attended by the development of free acid, which, if it accumulates in the material, even to a very trifling extent, greatly promotes decomposition.

Numerous experimental data have been collected with respect to the establishment and acceleration of decomposition in guncotton by free acid whilst exposed to light or elevated temperature. This acid is present either in the imperfectly purified material or has been developed by decomposition of guncotton or its organic impurities. Samples of guncotton which, by exposure to elevated temperatures or for considerable periods to strong daylight, had sustained changes resulting in a considerable development of acid, have afterwards been thoroughly purified by washing. When exposed to light for months, and in some instances for two or three years (up to the present time), they have undergone no further change, while corresponding samples confined in close vessels without being purified, have continued, in some instances, to undergo decomposition, and the original substance has been completely transformed into the products repeatedly spoken of.

Abel found that the guncotton regularly produced at Waltham Abbey contained a small amount of material soluble in ether-alcohol, an average amount of 1.62% in the guncotton which was made by treating cotton with 18 times its weight of mixed acid, and an average of 2.13% in the guncotton which was made by the use of 10 parts of acid. "The employment of the higher proportion of acid furnished results more nearly approaching perfection than those obtained when the guncotton was left in contact with a smaller proportion of the acid mixture. As far as can be judged at present, however, from the general properties of the products, the difference observed when the larger or the smaller proportion of acid is used, is not of sufficient importance to render necessary the consumption of the larger quantity of acid in the manufacture." Abel was able to carry out satisfactory combustion analyses, with the following average results:

Material soluble in ether-alcohol, C 30.50%; H 2.91%; N 11.85%;
Material insoluble in ether-alcohol, C 24.15%; H 2.46%; N 13.83%.

He concluded that the different analytical results which had been procured with different samples of guncotton resulted from the samples containing different amounts of the ether-alcohol soluble material, and judged that completely nitrated guncotton is the trinitrate of cellulose, $[C_6H_7O_2(ONO_2)_3]_n$, as had been first suggested by Crum. This substance contains theoretically 14.14% nitrogen.

1868. E. A. Brown, assistant to Abel, discovered [91] that dry compressed guncotton could be made to detonate very violently by the explosion of a fulminate detonator such as Nobel had already used for exploding nitroglycerin. Shortly afterwards he made the further important discovery that wet guncotton could be exploded by the explosion of a small quantity of dry guncotton (the principle of the booster). This made it possible to use large blocks of wet guncotton in naval mines with comparative safety.

Nitrocellulose (NC)

Cellulose occurs everywhere in the vegetable kingdom; it is wood fiber and cell wall, the structural material of all plants. Cotton fiber is practically pure cellulose, but cellulose of equal purity, satisfactory in all respects for the manufacture of explosives and smokeless powder, may be produced from wood. Cellulose and starch both yield glucose on hydrolysis, and the molecules of both these substances are made up of anhydroglucose units linked together.

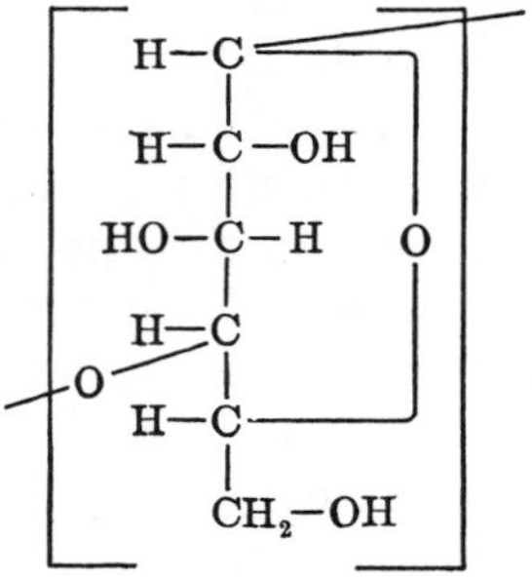

---

[91] Brit. Pat. 3115 (1868).

The two substances differ in the configuration of the number 1 carbon atom. In cellulose this atom has the β-configuration; 2000 or 3000 anhydroglucose units are linked together in long, straight, threadlike masses which are essentially one dimensional. In starch the number 1 carbon atom has the α-configuration which leads to spiral arrangements essentially three dimensional, and the molecule contains not more than 25 or 30 anhydroglucose units.

Cellulose contains 3 hydroxyl groups per anhydroglucose unit, and yields a trinitrate on complete nitration (14.14% N). An absolutely complete nitration is difficult to secure, but a product containing 13.75% nitrogen may be produced commercially. If the conditions of nitration, concentration of acid, temperature, and duration of the reaction, are less severe, less nitrogen is introduced, and products ranging all the way from a few per cent of nitrogen upward, and differing widely in solubilities and viscosities, may be secured. In the cellulose nitrates which contain less than enough nitrogen to correspond to the trinitrate, the nitrate groups are believed to be distributed at random among the three possible positions, and no definite structural formulas can be assigned to the materials. Nor is it to be supposed that a sample which may correspond in empirical composition to cellulose mononitrate or dinitrate really represents a single chemical individual.

Collodion is a nitrocellulose which is soluble in ether-alcohol and contains, according to the use for which it is destined, from 8%, more or less, of nitrogen to 12% or thereabouts. The name of *pyroxylin* is now generally applied to collodion of low nitrogen content intended for use in pharmacy, in the making of lacquers or of photographic film, or intended in general for industrial uses outside of the explosives industry. In 1847 Maynard discovered that nitrocellulose existed which was soluble in a mixture of ether and alcohol although it would not dissolve in either of these solvents taken singly.[92] The discovery soon led to the invention of collodion photography by Archer in 1851. Chardonnet's first patent [93] for artificial silk was granted in 1884. *Celluloid,* made by dissolving collodion nitrocellulose in camphor with the use of

[92] After the material is dissolved, the solution may be diluted either with alcohol or with ether without precipitating.

[93] French Pat. 165,345 (1884).

heat and pressure, was patented by J. W. and I. S. Hyatt[94] in 1870. Worden[95] states that collodion for the manufacture of celluloid is made by nitrating tissue paper with a mixed acid which contains nitric acid 35.4%, sulfuric acid 44.7%, and water 19.9%. Twenty-two pounds of acid are used per pound of paper. The nitration is carried out at 55° for 30 minutes, and the product contains 11.0–11.2% nitrogen. Ether-alcohol solutions of collodion, to which camphor and castor oil have been added in order that they may yield tough and flexible films on evaporation, are used in pharmacy for the application of medicaments to the skin in cases where prolonged action is desired. Two per cent of salicylic acid, for example, in such a mixture makes a "corn remover." Collodion for use with nitroglycerin to make blasting gelatin is generally of higher nitrogen content. Here the desideratum is that the jelly should be stiff, and the higher nitrogen content tends in that direction, but the collodion dissolves in the nitroglycerin more slowly, and the product becomes stiffer on prolonged storage, and less sensitive, and may cause misfires. The nitrogen content of collodion for use in the manufacture of blasting explosives is generally between 11.5 and 12.0%. The official definition in England of collodion for this purpose gives an upper limit of 12.3% nitrogen.

Two kinds of nitrocellulose were used in France at the time of the first World War, *coton-poudre No. 1* (CP₁), insoluble in ether-alcohol and containing about 13% nitrogen, and *coton-poudre No. 2* (CP₂), soluble in ether-alcohol and containing about 12% nitrogen.[96] CP₁ thus contained a little less nitrogen than the material which we are accustomed to call guncotton, and CP₂ contained a little more than the material which we are accustomed to call collodion. CP₁ and CP₂ were not respectively wholly insoluble and wholly soluble in ether-alcohol; their compositions were approximate, and CP₂ always contained a certain amount of material soluble in alcohol alone. A mixture of CP₁ and CP₂ colloided with ether-alcohol was used for making *pou-*

[94] U. S. Pat. 105,338 (1870).

[95] *J. Soc. Chem. Ind.,* **29**, 540 (1910).

[96] The French are accustomed to report their analyses of nitrocellulose, not as per cent nitrogen, but as cubic centimeters of NO (produced in the nitrometer and measured under standard conditions) per gram of sample. Per cent nitrogen times 15.96 equals number of cubic centimeters of NO per gram of nitrocellulose.

dre B. Either CP_1 with nitroglycerin and an acetone solvent or both with nitroglycerin and an ether-alcohol solvent were used for making ballistite, and both of them with nitroglycerin and with non-volatile solvents were used in attenuated ballistite. CP_2 was also used in France for the manufacture of blasting gelatin.

Mendeleev studied the nitration of cellulose during the years 1891 to 1895 in an effort to prepare a nitrocellulose which should have the largest content of nitrogen (and hence the greatest explosive power) compatible with complete solubility in ether-alcohol. He produced *pyrocellulose* containing 12.60% nitrogen. Russia adopted a military smokeless powder made from pyrocellulose by colloiding with ether-alcohol, and the United States in 1898 was using a similar powder in the Spanish-American War.

The word *guncotton* has about the same meaning in English and in American usage, namely, nitrocellulose containing 13% or more of nitrogen, usually 13.2–13.4%, insoluble in ether-alcohol and soluble in acetone and in ethyl acetate. One American manufacturer prefers to call guncotton *high-grade nitrocellulose*.

Preparation of Pyrocellulose. Equal volumes of sulfuric acid (1.84) and nitric acid (1.42) are mixed by pouring the sulfuric acid with stirring into the nitric acid, and the mixture is allowed to cool to room temperature. Five grams of absorbent cotton, previously dried at 100° for 2 hours, is thrust quickly into 150 cc. of this mixed acid and allowed to remain there for 30 minutes while it is stirred occasionally with a glass rod. The cotton is removed, freed as much as possible from acid by pressing against the side of the vessel, and introduced quickly into a large beaker of cold water where it is stirred about in such manner as to accomplish the prompt dilution of the acid with which it is saturated. The product is washed thoroughly in running water, and boiled for an hour with distilled water in a large beaker, then boiled three times with fresh portions of distilled water for a half hour each time. If the water from the last boiling shows the slightest trace of acidity to litmus paper, the pyrocellulose ought to be rinsed and boiled once more with distilled water. Finally, the excess of water is wrung out, and the pyrocellulose is dried in a paper tray for 48 hours at room temperature.

Pyrocellulose is made commercially from purified cotton *linters* or *hull shavings* or wood cellulose, most commonly by the mechanical dipper process. The thoroughly dry cellulose is introduced into the mixed acid contained in an iron or stainless steel nitrator which is equipped with two paddles revolving vertically in opposite directions and designed to thrust the cotton quickly under the surface of the acid. For 32 pounds of cellulose a charge of about 1500 pounds of mixed acid is used. This contains approximately 21% nitric acid, 63% sulfuric acid, and 16% water. It may contain also a small amount, say 0.5%, of nitrous acid, NO_2 or N_2O_4, which, however, is calculated as being equivalent to a like amount of water and is not reckoned as any part of the *nitrating total* of actual nitric and sulfuric acids. The sulfuric acid content of the nitrating acid is kept as constant as possible in practice; the nitric acid content may vary somewhat, less than 1%, however, for slightly more nitric acid is necessary in warm weather to offset the tendency toward denitration which exists at that time. At the start the acid has a temperature of about 30°, the introduction of the cellulose requires about 4 minutes, and the nitration is continued for 20 minutes longer while the mixture is stirred mechanically with the paddles and the temperature is kept between 30° and 34°. When the nitration is complete, a valve in the bottom of the nitrator is opened and the slurry is allowed to run into a centrifuge on the floor below. Here the crude nitrocellulose is separated quickly from the spent acid which is fortified for use again or, in part, goes to the acid recovery plant. Wringer fires are by no means uncommon, especially on damp days, for the air which is sucked through the acid-soaked material in the centrifuge gives up its moisture to the strong acid and dilutes it with the development of considerable heat. The nitrated product is forked through an orifice in the bottom of the wringer and falls into an immersion basin below, where it is *drowned* by being mixed rapidly with a swiftly moving stream of water. Thence it proceeds on its way down the *guncotton line* where it is *stabilized* or purified and then prepared for shipment or for use.

The crude nitrocellulose contains certain amounts of cellulose sulfate, of nitrate of oxycellulose, and possibly of some cellulose nitrate which is less stable than the ordinary, all of which are capable of being hydrolyzed by long-continued boiling with slightly acidified water. Guncotton requires a longer stabilizing boil than pyrocellulose. After the boiling the acid is washed off and removed from the nitrocellulose, yielding a product which is now stabilized because it contains neither free acid nor component materials which are prone to decompose with the formation of acid.

The *preliminary boiling* or *sour boiling* is carried out in large wooden tubs heated by means of steam. At the beginning the nitrocellulose is boiled with water which contains 0.25% to 0.50%

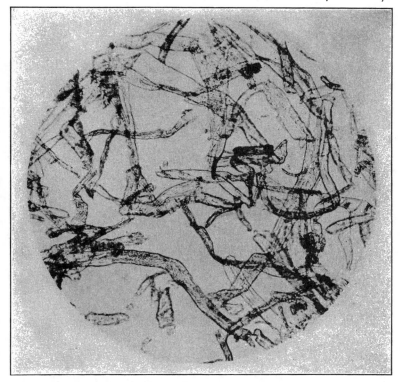

FIGURE 62. Nitrocellulose Fibers before Beating (132✕). (Courtesy Western Cartridge Company.)

of acidity calculated as sulfuric acid. The first boil lasts usually for 16 hours during which time the acidity of the solution increases. The increase is due largely to actual sulfuric acid. After 16 hours the steam is shut off, the solution is decanted from the nitrocellulose, the tub is filled with fresh water, and the material is boiled again for 8 hours. The boiling is repeated until each tubful has been boiled for 40 hours with at least 4 changes of water.

The hollow fibers still contain an acid solution within them. In order that this acid may be washed out, they are *pulped* or broken up into short lengths by means of apparatus like that which is used in the manufacture of paper. A Jordan mill cuts the fibers off rather sharply, leaving square ends, but a beater tears

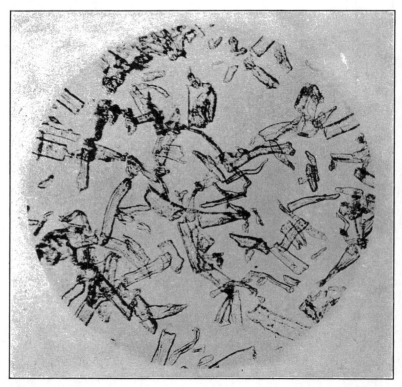

FIGURE 63. Nitrocellulose Fibers after Beating (132×). (Courtesy Western Cartridge Company.)

them, leaving ends which appear rough and shredded under the microscope and which result on the whole in the better opening up of the tubular fibers. The two machines are usually used in series. A weak solution of sodium carbonate is added during the pulping to neutralize the acid which is liberated. The pulping is continued until the desired fineness has been attained as shown by laboratory test.

The pulped fibers still retain acid adsorbed or occluded on their surface. This is removed by *poaching* the nitrocellulose, by boiling it again, first for 4 hours with fresh water with or without the addition of dilute sodium carbonate solution,[97] then for 2 hours with water without addition of soda, then twice with water for 1 hour each time. The material is then washed at least 8 times by thorough agitation with cold water, and by decantation each time of at least 40% of the liquid. After the washing, the material undergoes *screening*, where it passes through apertures 0.022 inch in width, *wringing*, whereby its moisture content is reduced to 26–28%, and finally *packing* for shipment or for storage in containers which are hermetically sealed.

Guncotton is made in substantially the same way as pyrocellulose except that a stronger mixed acid containing approximately 24% nitric acid, 67% sulfuric acid, and 9% water is used. Long-fiber high-grade guncotton is usually manufactured by the pot process and with the use of mixed acid which is nearly anhydrous. Iron pots are generally used. For the nitration of 4 pounds of dry cotton, 140 pounds of acid is introduced into the pot and the cotton is immersed in it, pressed down, and allowed to digest for 20 or 30 minutes. The contents of several pots are centrifuged at once, and the product is stabilized in the same way as pyrocellulose except that it is not pulped.

There can be no doubt that, in the standard method of stabilizing nitrocellulose, there are, among the results which the poaching accomplishes, at least some which would have been accomplished much earlier during the boiling if the material at that time had been pulped. This seems especially evident with respect to the hydrolysis of easily hydrolyzed material adjacent to the inner wall of the tubular fibers. Olsen,[98] discussing the standard method, has written, "The preliminary boiling tub treatment reduced the acidity of the fibers and of the interstitial material, but the pulping process, by macerating these fibers, has set free an additional amount of acid. It is, therefore, necessary to repurify the pyrocotton by boiling." He discovered that a marked reduction in time and in cost could be secured by carrying out the pulping operation prior to the hydrolyzing boils. If the pulping is done at

[97] Not more than 10 gallons of sodium carbonate solution (1 pound per gallon) for every 2000 pounds of nitrocellulose (dry weight).
[98] U. S. Pat. 1,798,270 (1931).

the outset, "less than half of the 16 hours sour boiling usually employed will suffice for obtaining the desired degree of purity when followed by alternating boils in fresh water and washes with cold fresh water, again less than half of the amount of boiling being sufficient." With less than 20 hours total time of purification, he obtained results as good as are ordinarily procured by the 52 hours of the standard method.

FIGURE 64. Boiling Tubs for Purification of Nitrocellulose.

Olsen's quick stabilization process [99] is the result of further thinking along this same line and represents an ingenious application of a simple principle of colloid chemistry. After the nitrocellulose has been thoroughly pulped, and after the easily decomposed cellulose sulfate, etc., have been hydrolyzed, there remains only the necessity for removing the acid which clings to the fiber. The acid, however, is adsorbed on the nitrocellulose, or bound to it, in such manner that it is not easily washed away by water or even by dilute soda solution; many boilings and washings are necessary to remove it. Olsen has found that the acid is removed rapidly and completely if the nitrocellulose is digested or washed with a solution of some substance which is adsorbed by nitro-

[99] U. S. Pat. 1,893,677 (1933).

cellulose with greater avidity than the acid is adsorbed, that is, with a solution of some substance which has, as he says, a greater *adhesion tension* for nitrocellulose than the acid has. Such substances are aniline red, Bismarck brown, methyl orange,

FIGURE 65. Fred Olsen. Has done important work on cellulose and has made many improvements in detonating explosives, high explosives, and smokeless powder; in particular, has invented processes for the quick stabilization of nitrocellulose and for the production of ball-grain powder. Chief of Chemical Research, Aetna Explosives Company, 1917-1919; Chemical Adviser, Picatinny Arsenal, 1919-1928; Technical Director, Western Cartridge Company, 1929—.

m-phenylenediamine, urea, substituted ureas such as diethyldiphenylurea, and diphenylamine. A 0.5% solution of urea in water may be used. A half-hour washing with a 0.5% solution of diphenylamine in alcohol was more effective in producing stability than 20 hours of boiling with water. A solution of 0.1 gram of Bismarck brown in 300 cc. of water gave better stabilization of 30 grams of nitrocellulose in 1 hour than 10 boilings of 1 hour each with separate 300-cc. portions of water.

Nitrocellulose, like all other nitric esters with the possible exception of PETN, is intrinsically unstable, even at ordinary temperatures. Yet the decomposition of a thoroughly purified sample is remarkably slow. Koehler and Marqueyrol [100] have made a careful study of the decomposition of nitrocellulose at various temperatures in the vacuum of a mercury pump. They found that it evolved gas at the rate of about 0.7 cc. per gram per day at 100°, 0.01 cc. per gram per day at 75°, and 0.0001 cc. per gram per day at 40°.

A sample of CP_1 was freed from carbonate by digestion with carbonated water and subsequent washing; it was dried thoroughly, and 35.152 grams of the material (analyzing 211.2 cc. NO per gram) was heated in vacuum at 75°. The results are summarized in the following table, where all gas volumes have been reduced to 0° and 760 mm. The residual gas, insoluble both

Duration of Heating at 75°	Total Volume, Cubic Centimeters	Cubic Centimeters per Gram per Day	Composition of Gas, %		
			NO	CO_2	Residue
1st period (5 days)	2.25	0.0128	62.5	16.7	20.8
2nd period (56 days)	17.29	0.0088	63.2	19.5	17.3
3rd period (56 days)	18.25	0.00927	60.8	21.5	17.6
4th period (56 days)	18.34	0.0080	65.5	18.0	16.5
5th period (56 days)	18.19	0.0079	60.0	20.7	19.6
6th period (56 days)	18.3	0.0084	61.2	20.4	18.3

in ferrous sulfate and in caustic soda solution, was analyzed and was found to consist approximately of 46% carbon monoxide, 18% nitrous oxide, 35% nitrogen, and a trace of hydrocarbons. After 309 days of heating at 75°, the temperature of the oven was reduced, and the same sample of nitrocellulose was heated in vacuum at 40° for 221 days. During this time it evolved a total of 0.697 cc. of gas or 0.0001154 cc. per gram per day. The same sample was then heated in vacuum at 100°, as follows.

[100] *Mém. poudres*, **18**, 101, 106 (1921).

Duration of Heating at 100°	Total Volume, Cubic Centimeters	Cubic Centimeters per Gram per Day	Composition of Gas, %		
			NO	CO₂	Residue
1st period (30 hrs.)	29.09	0.662	51.9	24.1	24.0
2nd period (8.5 hrs.)	8.57	0.689 ⎱	68.1	17.6	14.3
3rd period (9 hrs.)	8.09	0.614 ⎰			

The residual gas, neither NO nor CO₂, was found to contain about 64% of carbon monoxide, the remainder being nitrous oxide and nitrogen with a trace of hydrocarbons. The nitrocellulose left at the end of the experiment weighed 34.716 grams corresponding to a loss of 1.24% of the weight of the original material. It gave on analysis 209.9 cc. NO per gram corresponding to a denitration per gram of 2.2 cc.

The gases from the decomposition of nitrocellulose in vacuum contain nothing which attacks nitrocellulose. If the decomposition occurs in air, the nitric oxide which is first produced combines with oxygen to form nitrogen dioxide, and the red fumes, which are acidic in the presence of moisture, attack the nitrocellulose and promote its further decomposition. The decomposition then, if it occurs in the presence of air or oxygen, is self-catalyzed. The amount of nitric oxide which is produced if the decomposition occurs in the absence of air, or the amount of nitrogen dioxide which is produced in the first instance if the decomposition occurs in the presence of air, is a function solely of the mass of the sample. The extent to which the red fumes attack the nitrocellulose depends, on the other hand, upon the concentration of the gases and upon the area of the surface of the sample which is accessible to their attack. The greater the density of loading of the sample, the greater will be the concentration of the red fumes. For the same density of loading, the finer the state of subdivision of the sample, the greater will be the surface. Pellets of compressed nitrocellulose, heated in the air, decompose more rapidly than the same nitrocellulose in a fluffier condition. The pellets give a poorer heat test (see below) but obviously consist of material which has the same stability. Likewise, nitrocellulose which has been dissolved in ether-alcohol and precipitated by the addition of water, decomposes in the air more rapidly than the original, bulkier material. Straight nitrocellulose powder always gives a better heat test than the nitrocellulose from which it was made. If small grains and large grains of smokeless powder are made from the same nitrocellulose, the large grains will give the better heat test.

In this country the most common heat tests which are made regularly upon nitrocellulose and smokeless powder are the 65.5° KI starch test and the 134.5° methyl violet test. In the former of these, five several portions of the material under test, differing in their moisture content from nearly dry to thoroughly dry, are heated in test tubes in a bath warmed by the vapors of boiling methyl alcohol. Within each tube, a strip of potassium iodide starch paper, spotted with a 50% aqueous solution of glycerin, hangs from a hook of platinum wire a short distance above the sample, the hook itself being supported from a glass rod through a cork stopper. The tubes are examined constantly, and the time needed for the first appearance of any color on the test paper in any one of the tubes is reported.

In the 134.5° methyl violet test, heavy glass test tubes about a foot long are used. They are closed loosely at their upper ends with perforated or notched cork stoppers, and are heated for almost their whole length in a bath which is warmed by the vapors of boiling xylene. Two tubes are used. The samples occupy the lower 2 inches of the tubes, strips of methyl violet paper are inserted and pushed down until their lower ends are about 1 inch above the samples, the tubes are heated and examined every 5 minutes, and the times are noted which are necessary for the test papers to be turned completely to a salmon-pink color, for the first appearance of red fumes, and for explosion. The explosion usually manifests itself by the audible popping of the cork from the tube, but causes no other damage. A test similar to this one, but operated at 120°, using blue litmus paper and reporting the time necessary for the paper to be reddened completely, is sometimes used.

In the Bergmann-Junk test the number of cubic centimeters of nitrogen dioxide produced by heating a 5-gram sample for 5 hours at 132° is reported. The determination was originally made by absorbing the gas in ferrous sulfate solution, liberating the nitric oxide by warming, and measuring its volume. A method based upon the absorption of the gas in caustic soda solution and the titration of its acidity is now often used instead.

There are many other variations of the heat test.[101] They are sometimes called *stability tests*, but most of them, it will be noted, involve the self-catalyzed decomposition of the sample in an atmosphere of air or of red fumes. They indicate the comparative stability only of materials which are physically alike. True indications of the stability of nitric esters are to be secured only by studying the decomposition of the substances in vacuum. For this purpose the 120° vacuum stability test is most generally preferred.

Ash in nitrocellulose is determined by gelatinizing the sample with acetone which contains 5% of castor oil, setting fire to the colloid, allowing it to burn tranquilly, and igniting the charred residue to constancy of weight. It is sometimes determined as sulfate by dissolving the sample in pure concentrated sulfuric acid and igniting to constant weight.

Nitrogen in nitrocellulose is determined by means of the *nitrometer*, an instrument of great usefulness to the chemist who is working with nitric esters or with nitroamines.

Determination of Nitrogen

Nitric acid and organic and inorganic nitrates, and in general all substances which contain free nitric acid or yield nitric acid when they are treated with concentrated sulfuric acid, are analyzed by means of the nitrometer. The method depends upon the measurement of the volume of the nitric oxide which is produced when concentrated sulfuric acid acts upon the sample in the presence of mercury. It is satisfactory also for the determination of nitro group nitrogen in certain nitroamines, in nitroguanidine and in tetryl but not in methylnitramine. It is not satisfactory in the presence of mononitro aromatic compounds or of other substances which are nitrated readily by a solution of nitric acid in concentrated sulfuric acid.

[101] U. S. War Department Technical Manual TM 9–2900 and the U. S. Bureau of Mines *Bulletins* on the analysis of explosives describe the standard heat tests in detail. "Explosives, Matches, and Fireworks" by Joseph Reilly, New York, D. Van Nostrand Company, Inc., 1938, pp. 71–83, describes about 40 different heat tests.

Cold concentrated sulfuric acid does not attack mercury. Cold nitric acid acts upon mercury to form mercurous nitrate with the evolution of nitric oxide. If concentrated sulfuric acid is present, mercurous nitrate cannot form, and the nitric acid is converted by the mercury quantitatively into nitric oxide. The method appears to have been .used for the first time by Walter Crum [102] who applied it at an early date to the analysis of guncotton.

FIGURES 66 and 67. Georg Lunge and His Nitrometer. Obverse and reverse of commemorative bronze plaquette by Hans Frei in celebration of Lunge's seventieth birthday.

He introduced .the sample of guncotton into a eudiometer filled with mercury and inverted in that liquid, and carried out the reaction and measured the gas volume in the same eudiometer. Since he was unable to separate the guncotton from the air entangled with it, the measured gas volume was too large. The true volume of nitric oxide was determined by admitting a solution of ferrous sulfate to the eudiometer and noting the volume of gas which was absorbed.

The Lunge nitrometer is so designed that the nitrate or nitric ester is dissolved first in concentrated sulfuric acid and the solution, without entrained gas, is afterwards admitted to the re-

[102] *Loc. cit.*

action vessel. In the usual form of the instrument as used in Europe, the gas from the reaction is measured in cubic centimeters at atmospheric pressure, the barometer and the thermometer are read, and the weight of the nitrogen in the nitric oxide and the percentage of nitrogen in the sample are calculated.

In the extremely ingenious DuPont nitrometer, a 1-gram sample is used for the analysis, and the gas is collected in a measuring tube which has been graduated to read, at a certain temperature and pressure, the correct percentage of nitrogen in the 1-gram sample. By means of a compensating bulb and leveling device, the gas in the measuring tube is brought to the volume which it would occupy if it were confined at the temperature and pressure at which the graduations are correct, and the percentage of nitrogen is then read off directly. The DuPont nitrometer [103] was invented by Francis I. DuPont about 1896. It quickly came into general use in the United States, and represents the form of the nitrometer which is preferred and generally used in this country. Lunge in 1901 claimed that it differs in no significant respect from the "gasvolumeter" or "five-part nitrometer" [104] which he had described in 1890.

Calibration and Use of the DuPont Nitrometer. The five essential parts of the DuPont nitrometer are illustrated in Figure 68. The graduations on the measuring bulb correspond to dry nitric oxide measured at 20° and 760 mm., which nitric oxide contains the indicated number of centigrams of nitrogen. Thus, the point marked 10 indicates the volume which would be occupied under the standard conditions of temperature and pressure by the quantity of dry nitric oxide which contains 0.10 gram of nitrogen, that is, by the nitric oxide produced in the nitrometer reaction from a 1-gram sample of nitrate containing 10% nitrogen. The point marked 12 corresponds to 12/10 of this volume, that marked 14 to 14/10, and so on. And the tube reads correctly the per cent of nitrogen in a 1-gram sample provided the gas is measured at 20° and 760 mm.

In setting up the instrument, dry air is introduced into the compensating bulb and the outlet at the upper end of the bulb is sealed. Dry air is introduced into the measuring bulb, the outlet is connected to a sulfuric acid manometer, and the mercury reservoir and the compensating bulb are raised or lowered until the portions of air confined

[103] Pitman, *J. Soc. Chem. Ind.*, **19**, 982 (1900).
[104] *Ibid.*, **9**, 547 (1890); **20**, 100 (1901).

in both bulbs are at atmospheric pressure. The stopcock is closed, the volume in the measuring bulb is read, thermometer and barometer are noted, the volume which the air in the measuring bulb would occupy at 20° and 760 mm. is calculated, and the mercury reservoir and the bulbs are adjusted until the air in the measuring bulb occupies this calculated volume and until the air in the compensating bulb is at exactly the same pressure as that in the measuring bulb. A glass tube

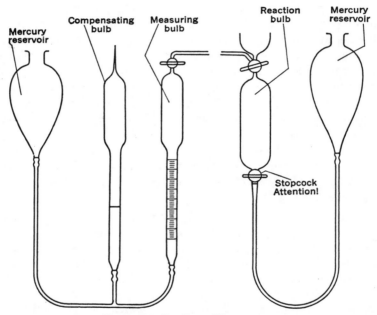

FIGURE 68. Du Pont Nitrometer.

bent twice at right angles and containing some water is used for leveling the mercury in the two bulbs. The position of the mercury in the compensating bulb is now marked by means of a strip of paper glued to the glass. Whenever in the future the gas in the compensating bulb is again confined in this same volume, and whenever the nitric oxide in the measuring bulb is confined at the same pressure as the gas in the compensating bulb, then the nitric oxide will occupy the volume which it would occupy if confined at 20° and 760 mm., and, if a 1-gram sample was taken for the analysis, the reading will indicate correctly the nitrogen content. If a sample larger or smaller than 1 gram was taken, then the reading is to be corrected accordingly.

At the beginning of an analysis, the reaction bulb and the measuring bulb and the capillary tubes at the tops of the bulbs are completely filled with mercury. A sample of about 1 gram of nitrocellulose is weighed in a small weighing bottle, dried for an hour and a half at 100°, cooled in a desiccator, and weighed accurately. A little 95% sulfuric acid is poured onto the nitrocellulose and the whole is washed into the reaction bulb. The weighing bottle is rinsed out with several small portions of sulfuric acid, the same acid is used for rinsing the cup and is finally introduced into the reaction bulb, until altogether about 20 cc. of acid has been used, care being taken that no air is introduced. The mercury reservoir is lowered to give a reduced pressure in the reaction bulb and the bulb is shaken gently, *the stopcock at its bottom being open,* until the generation of gas has practically ceased. The bulb is then raised until the level of the mercury drops nearly to its lower shoulder, the stopcock is closed, and the bulb is shaken vigorously for 3 minutes. The cock is opened and the apparatus is allowed to stand for several minutes. The mercury level is then adjusted as before, the cock is closed, and the shaking is repeated for another 3 minutes. Finally the gas is transferred to the measuring bulb and allowed to stand for about 20 minutes. The measuring bulb and the compensating bulb are then adjusted in such fashion that the mercury in both stands at the same level and that the mercury in the compensating bulb stands at the point indicated by the paper strip. The volume in the measuring bulb is then read. After each determination the reaction bulb is rinsed out twice with concentrated sulfuric acid.

In practice it is convenient to standardize the nitrometer from time to time by means of a sample of pure potassium nitrate (13.85% N) or of nitrocellulose of known nitrogen content.

The nitrometer is dangerous to one who does not understand it fully. The closing at the wrong time of the stopcock at the bottom of the reaction bulb may result in the explosion of that vessel and the throwing about of glass and of acid.

Nitrostarch

Nitrostarch [105] is manufactured and used in the United States, but has not found favor in other countries. In all the early attempts to manufacture nitrostarch, the starch was dissolved in strong nitric acid and the nitric ester was precipitated by mixing the solution with sulfuric acid or with the spent acid from some

[105] The article by Urbanski and Häckel in *Z. ges. Schiess- u. Sprengstoffw.,* **30,** 98 (1935), is accompanied by an extensive bibliography.

other nitration, as from the nitration of glycerin. The product resembled the xyloïdine of Braconnot, showed a very poor stability, and could not be stored or handled safely in the dry condition. The pulverulent, dusty form of the dry material probably also contributed to the disrepute into which it fell in Europe. In this country starch is nitrated with mixed acid in which it does not dissolve, and the product retains the appearance of ordinary starch, as guncotton retains the appearance of cotton.

Cassava or tapioca starch was preferred at first, for it was claimed that it contained less fat than corn starch and that the

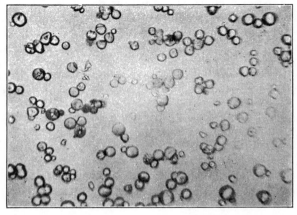

FIGURE 69. Nitrostarch Granules (about 30×). (Courtesy Trojan Powder Company.)

granules, being smaller than those of corn and potato starch, permitted a more uniform nitration and a more efficient purification. Since 1917 corn starch has been used in this country. The starch is first freed from fats and from pectic acid by washing with dilute caustic soda or ammonia solution and then with water, and it is dried until it contains less than 0.5% of moisture. In one process which produced a nitrostarch containing 12.75% nitrogen, a mixed acid containing 38% of nitric acid and 62% of sulfuric acid was used, 800 pounds of the acid in a single nitrator for 200 pounds of starch. The initial temperature of the acid was 32°, the mixture was agitated by a mechanical stirrer having a downward pitch, and the temperature during the nitration was kept between 38° and 40°. At the end of the nitration the contents of the nitrator was drowned in a small tub of water. The product was purified entirely by cold-water washings, without boiling. Ammonia was used to neutralize the acidity during the preliminary washing, and it is probable that this use of ammonia determined the stability of the product, perhaps because ammonia was preferentially adsorbed, instead of acid, by the material of the nitrostarch granules. The product was dried at 35–40°.

Nitrostarch gives no color with iodine. It is insoluble in water and does not gelatinize to form a paste as starch does when it is boiled with water. It is not notably hygroscopic, but may take up 1 or 2% of moisture from a damp atmosphere. It is soluble in acetone. The varieties of nitrostarch which are soluble in etheralcohol contain about the same amounts of nitrogen as the varieties of nitrocellulose which dissolve in that mixed solvent. Nitrostarch does not form a good film or tough colloid as nitrocellulose does.

During the first World War a *Trojan explosive* which contained nitrostarch was used in trench mortar shells and in hand and rifle grenades.[106] Its composition was as follows.

	Not Less Than	Not More Than
Nitrostarch	23.0%	27.0%
Ammonium nitrate	31.0	35.0
Sodium nitrate	36.0	40.0
Charcoal	1.5	2.5
Heavy hydrocarbons	0.5	1.5
Anti-acid	0.5	1.5
Diphenylamine	0.2	0.4
Moisture	....	1.2

All the *dope materials* were first ground to the desired fineness and dried, and then turned over in a large mixing barrel while the dry nitrostarch was added. Trench mortar shells were loaded by *stemming,* but the explosive was *jarred* into the grenades through small funnel-shaped openings. Another nitrostarch explosive, which was used only in grenades, was called *Grenite* and consisted almost entirely of nitrostarch (about 97%) with small amounts (about 1.5% each) of petroleum oil and of gum arabic. It was made by spraying the dry materials with a solution of the

106 U. S. War Department Technical Manual TM 9–2900, p. 109.

binder while the mixture was stirred in a rotary mixer. The resulting granules were dried and screened, and yielded a free-running explosive which could be loaded easily by machine.

Three United States patents [107] granted in 1916 to Bronstein and Waller describe several nitrostarch blasting explosives, of which the following table reports typical examples. In actual use,

	I	II	III	IV	V	VI
Nitrostarch	30.0%	39.0%	30.0% *	40.0%	40.0%	40.0%
Ammonium nitrate TNT mixture	15.0	20.0	...	...	...	20.0
Sodium nitrate	46.8	37.25	58.0	37.7	34.7	17.7
Barium nitrate	...	...	...	20.0	20.0	20.0
Carbonaceous material	3.0	...	5.0	...	...	...
Paraffin oil	0.7	0.75	0.5	0.8	0.8	0.8
Sulfur	3.0	2.0	5.0	...	3.0	...
Calcium carbonate	1.5	1.0	1.5	1.5	1.5	1.5

these explosives would also contain a small amount of some stabilizer, say 0.2% of diphenylamine or of urea.

Utilization of Formaldehyde

At the time of the first World War the methyl alcohol which was needed for the preparation of tetryl was procured from the distillation of wood. It was expensive and limited in amount. Formaldehyde was produced then, as it is now, by the oxidation of methyl alcohol, and a demand for it was a demand upon the wood-distillation industry. Formaldehyde was the raw material from which methylamine was produced commercially, and the resulting methylamine could be used for the preparation of tetryl by the alternative method from dinitrochlorobenzene. It was also the raw material from which certain useful explosives could be prepared, but its high price and its origin in the wood-distillation industry deprived the explosives in question of all but an academic interest. With the commercial production of synthetic methyl alcohol, the same explosives are now procurable from a raw material which is available in an amount limited only by the will of the manufacturers to produce it.

Carbon monoxide and hydrogen, heated under pressure in the presence of a suitable catalyst, combine to form methyl alcohol. A mixture of zinc oxide and chromium oxide has been used as a

[107] U. S. Pats. 1,188,244, 1,188,245, 1,188,246 (1916).

catalyst for the purpose. Carbon monoxide and hydrogen (equimolecular amounts of each) are produced as *water gas* when steam is passed over hot coke.

$$C + H_2O \longrightarrow CO + H_2$$

Additional hydrogen, from the action of iron on steam or from the electrolysis of water, is added to the water gas to provide the mixture which is needed for the synthesis of methyl alcohol.

$$CO + 2H_2 \longrightarrow CH_3-OH$$

It is evident that carbon dioxide may be used instead of the monoxide if a correspondingly larger amount of hydrogen is also used.

$$CO_2 + 3H_2 \longrightarrow CH_3-OH + H_2O$$

Methyl alcohol in fact is made in this manner from the carbon dioxide which results from certain industrial fermentations. When methyl alcohol vapor is mixed with air and passed over an initially heated catalyst of metallic copper or silver gauze, oxidation occurs, sufficient heat is evolved to maintain the catalyst at a bright red, and formaldehyde is formed.

$$2CH_3-OH + O_2 \longrightarrow 2H_2O + 2 \begin{matrix} H \\ \diagdown \\ \diagup \\ H \end{matrix} C=O$$

Of the several explosives which are preparable from formaldehyde, two are the most powerful and brisant of the solid high explosives which are suitable for military use. One of these, *cyclotrimethylenetrinitramine* or *cyclonite,* is a nitroamine and is discussed in the chapter which is devoted to those substances. The other, *pentaerythrite tetranitrate* or PETN, is a nitric ester. Both may be prepared from coke and air.

Formaldehyde enters readily into combination with substances which add to its unsaturated carbonyl group. If a substance containing an active hydrogen adds to formaldehyde or condenses with it, the active hydrogen attaching itself to the oxygen of the formaldehyde and the rest of the molecule attaching itself to the carbon, the result is that the position originally occupied by the active hydrogen is now occupied by a —CH_2—OH or methylol group. Hydrogens which are active in condensation reactions are those which are α- to a carbonyl, a nitro, or a cyano group, etc., that is, they are attached to a carbon atom to which a carbonyl, a nitro, or a cyano group is also attached and are in general the hydrogen atoms which are involved in the phenomena of tautomerism. The condensation of formaldehyde with acetaldehyde, with nitromethane, with cyclopentanone, and with cyclohexanone thus leads to polyhydric primary alcohols the nitric esters of which are useful explosives.

Pentaerythrite Tetranitrate (PETN, penta, niperyth, penthrit)

Four equivalents of formaldehyde in warm aqueous solution in the presence of calcium hydroxide react with one equivalent of acetaldehyde to form pentaerythrite. Three of the four react with the three α-hydrogens of the acetaldehyde, the fourth acts as a reducing agent, converts the —CHO group to —CH$_2$—OH, and is itself oxidized to formic acid.

$$3\ \underset{H}{\overset{H}{\diagdown}}C{=}O + CH_3{-}CHO$$

$$H{-}COOH + HO{-}CH_2{-}\underset{\underset{OH}{|}}{\overset{\overset{CH_2OH}{|}}{C}}{-}CH_2{-}OH \quad PE$$

$$\left[HO{-}CH_2{-}\underset{\underset{OH}{|}}{\overset{\overset{CH_2OH}{|}}{C}}{-}CHO \right] + \underset{H}{\overset{H}{\diagdown}}C{=}O \qquad NO_2{-}O{-}CH_2{-}\underset{\underset{ONO_2}{|}}{\overset{\overset{ONO_2\;CH_2}{|}}{C}}{-}CH_2{-}ONO_2 \quad PETN$$

The name, pentaerythrite, indicates that the substance contains five carbon atoms and (like erythrite) four hydroxyl groups. In commercial practice [108] the reaction is carried out at 65–70°. After 2 hours at this temperature, the calcium is precipitated by means of sulfuric acid, the mixture is filtered, and the filtrate is concentrated and crystallized by evaporation in vacuum. Penta-

---

[108] For the laboratory preparation see *Org. Syntheses*, **4**, 53 (1912), John Wiley & Sons, New York. Also Tollens and Wigand, *Ann.*, **265**, 316 (1891); Rave and Tollens, *ibid.*, **276**, 58 (1893); Stettbacher, *Z. ges. Schiess- u. Sprengstoffw.*, **11**, 182 (1916).

erythrite crystallizes from water in white tetragonal crystals, m.p. 253°. One part requires 18 parts of water at 15° for its solution.

PETN may be prepared, according to Naoúm,[109] by adding 100 grams of finely powdered pentaerythrite to 400 cc. of nitric acid (1.52) while the temperature is maintained between 25° and 30° by efficient cooling. Toward the end of the nitration a certain amount of the tetranitrate crystallizes out. The separation of the product is completed by the gradual addition of 400 cc. of concentrated sulfuric acid (1.84) while the stirring and cooling are continued. The mixture is not drowned, but the crude PETN (85–90% of the theory) is filtered off directly, and washed first with 50% sulfuric acid and then with water. It still contains some occluded acid and is purified, according to Naoúm, by dissolving in hot acetone to which a little ammonium carbonate is added, and filtering the hot solution into twice its volume of 90% alcohol by which the PETN is precipitated in fine needles.

Pentaerythrite may also be nitrated satisfactorily, and probably in better yield, without the use of sulfuric acid and with the use of nitric acid from which the nitrous acid has been removed.

Preparation of Pentaerythrite Tetranitrate. Four hundred cc. of strong *white* nitric acid—prepared by adding a little urea to fuming nitric acid, warming, and blowing dry air through it until it is completely decolorized—is cooled in a 600-cc. beaker in a freezing mixture of ice and salt. One hundred grams of pentaerythrite, ground to pass a 50-mesh sieve, is added to the acid a little at a time with efficient stirring while the temperature is kept below 5°. After all has been added, the stirring and the cooling are continued for 15 minutes longer. The mixture is then drowned in about 3 liters of cracked ice and water. The crude product, amounting to about 221 grams or 95% of the theory, is filtered off, washed free from acid, digested for an hour with a liter of hot 0.5% sodium carbonate solution, again filtered off and washed, dried, and finally recrystallized from acetone. A good commercial sample of PETN melts at 138.0–138.5°. The pure material melts at 140.5–141.0°, short prismatic needles, insoluble in water, difficultly soluble in alcohol and ether.

Pentaerythrite tetranitrate is the most stable and the least reactive of the explosive nitric esters. It shows no trace of decomposition if stored for a very long time at 100°. While nitrocellulose

[109] *Op. cit.,* p. 244.

is destroyed within a few minutes by boiling with a 2.5% solution of caustic soda, PETN requires several hours for its complete decomposition. Ammonium sulfide solution attacks PETN slowly at 50°, and a boiling solution of ferrous chloride decomposes it fairly rapidly. It does not reduce Fehling's solution even on boiling, and differs in this respect from erythrite tetranitrate.

PETN does not take fire from the spit of a fuse. If a small quantity is submitted to the action of a flame, it melts and takes fire and burns quietly with a slightly luminous flame without smoke. Above 100° it begins to show appreciable volatility, and at 140–145°, or at temperatures slightly above its melting point, it shows red fumes within half an hour. It inflames spontaneously at about 210°. It is relatively insensitive to friction but makes a loud crackling when rubbed in a rough porcelain mortar. It may be exploded readily by pounding with a carpenter's hammer on a concrete floor. In the drop test it is detonated by a 20-cm. drop of a 2-kilogram weight, sometimes by a drop of 10 or 15 cm.

Naoúm [110] reports that 10 grams of PETN in the Trauzl test with sand tamping gave a net expansion of about 500 cc., with water tamping one of 560 cc. The same investigator [111] found a velocity of detonation of 5330 meters per second for the material, only slightly compressed, at a density of loading of 0.85 in an iron pipe 25 mm. in internal diameter. For PETN compressed to a density of 1.62 Kast [112] found a velocity of detonation of 8000 meters per second.

PETN is extraordinarily sensitive to initiation. It is detonated by 0.01 gram of lead azide, whereas tetryl requires 0.025 gram of lead azide for its certain detonation. This sensitivity and its great brisance combine to make PETN exceptionally serviceable in compound detonators.

Under high pressure powdered PETN agglomerates to a mass which has the appearance of porcelain, but which, when broken up into grains, is a very powerful smokeless powder functioning satisfactorily with the primers which are commonly used in small arms ammunition. The powder is hot and unduly erosive, but cooler powders have been prepared by incorporating and compressing PETN in binary or in ternary mixtures with TNT,

110 *Ibid.*, p. 246.
111 *Ibid.*, p. 247.
112 *Z. angew. Chem.*, **36**, 74 (1923).

nitroguanidine, and guanidine picrate. A mixture of PETN with guanidine picrate is less sensitive to heat and to shock than ordinary colloided smokeless powder, and is stable at all temperatures which are likely to be encountered. PETN does not colloid with nitrocellulose. It dissolves readily in warm trinitrotoluene, and mixtures may be prepared which contain 65% or more of PETN. The richer mixtures may be used as propellent powders. The less-rich mixtures are brisant and powerful high explosives comparable in their behavior and effects to TNB.

Stettbacher [113] in 1931 described several dynamite-like explosives which contained both PETN and nitroglycerin. He called them by the general name of *Penthrinit,* and described simple penthrinit, *gelatin penthrinit,* and *ammonpenthrinit.* Naoúm [114] later in the same year reported comparative tests of ammonpenthrinit and gelatin dynamite, as follows.

Composition	AMMONPENTHRINIT	GELATIN DYNAMITE
PETN	37%	....
Nitroglycerin	10%	63%
Collodion nitrocotton	....	2%
Dinitrotoluene	5%	....
Wood meal	....	5%
Ammonium nitrate	48%	30%
Trauzl test (average)	430 cc.	465 cc.
Velocity of detonation (average)	6600 meters per sec.	7025 meters per sec.
At density of loading	1.36	1.47

A Swiss patent of 1932 to Stettbacher [115] covers the conversion of PETN into a plastic mass by means of 10–30% of a fluid nitric ester such as nitroglycerin or nitroglycol. It states that a mixture of 80% PETN and 20% nitroglycerin is a plastic mass, density 1.65, which does not separate into its components and which is suitable for loading shells and detonators. For the latter purpose it is initiated with 0.04 gram of lead azide.

Dipentaerythrite Hexanitrate (Dipenta)

The formation of a certain amount of dipentaerythrite is unavoidable during the preparation of pentaerythrite. It is nitrated

[113] *Z. ges. Schiess- u. Sprengstoffw.,* **26,** 8, 39 (1931).
[114] *Ibid.,* **26,** 42 (1931).
[115] Swiss Pat. 137,476 (1932).

along with the latter substance, and, unless a special purification is made, remains in the PETN where its presence is undesirable because of its lower stability.

$$
\begin{array}{c}
\text{OH} \\
| \\
\text{CH}_2 \\
| \\
\text{HO-CH}_2\text{-C-CH}_2\text{-[OH} \quad \text{H]O-CH}_2\text{-C-CH}_2\text{-OH} \\
| \\
\text{CH}_2 \\
| \\
\text{OH}
\end{array}
$$

$$
\begin{array}{c}
\text{OH} \quad\quad \text{OH} \\
| \quad\quad\quad | \\
\text{CH}_2 \quad\quad \text{CH}_2 \\
| \quad\quad\quad | \\
\text{HO-CH}_2\text{-C-CH}_2\text{-O-CH}_2\text{-C-CH}_2\text{-OH} \\
| \quad\quad\quad | \\
\text{CH}_2 \quad\quad \text{CH}_2 \\
| \quad\quad\quad | \\
\text{OH} \quad\quad \text{OH}
\end{array}
$$

Dipentaerythrite

$$
\begin{array}{c}
\text{ONO}_2 \quad\quad \text{ONO}_2 \\
| \quad\quad\quad | \\
\text{CH}_2 \quad\quad \text{CH}_2 \\
| \quad\quad\quad | \\
\text{NO}_2\text{-O-CH}_2\text{-C-CH}_2\text{-O-CH}_2\text{-C-CH}_2\text{-ONO}_2 \\
| \quad\quad\quad | \\
\text{CH}_2 \quad\quad \text{CH}_2 \\
| \quad\quad\quad | \\
\text{ONO}_2 \quad\quad \text{ONO}_2
\end{array}
$$

Dipentaerythrite hexanitrate

Dipentaerythrite hexanitrate [116] is procured in the pure state by the fractional crystallization from moist acetone of the crude PETN which precipitates when the nitration mixture is drowned in water, white crystals, m.p. 72°. The crystals have a specific gravity of 1.630 at 15°, after being fused and solidified 1.613 at 15°. The substance is less sensitive to friction, less sensitive to the mechanical shock of the drop test, and less sensitive to temperature than PETN, but it is less stable and decomposes much more rapidly at 100°.

Brün [117] reports measurements by the Dautriche method of the

[116] Friederich and Brün, *Ber.*, **63**, 2861 (1930); Brün, *Z. ges. Schiess- u. Sprengstoffw.*, **27**, 71, 125, 156 (1932).
[117] *Ibid.*, **27**, 126 (1932).

velocities of detonation of several explosives loaded in copper tubes 10 mm. in diameter and compressed under a pressure of

EXPLOSIVE	DENSITY	VELOCITY OF DETONATION, METERS PER SECOND
Dipentaerythrite hexanitrate	{ 1.589	 7370
	{ 1.589	 7450
Pentaerythrite tetranitrate	{ 1.712	 8340
	{ 1.712	 8340
Tetryl	{ 1.682	 7530
	{ 1.682	 7440
Trinitrotoluene	{ 1.615	 7000
	{ 1.615	 7000

2500 kilograms per square centimeter. He also reports that a 10-gram sample of dipentaerythrite hexanitrate in the Trauzl test gave a net expansion of 283 cc. (average of 2), and PETN under the same conditions gave a net expansion of 378 cc. (average of 3).

Trimethylolnitromethane Trinitrate (Nitroisobutanetriol trinitrate, nitroisobutylglycerin trinitrate, nib-glycerin trinitrate)[118]

This explosive was first described in 1912 by Hofwimmer[119] who prepared it by the condensation of three molecules of formaldehyde with one of nitromethane in the presence of potassium bicarbonate, and by the subsequent nitration of the product.

[118] The first two of these names are scientifically correct. The third is not correct but is used widely. The trihydric alcohol from which the nitric ester is derived is not an isobutylglycerin. In the abbreviated form of this name, the syllable, nib, stands for nitro-iso-butyl and is to be pronounced, not spelled out like TNT and PETN.

[119] *Z. ges. Schiess- u. Sprengstoffw.*, **7**, 43 (1912). Brit. Pat. **6447** (1924). Stettbacher, *Nitrocellulose*, **5**, 159, 181, 203 (1935).

At a time when the only practicable methods for the preparation of nitromethane were the interaction of methyl iodide with silver nitrite and the Kolbe reaction from chloracetic acid, the explosive was far too expensive to merit consideration. The present cheap and large scale production of nitromethane by the vapor-phase nitration of methane and of ethane has altered the situation profoundly. Trimethylolnitromethane trinitrate is an explosive which can now be produced from coke, air, and natural gas. Nitromethane too has other interest for the manufacturer of explosives. It may be used as a component of liquid explosives, and it yields on reduction methylamine which is needed for the preparation of tetryl.

The crude trimethylolnitromethane from the condensation commonly contains a small amount of mono- and dimethylolnitromethane from reactions involving one and two molecules of formaldehyde respectively. It is recrystallized from water to a melting point of 150°, and is then nitrated. Stettbacher reports that the pure substance after many recrystallizations melts at 164–165°. The nitration is carried out either with the same mixed acid as is used for the nitration of glycerin (40% nitric acid, 60% sulfuric acid) or with very strong nitric acid, specific gravity 1.52. If the trihydric alcohol has been purified before nitration, there is but little tendency for the nitrate to form emulsions during the washing, and the operation is carried out in the same way as with nitroglycerin. In the laboratory preparation, the nitric ester is taken up in ether, neutralized with ammonium carbonate, dried with anhydrous sodium sulfate, and freed from solvent in a vacuum desiccator.

The explosive is procured as a yellow oil, more viscous than nitroglycerin, density 1.68 at ordinary temperature. It has but little tendency to crystallize at low temperatures. A freezing point of −35° has been reported. It is very readily soluble in ether and in acetone, readily soluble in alcohol, in benzene, and in chloroform, and insoluble in ligroin. It is less soluble in water and less volatile than nitroglycerin. Because it is less volatile, it is slower to cause headaches, and for the same reason the headaches are slower to go away. It is distinctly inferior to nitroglycerin as a gelatinizing agent for collodion nitrocotton. The nitro group attached directly to an aliphatic carbon atom appears to have an unfavorable effect on stability, for trimethylolnitromethane trinitrate gives a poorer potassium iodide 65.5° heat test than nitroglycerin. Naoúm [120] reports the data which are tabulated below.

	TRIMETHYLOL-NITROMETHANE TRINITRATE	NITRO-GLYCERIN
Trauzl test: 75% kieselguhr dynamite	325 cc.	305 cc.
93% blasting gelatin	580 cc.	600 cc.
Drop test, 2-kilogram weight	6 cm.	2 cm.

Nitropentanone and Related Substances

Cyclopentanone and cyclohexanone contain four active hydrogen atoms and condense with formaldehyde to form substances which contain four —CH$_2$—OH groups. The latter may be converted directly into explosive tetranitrates or they may be reduced, the carbonyl groups yielding secondary alcohol groups, and the products then may be nitrated to pentanitrates.

Tetramethylolcyclopentanone tetranitrate

Tetramethylolcyclopentanol pentanitrate

The explosives derived in this way from cyclopentanone and cyclohexanone were patented in 1929 by Friederich and Flick.[121] They

[120] *Op. cit.*, p. 241.
[121] Ger. Pat. 509,118 (1929).

are less sensitive to mechanical shock than PETN, and three out of four of them have conveniently low melting points which permit them to be loaded by pouring. *Tetramethylolcyclopentanone tetranitrate,* called *nitropentanone* for short, melts at 74°. *Tetramethylolcyclopentanol pentanitrate* is called *nitropentanol* and melts at 92°. *Tetramethylolcyclohexanone tetranitrate,* m.p. 66°, is called *nitrohexanone,* and *tetramethylolcyclohexanol pentanitrate,* m.p. 122.5°, *nitrohexanol.* They are less brisant than PETN. Wöhler and Roth [122] have measured their velocities of detonation at various densities of loading, as follows.

EXPLOSIVE	DENSITY OF LOADING	VELOCITY OF DETONATION, METERS PER SECOND
Nitropentanone	1.59	7940
	1.44	7170
	1.30	6020
	1.13	4630
Nitropentanol	1.57	7360
	1.51	7050
	1.29	6100
	1.11	5940
	1.01	5800
	0.91	5100
	0.75	5060
Nitrohexanone	1.51	7740
	1.42	7000
	1.25	5710
Nitrohexanol	1.44	7670
	1.28	6800
	1.00	5820
	0.81	5470

[122] *Z. ges. Schiess- u. Sprengstoffw.,* **29,** 332–333 (1934).

CHAPTER VI

SMOKELESS POWDER

An account of smokeless powder is, in its main outlines, an account of the various means which have been used to regulate the temperature and the rate of the burning of nitrocellulose. After the degree of nitration of the nitrocellulose, other factors which influence the character of the powder are the state of aggregation of the nitrocellulose, whether colloided or in shreds, the size and shape of the powder grains, and the nature of the materials other than nitrocellulose which enter into its composition.

Bulk Powder

The first successful smokeless powder appears to have been made by Captain Schultze of the Prussian Artillery in 1864. At first he seems only to have impregnated little grains of wood with potassium nitrate, but afterwards he purified the wood by washing, boiling, and bleaching, then nitrated it, purified the nitrated product by a method similar to that which had been used by von Lenk, and finally impregnated the grains with potassium nitrate alone or with a mixture of that salt and barium nitrate.[1] The physical structure of the wood and the fact that it contained material which was not cellulose both tended to make the nitrated product burn more slowly than guncotton. The added nitrates further reduced the rate of burning, but Schultze's powder was still too rapid for use in rifles. It found immediate favor for use in shot guns. It was manufactured in Austria by a firm which in 1870 and 1871 took out patents covering the partial gelatinization of the powder by treatment with a mixture of ether and alcohol. The improved powder was manufactured between 1872 and 1875 under the name of *Collodin*, but the Austrian gov-

[1] Brit. Pat. 900 (1864).

ernment stopped its manufacture on the grounds that it infringed the government's gunpowder monopoly. A company was formed in England in 1868 to exploit Schultze's invention, a factory was established at Eyeworth in the New Forest in 1869, and the

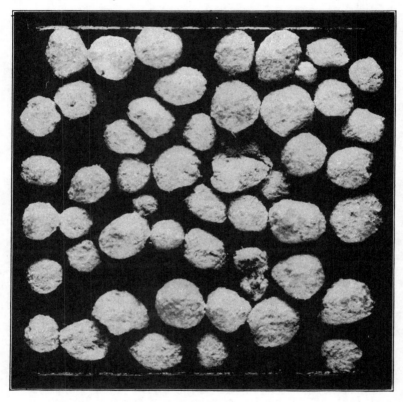

FIGURE 70. Shreddy Grains of Bulk Powder (25×). (Courtesy Western Cartridge Company.)

methods of manufacture were later improved by Griffiths and achieved great success. In 1883 Schultze entered into a partnership in Germany and started a factory at Hetzbach in Hesse-Darmstadt.

The next successful smokeless powder was invented [2] at the works of the Explosives Company at Stowmarket in England. It

[2] Brit. Pat. 619 (1882) to Walter F. Reid and D. Johnson.

was called E. C. powder (Explosives Company), and consisted of nitrocotton mixed with potassium and barium nitrates with the addition of coloring matter and small amounts of other organic material. It was made into grains which were hardened by being partially gelatinized with ether-alcohol. A separate company was organized to develop the invention, and the manufacture was started at Green Street Green, near Dartford, in Kent.

Schultze powder and E. C. powder are known as *bulk sporting* powders, either because they are loaded by bulk or because, for the same bulk, they have about the same power as black powder. Bulk powders burn quickly. They are used in shot guns, in hand grenades, in blank cartridges, and occasionally in the igniter charges which set fire to the dense colloided propellent powder which is used in artillery.

Bulk powders are made in considerable variety, but they consist always of nitrocellulose fibers which are stuck together but are not completely colloided. Some contain little else but nitrocellulose; others contain, in addition to potassium and barium nitrates, camphor, vaseline, paraffin, lampblack, starch, dextrine, potassium dichromate or other oxidizing or deterrent salts, and diphenylamine for stabilization, and are colored in a variety of brilliant hues by means of coal-tar dyes. In the United States bulk powders are manufactured by one or the other of two processes, either one of which, however, may be modified considerably; the materials are incorporated under wooden wheels, grained, and partially gelatinized, or the grains are formed in a still where a water suspension of pulped nitrocellulose is stirred and heated with a second liquid, a solvent for nitrocellulose which is volatile and immiscible with water.

Three typical bulk powders are made up according to the approximate formulas tabulated below. The nitrogen content of

Nitrocellulose	84.0	87.0	89.0
% N in nitrocellulose	13.15	12.90	12.90
Potassium nitrate	7.5	6.0	6.0
Barium nitrate	7.5	2.0	3.0
Starch	....	....	1.0
Paraffin oil	....	4.0	....
Diphenylamine	1.0	1.0	1.0

the nitrocellulose is an average secured by mixing pyrocellulose and guncotton. A batch usually amounts to 200 pounds, 100 pounds of water is added and about 90 grams of rosaniline or some other, generally bright-colored, water-soluble dyestuff, and the charge is incorporated by milling for about 45 minutes in a wheel mill which is built like a black-powder mill but is smaller and has light wooden wheels. The charge is then run through a mechanical rubber, which consists of wooden blocks rubbing with a reciprocating motion on a perforated zinc plate; the larger lumps are broken up and the material is put into proper condition for granulating. For this purpose about 50 pounds is placed in a copper pan or "sweetie barrel" which is revolving in a vat of hot water and is heated by that means. The pan rotates fairly rapidly, say at about 15 r.p.m., and carries the powder up along its sloping side to a point where it is scraped off by suitably arranged wooden scrapers and falls back again. It thus receives a rolling motion which has the effect of granulating the powder into spherical grains. The operation requires about 40 minutes, and its completion is indicated by the failure of the powder to carry up on the pan because of the loss of moisture.

After it has been granulated, the powder is given a preliminary screening with a 12-mesh sieve. The material which is retained on the sieve is returned to the wheel mill. That which passes through is hardened. It is put into a horizontal revolving cylinder and a mixed solvent, consisting of about 1 part of acetone and 6 parts of alcohol, is added in the proportion of 1 gallon of solvent to 15 pounds of powder. Acetone dissolves nitrocellulose, alcohol does not; the mixed solvent swells and softens the fibers and makes them stick together. The cylinder is rotated, while hot air is blown through, until the solvent has been volatilized. During this process the temperature is allowed to rise as high as 50° or 55°. The product, which consists of grains now more or less completely agglutinated, is given a final screening. In a typical case, the portion passed by a 12-mesh sieve and retained by a 50-mesh sieve is taken; it is given a final drying and is ready for use.

In a typical example of the still process for the manufacture of bulk sporting powder, 500 pounds of pulped nitrocellulose (12.60% N) is placed in a vertical cast-iron still along with 700 gallons of water containing 2% of potassium nitrate and 6% of barium nitrate dissolved in it. The material is mixed thoroughly

FIGURE 71. Sweetie Barrel. (Courtesy Western Cartridge Company.) The moist and mixed ingredients of bulk powder, tumbled in this apparatus, take on the form of grains. Similar equipment is used for sugar-coating pills and for applying a deterrent coating or a graphite glaze to grains of colloided smokeless powder.

and agitated actively by mechanical stirrers while 145 gallons of mixed solvent (2 parts butyl acetate, 3 parts benzene) containing about 3 pounds of diphenylamine dissolved in it is pumped in. The stirring is violent enough to break the solvent phase up into many small droplets, and around each droplet a globular cluster of nitrocellulose shreds builds up. The mixture is stirred continuously and distilled in vacuum at a temperature of about 30°. The distillate is collected in a separating device in such manner that the solvent is drawn off while the water is returned to the still. At the end of the process the contents of the still consists of water with potassium and barium nitrates in solution along with granules of the wet but otherwise finished powder. The individual grains of the powder are broken apart by a very violent stirring, filtered off in a centrifuge, and dried. The finished powder contains about 1 or 1.5% of potassium nitrate and about 3.5% of barium nitrate.

Early History of Colloided Powders

1884. The first smokeless powder which was satisfactory for use in rifled guns was the dense, colloided *poudre B*,[3] invented by the French physicist, Paul Vieille, and adopted immediately for the use of the French army and navy. It was made by treating a mixture of soluble and insoluble nitrocotton with ether-alcohol, kneading to form a stiff jelly, rolling into thin sheets, cutting into squares and drying, or, in later practice, extruding through a die in the form of a strip, cutting to length, and drying. The results of the first proof firing of this powder, made with a 65-mm. cannon, were communicated to the Minister of Armaments on December 23, 1884.

> It was then established that the new processes would permit the ballistic effect of black powder to be secured with the same pressure and with the charge reduced to about a third, and that the power of the arms could be increased notably, with a slight reduction of the charge, while still keeping to the ordinary pressures. The standard powder for the model 1886 rifle was determined in the early months of the year 1885. . . . The standard powder made possible an increase of velocity of 100 meters per second for the same pressures.

[3] *Poudre blanche,* white powder in contradistinction to *poudre N, poudre noire,* black powder.

. . . This substitution has had the foreseen consequence of suppressing the smoke from the shooting.[4]

The author of the note in the *Mémorial des poudres* in which the above-quoted public announcement was made concerning the new powder was so impressed by the importance of the invention that he concludes the note by saying:

It results from this that the adaptation to firearms of any other explosive known at the present time would be able to bring to the armament only a perfectioning of detail, and that a new progress, comparable to that which has been realized recently, cannot be made except by the discovery of explosives of a type entirely different from those which chemistry today puts at our disposition.

French powder for the military rifle consists of small square flakes lightly glazed with graphite. The glazing serves to retard slightly the rate of burning of the surface layer, and, more important, it serves to make the powder electrically conducting and to prevent the accumulation of a static charge during the blending of small lots of the powder into a single, ballistically uniform large lot. For guns the powder consists of unglazed strips. The squares and strips, ignited over their entire surfaces, burn for lengths of time which depend upon their thicknesses, and they retain, during the burning, surfaces which change but little in area until at the end the grains are completely consumed.

1888.   The second successful dense smokeless powder was the *ballistite* which was invented by Alfred Nobel.[5] This was a stiff gelatinous mixture of nitroglycerin and soluble nitrocellulose in proportions varying between 1 to 2 and 2 to 1, prepared with the use of a solvent which was later removed and recovered. Nobel appears to have been led to the invention by thinking about celluloid, for the patent specification states that the substitution of almost all the camphor in celluloid by nitroglycerin yields a material which is suitable for use as a propellant. In the method of manufacture first proposed, camphor was dissolved in nitroglycerin, benzene was added, and then dry, pulped, soluble nitrocellulose; the mixture was kneaded, the benzene was allowed to evaporate, and the material was rolled between warm rollers

[4] *Mém. poudres,* **3,** 11–12 (1890).
[5] Brit. Pat. 1471 (1888).

to make it completely homogeneous. It was rolled into thin sheets which were cut with a knife or scissors into the desired shape and size. The use of nitrostarch instead of part of the nitrocellulose, and the addition of pulverized chlorate or picrate in various proportions, were also mentioned in the patent.

FIGURE 72. Paul Vieille (1854-1934). Inventor of *poudre B*, the first progressive-burning smokeless powder, 1884. Author of classic researches on erosion. Secretary and later, as successor to Berthelot, President of the French Powder and Explosives Commission.

1889. Nobel soon discovered [6] that the use of soluble nitrocellulose made it possible to manufacture ballistite without using camphor or any other solvent. The nitroglycerin and soluble nitrocellulose were brought together under water. As soon as the nitroglycerin had been absorbed by the nitrocellulose, the mass was heated to 80° to complete the gelatinization, and was then rolled and cut up in the usual way. In an alternative process the gelatinization was hastened by using more nitroglycerin than was

[6] Brit. Pat. 9361 (1889).

desired in the powder, and the excess was removed by means of 75% methyl alcohol by which it was extracted while the nitrocellulose was unaffected by that solvent.

1889. Lundholm and Sayers [7] devised a better process of incorporating the materials. The nitroglycerin and the soluble nitrocellulose were brought together under hot water and stirred by means of compressed air. The nitroglycerin presently gelatinized, or dissolved in, the nitrocellulose. The doughlike mass was removed, and passed between rollers heated to 50° or 60° whereby the water was pressed out. The sheet was folded over and passed through the rolls again, and the process was repeated until a uniform colloid resulted. It was rolled to the desired thickness and cut into squares which were generally glazed with graphite and finally blended.

1889. At about the time that Vieille was developing *poudre B*, the British government appointed a committee to investigate and report upon a smokeless powder for the use of the British service. Samples of ballistite and other smokeless powders were procured, the patent specifications relative to them were studied, and the decision was reached to use a powder which differed from Nobel's ballistite in being made from insoluble nitrocellulose containing more nitrogen than the soluble material which he used. The guncotton and nitroglycerin were incorporated together by means of acetone, mineral jelly (vaseline) was added, the colloid was pressed through dies into the form of cords of circular or oval cross section, and the acetone was evaporated off. The product was called *cordite*. The experimental work in connection with its development was done mostly in Abel's laboratory, and mostly by Kellner who later succeeded Abel as War Department chemist. Patents [8] in the names of Abel and Dewar, members of the committee, were taken out on behalf of the government in 1889, and later in the same year the manufacture of cordite was commenced at the royal gunpowder factory at Waltham Abbey.

The mineral jelly was added to cordite originally with the idea that it would lubricate the barrel of the gun, but it seems to have no such effect. Actually it is consumed during the combustion. Because of it the powder gases contain a larger number of mols

[7] Brit. Pat. 10,376 (1889).
[8] Brit. Pats. 5614, 11,664 (1889).

at a lower temperature, and produce, with less erosion, substantially the same ballistic effect as the same weight of powder made up without mineral jelly. The original cordite Mk. I. contained guncotton 37%, nitroglycerin 58%, and mineral jelly 5%. This produced such serious erosion of the guns in the British South African war that the composition was modified; the relative amount of nitroglycerin was reduced for the purpose of making it cooler. Cordite M. D. (modified) consists of guncotton 65%, nitroglycerin 30%, and mineral jelly 5%.

Mineral jelly in cordite has a distinct stabilizing action. The material is known to take up nitric oxide in the nitrometer and to cause a falsely low nitrogen analysis if it is present in the material which is being analyzed.[9]

Any distinction between cordite and ballistite which is based upon the methods by which the materials are manufactured is now no longer valid. Certain cordites are made without the use of a volatile solvent. Ballistites are made from soluble and from insoluble nitrocellulose, with and without the use of acetone, ethyl acetate, or other volatile solvent. Cordite is the name of the propellant which is used by the British armed forces. Ballistite, generally in flakes, sometimes in cords and in single-perforated tubes, is the preferred military powder of Italy, Germany, and the Scandinavian countries.

1891. Charles E. Munroe commenced investigations of smokeless powder at the Naval Torpedo Station, Newport, Rhode Island, about 1886, and about 1891 invented *indurite*. This was made from guncotton, freed from lower nitrates by washing with methyl alcohol, and colloided with nitrobenzene. The colloid was rolled to the desired thickness and cut into squares or strips which were hardened or indurated by the action of hot water or steam. Most of the nitrobenzene was distilled out by this treatment, and the colloid was left as a very hard and tough mass. Indurite was manufactured which gave satisfactory tests in guns ranging in caliber from the one-pounder to the six inch.[10]

1895–1897. After Munroe's resignation from the Torpedo Station, Lieutenant John B. Bernadou, U. S. Navy, took up the

[9] *U. S. Bur. Mines Bull.* 96, "The Analysis of Permissible Explosives," by C. G. Storm, Washington, 1916, p. 44.

[10] Ballistic tests are reported in Munroe's interesting article on "The Development of Smokeless Powder," *J. Am. Chem. Soc.,* **18,** 819–846 (1896).

work on smokeless powder and in 1895 patented a powder consisting of a mixture of guncotton, collodion cotton, and potassium nitrate, colloided with acetone, and in 1897 an improved powder made from nitrocellulose alone colloided with ether-alcohol. The nitrocellulose first used contained approximately 12.45% nitrogen, but this was later replaced by pyrocellulose, 12.60% nitrogen. The powder was made in multiperforated cylindrical grains, and was substantially the same as was used by the United States in the first World War. Patents covering various improvements in the manufacture of pyrocellulose powder were taken out in the names of Lieutenant Bernadou and Captain Converse, U. S. Navy, and were licensed or sold to private interests, the United States government retaining the right to manufacture under these patents powder for its own use.

1900–1907.   About 1900 the Navy Department built the Naval Powder Factory at Indian Head, Maryland. The plant was capable of producing several thousand pounds of smokeless powder per day, and was enlarged during the course of a few years to a capacity of about 10,000 pounds daily. About 1907 the Ordnance Department, U. S. Army, built at Picatinny Arsenal, Dover, New Jersey, a powder plant with a capacity of several thousand pounds per day.

Classification of Colloided Nitrocellulose Powders

American pyrocellulose powder and French *poudre B* are *straight nitrocellulose* or *single-base* powders. They are made by the use of a volatile solvent, generally ether-alcohol, which solvent is removed wholly or in large part during the process of manufacture. They are the simplest of colloided powders, the pyrocellulose powder being really the simpler of the two, for it is made from one single kind of nitrocellulose. Modified forms of these powders are made by incorporating into the colloid nonvolatile solvents (i.e. solvents which remain in the finished powder) which may be either explosive or non-explosive or by distributing throughout the colloid as a separate phase materials, either explosive or non-explosive, which affect the rate or the temperature of the burning or the strength of the powder. Aromatic nitro compounds, such as DNT, TNX oil, etc., dissolve nitrocellulose or are dissolved by it, and thus constitute themselves non-volatile solvents, but they are also explosives in their own right, and a nitrocellulose powder which contains one of them might, it would seem, be designated with propriety as a *double-base powder*. This, however, is not in accordance with prevailing usage. The name of double-base powder is reserved for such powders as ballistite and cordite which contain nitrocellulose and nitroglycerin (or perhaps some substitute for nitroglycerin such as nitroglycol). Double-base powders are made both with and without volatile solvent, and are also capable of being modified in all of the ways in which a single base powder may be modified. We have, therefore, colloided powder of various kinds, as follows.

   I. Nitrocellulose powder without nitroglycerin
    a. with volatile solvent,
    b. with non-explosive non-volatile solvent,
    c. with explosive non-volatile solvent,
    d. with non-explosive non-volatile non-solvent,
    e. with explosive non-volatile non-solvent.
  II. Nitrocellulose powder with nitroglycerin
    a. with volatile solvent,
    b. with non-explosive non-volatile solvent,
    c. with explosive non-volatile solvent,
    d. with non-explosive non-volatile non-solvent,
    e. with explosive non-volatile non-solvent.
 III. Coated and laminated powders the grains of which are non-homogeneous combinations of the powders above classified.

This classification is offered, not in any belief that it clarifies a matter which is otherwise difficult to understand, but because it directs attention to the various possibilities and displays their relationships to one another. Some of the possibilities correspond to powders which are or have been used in this country or in Europe, and which are sufficiently described for our present purpose if they are mentioned specifically. Others will be discussed at greater length in the sections, below, which are concerned with the absorption of moisture, with gelatinizing agents, and with flashless charges and flashless powder. All the possibilities are actually exploited, though not always separately.

Cordite MD, it may be noted, is a double base powder made with volatile solvent and containing a non-volatile, non-explosive non-solvent, namely mineral jelly, and is classified in class II *a d*, while a flashless ballistite of class II *b c* is made by incorporating centralite and DNX oil with nitroglycerin and nitrocellulose, and one of class II *b e* by mixing centralite and nitroguanidine with nitroglycerin and nitrocellulose. The nitroguanidine does not dissolve in the colloid but is distributed through it in a state of fine subdivision. Ten or 15 parts of nitroguanidine incorporated with 90 or 85 parts of pyrocellulose colloided with ether-alcohol gives a mixture which may be extruded through dies and yields a powder (I *a e*) which is flashless. PETN is another substance, insoluble in nitrocellulose colloids, which in the state of a fine powder may be incorporated in single-base or in double-base mixtures to yield powders (I *a e* and II *a e*) which are hotter and more powerful than otherwise.

Manufacture of Single-Base Powder

The operations in the manufacture of smokeless powder from pyrocellulose, briefly, are as follows.

1. *Dehydrating.* The pulped pyrocellulose contains about 25% moisture when it arrives at the smokeless powder plant. Most of this is squeezed out by pressing with a moderate pressure, say 250 pounds per square inch, for a few moments. The pressure is then released, alcohol in an amount at least equal to the dry weight of the pyrocellulose is forced into the mass by means of a pump, and the pressure is increased to about 3500 pounds per square inch. The process is managed in such fashion that the resulting cylindrical block consists of pyrocellulose moistened with exactly the amount of alcohol which is needed for the formation of the colloid. The requisite amount of ether is added later. The solvent consists altogether of 1 part by weight of alcohol and 2 parts of ether, 105 pounds of the mixed solvent for every 100 pounds of pyrocellulose if the colloid is to be made into 0.30-caliber powder, 100 parts if into powder of which the web thickness is approximately 0.025 inch, and 85 parts for powder having a web thickness of 0.185 inch. The block is received in a cannister of vulcanized fiber and is covered over in order that loss of solvent by evaporation may be reduced to a minimum. From this point on, in fact, the material is kept and is moved from one operation to another in covered cannisters at all times except when it is being worked.

FIGURE 73. Smokeless Powder Manufacture. (Courtesy E. I. du Pont de Nemours and Company, Inc.) Dehydrating Press. The nitrocellulose comes from the dehydrating press in the form of a cylindrical block, impregnated with alcohol, ready for the mixer where ether is added and where it is colloided.

FIGURE 74. Smokeless Powder Manufacture. (Courtesy E. I. du Pont de Nemours and Company, Inc.) Smokeless Powder Mixer—open to show the crumbly, partially colloided material. In use, the apparatus is closed tightly to prevent the loss of volatile solvent.

2. *Mixing or incorporating.* The compressed block from the dehydrating press is broken up by hand against the blades of the mixing machine. This is similar to the bread-mixing machines which are used in large commercial bakeries, and consists of a water-cooled steel box in which two shafts carrying curved blades rotate in opposite directions and effectively knead the material. The ether is added rapidly and mixed in as fast as possible. Diphenylamine sufficient to constitute 0.9–1.1% of the weight of the finished powder is previously dissolved in the ether, and is thus distributed uniformly throughout the colloid. The incorporated material has an appearance similar to that of a mass of brown sugar which has been churned; it is soft enough to be deformed between the fingers, and, when squeezed, welds together in the form of a film or colloid.

3. *Pressing.* The loose and not yet completely colloided material is pressed into a compact cylindrical mass by means of a pressure of about 3500 pounds per square inch in the *preliminary blocking press*. The *preliminary block* is then placed in the *macaroni press* where it is pressed or strained through 1 12-mesh steel plate, 2 sheets of 24-mesh and 1 sheet of 36-mesh steel wire screen, and through the perforations in a heavy plate of brass from which it emerges in wormlike pieces resembling macaroni. A pressure of 3000 to 3500 pounds per square inch is commonly used. The material drops directly into the cylinder of the *final blocking press*, where it is squeezed into a compact cylindrical block of the right size to fit the graining press. A pressure of about 3500 pounds per square inch is maintained for 1 or 2 minutes, and completes the colloiding of the pyrocellulose. The *final block* is dense, tough, elastic, light brown or amber colored, and translucent.

4. *Graining and cutting.* The colloid is forced by an hydraulic press through dies by which it is formed into single-perforated or into multiperforated tubes. For the formation of a single-perforated tube, the plastic mass is forced in the die into the space which surrounds a centrally fixed steel wire; it is then squeezed past the wire through a circular hole and emerges in the form of a tube. For the formation of a multiperforated tube, 7 such wires are accurately spaced within the die. A pressure of 2500 to 3800 pounds per square inch is used. For small arms powder the head of the press may contain as many as 36 dies, for large-

FIGURE 75. Smokeless Powder Manufacture. (Courtesy E. I. du Pont de Nemours and Company, Inc.) Blocking Press.

caliber powder, such as that for the 16-inch gun, it usually contains only one. The cord or rope of powder as it comes from the press is passed over pulleys or through troughs to a rotary cutter where it is cut into short cylinders about 2.1 to 2.5 times as long as their diameters or it is coiled up in a fiber cannister in which it is taken to another room for cutting. In France the colloid is

FIGURE 76. Smokeless Powder Manufacture. (Courtesy E. I. du Pont de Nemours and Company, Inc.) Finishing Press. The colloid is extruded in the form of a perforated cylinder which is later cut into pieces or grains.

pressed through slots from which it emerges in the form of ribbons which are cut into strips of a length convenient for loading into the gun for which the powder is intended.

5. *Solvent recovery.* The *green powder* contains a large amount of ether and alcohol which presents a twofold problem: (1) the recovery of as much of the valuable volatile solvent as is economically feasible, and (2) the removal of the solvent to such an extent that the finished powder will not be disposed either to give off or to take up much volatile matter or moisture under changing atmospheric conditions. For the recovery of the solvent, the pow-

der is put into a closed system and warm air at 55–65° is circulated through it; the air takes up the alcohol and ether from the powder and deposits much of it again when it is passed through a condenser. It is then heated and again passed through the powder. In some European plants the air, after refrigeration, is passed upward through an absorption tower down which cresol or other suitable liquid is trickling. This removes the ether which was not condensed out by the cooling, and the ether is recovered from it by distillation. The whole process of solvent recovery requires careful control, for the colloid on drying tends to form a skin on its surface (the way a pot of glue does when drying) and the skin tends to prevent the escape of volatile matter from the interior of the powder grain.

6. *Water-drying*. Powder is now most commonly dried by the rapid water-drying process whereby the formation of a skin upon its surface is prevented and certain other advantages are gained. Water at 65° is circulated throughout the powder. The water causes the production of microscopic cracks and pores through which the alcohol and ether escape more freely. These substances leave the powder to dissolve in the water, and then the ether in particular evaporates out from the water. When the volatile solvent content of the powder is sufficiently reduced, the powder grains are taken out and the water with which they are superficially covered is removed in a dry-house or in a continuous dryer at 55–65°. The finished powder contains 3.0 to 7.5% of volatile solvent in the interior of the grain, the amount depending upon the thickness of the web, and 0.9 to 1.4% of *external moisture*, mostly water actually resident in the cracks or pores of the surface. The amount of moisture which the powder thus holds upon its surface is an important factor in maintaining its ballistic stability under varying atmospheric conditions. The amount ought to be such that there is no great tendency for the moisture to evaporate off in dry weather, and such also that there is no great tendency for the powder to take up moisture in damp weather. The importance of surface moisture is so considerable that the French powder makers, long before there was any thought of using warm water to dry the powder, were accustomed to submit it to a *trempage* or tempering by immersion in water for several days. Later, periods of air-drying were alternated

with periods of *trempage* in warm water at temperatures sometimes as high as 80°.

Powder for small arms is generally *glazed* with graphite, by
which treatment its attitude toward the loss and absorption of
moisture is improved, and by which also it is made electrically
conducting so that it can be *blended* without danger from static

FIGURE 77. Smokeless Powder Blending Tower. The powder is blended
by being made to flow through troughs and bins. Lots as large as 50,000
pounds of rifle powder and 125,000 pounds of cannon powder have been
blended in this tower.

electricity and loaded satisfactorily by a volumetric method. The
powder is blended in order that large lots can be made up which
will be ballistically uniform, and hence that the proof firing, the
operations of loading, and the calculations of the artilleryman
may all be either simplified in kind or reduced in amount. Powder
in short cylindrical grains, such as is used in the United States,
is particularly easy to blend, but the blending of strips, or of
long tubes or cords, is obviously difficult or impracticable. The
finished powder is stored and shipped in airtight boxes which
contain 110–150 pounds.

Stabilizers

The spontaneous decomposition of nitrocellulose in the air produces nitrous and nitric acids which promote a further decomposition. If these products however are removed continuously, the uncatalyzed decomposition is extremely slow, and smokeless pow-

FIGURE 78. Bernhart Troxler. (Greystone Studios, Inc.) Introduced many innovations into the manufacture of smokeless powder and improved the design of equipment in such manner as to increase production while reducing the hazard—the steam-air-dry process for double-base powder, methods of coating, apparatus for solvent recovery, water drying, and air drying of single-base powder without transferring the powder during the three operations. His whole professional life has been devoted to smokeless powder, with the Laflin and Rand Powder Company until 1913, and afterwards with the Hercules Powder Company from the time when that company was organized and built its first smokeless powder line.

der may be stabilized by the addition to it of a substance which reacts with these acids and removes them, provided neither the substance itself nor the products of its reaction with the acids attacks the nitrocellulose.

Vieille suggested the use of amyl alcohol as a stabilizer, and powder containing this material was used in France until 1911 when, in consequence of the disastrous explosion of the battleship *Jena* in 1907 and of the battleship *Liberté* in 1911, both ascribed to the spontaneous inflammation of the powder, and in consequence of the researches of Marqueyrol, its use was discontinued entirely. Indeed no powder containing amyl alcohol was manufactured in France after October, 1910. Freshly manufactured *poudre BAm* smelled of amyl alcohol; the alcohol was converted by the products of the decomposition into the nitrous and nitric esters, and these soon broke down to produce red fumes anew and evil-smelling valerianic acid. The presence of the latter in the powder was easily detected, and was taken as evidence that the powder had become unstable. The Italians early used aniline as a stabilizer for their military ballistite. This forms nitro derivatives of aniline and of phenol, but it attacks nitrocellulose and is now no longer used. As early as 1909, diphenylamine was being used in the United States, in France, and in Germany, and, at the present time, it is the most widely used stabilizer in smokeless powder. The *centralites* (see below) also have a stabilizing action in smokeless powder but are used primarily as non-volatile solvents and deterrent coatings.

Calcium carbonate, either powdered limestone or precipitated chalk, is used as an anti-acid in dynamite where it serves as a satisfactory stabilizer. Urea is used in dynamite and in celluloid. It reacts with nitrous acid to produce nitrogen and carbon dioxide, and is unsuitable for use in smokeless powder because the gas bubbles destroy the homogeneity of the colloid and affect the rate of burning. The small gas bubbles however commend it for use in celluloid, for they produce an appearance of whiteness and counteract the yellowing of age.

In addition to the ability of certain substances to combine with the products of the decomposition of nitrocellulose, it is possible that the same or other substances may have a positive or a negative catalytic effect and may hasten or retard the decomposition by their presence. But it has not yet been made clear what types of chemical substance hasten the decomposition or why they do so. Nitrogen dioxide hastens it. Pyridine hastens it, and a powder containing 2 or 3% of pyridine will inflame spontaneously if heated for half an hour at 110°. Powders containing tetryl are very unstable, while those containing 10% of trinitronaphthalene (which does not react with the products of decomposition) are as stable as those containing 2% of diphenylamine (which does react).

In a series of researches extending over a period of 15 years Marqueyrol [11] has determined the effect of various substances, particularly naphthalene, mononitronaphthalene, diphenylbenzamide, carbazol, diphenylamine, and diphenylnitrosamine, upon the stability of smokeless powder at 110°, 75°, 60°, and 40°. Samples of *poudre BF* [12] were made up containing different amounts of each stabilizer, and were subjected to dry heat in open vessels and in vessels closed with cork stoppers. Samples were removed from time to time and the nitrogen content of their nitrocellulose was determined. A sample of the powder was taken up in a solvent, and precipitated in a granular state; the precipitate was washed with cold chloroform until a fresh portion of chloroform was no longer colored by 18 hours contact with it, and was dried, and analyzed by the nitrometer. It was necessary to isolate the pure nitrocellulose and to separate it from the stabilizer, for the reason that otherwise the stabilizer would be nitrated in the nitrometer and a low result for nitrogen would be secured. A selected portion of Marqueyrol's results, from experiments carried out by heating in open vessels, are shown in the tables on

Days of heating at 40°....	0	387	843	1174	2991	3945	4016
Analysis:							
no stabilizer	201.8	....	199.5	*147.8*	....	....	....
2% amyl alcohol	202.2	....	198.7	200.6	199.2	....	172.9
8% " "	201.4	....	199.2	200.8	198.9	....	198.2
1% diphenylamine	201.3	....	199.5	201.0	200.9	....	201.0
2% "	199.5	....	198.2	199.2	199.4	....	200.2
5% "	200.1	201.2	....	....	....	197.6	....
10% "	200.1	199.0	198.2	....	....	....	....

Days of heating at 60°	0	146	295	347	1059	2267	3935
Analysis:							
no stabilizer	201.8	*146.5*	....	....	....	....	....
2% amyl alcohol	202.2	197.4	....	*147.2*	....	....	....
8% " "	201.4	197.3	....	198.3	159.5	....	....
1% diphenylamine	201.3	197.6	....	200.0	....	....	....
2% "	199.5	196.1	....	198.3	....	....	....
5% "	200.1	....	196.0	....	....	....	185.7
10% "	200.1	....	192.3	....	....	....	173.0

[11] *Mém. poudres,* **23,** 158 (1928).
[12] F = *fusil,* rifle.

Days of heating at 75°....	0	86	231	312	516	652	667
Analysis:							
2% amyl alcohol	203.1	*191.4*	....	....	....	....	....
1% diphenylamine	201.3	196.0	....	198.0	190.9	....	....
2% "	199.5	194.7	....	198.1	....	192.1	....
5% "	200.1	....	192.8	....	....	....	186.2
10% "	200.1	....	184.2	....	....	....	175.9

Days of heating at 75°....	0	55	146	312	419	493	749
Analysis:							
1% diphenylnitrosamine	200.4	197.9	198.4	199.2	197.5	198.2	194.0
2% "	200.0	201.5	198.3	198.5	197.6	171.5	....
10% "	201.5	195.3	194.1	193.0	190.6	187.3	184.3

Days of heating at 75°....	0	60	85	108	197	377	633
Analysis:							
2% amyl alcohol	200.9	198.9	*196.9*	....	....	....	....
1.25% carbazol	200.6	199.4	....	199.1	*182.8*	....	....
10% "	200.3	198.7	....	197.7	198.2	193.0	190.2

Days of heating at 75°....	0	31	50	62	87	227	556
Analysis:							
1.5% diphenylbenzamide	200.2	199.3	*186.1*	....	....	....	....
10% "	200.2	198.1	....	....	200.3	....	....
1.5% mononitronaphthalene	202.4	199.8	....	....	....	....	....
10% mononitronaphthalene	202.0	198.1	....	*193.0*	....	....	....
1.5% naphthalene	202.3	200.2	194.8	....	....	....	....
10% "	201.8	199.2	....	....	....	199.1	200.2

pages 309–310, where the numbers representing *analyses* indicate the nitrogen contents as usually reported in France, namely, cubic centimeters of nitric oxide per gram of nitrocellulose. When the numbers are printed in italics, the samples which were taken for analysis were actively giving off red nitrous fumes.

Diphenylnitrosamine, which is always present in powders made from diphenylamine, is decomposed at 110°, and that temperature therefore is not a suitable one for a study of the stability of smokeless powder. At 75° diphenylnitrosamine attacks nitrocellulose less rapidly than diphenylamine itself, but this is not true at lower temperatures (40° and 60°) at which there is no appreciable difference between the two substances. Carbazol at 110° is an excellent stabilizer but at 60° and 75° is so poor as to deserve no further consideration. Ten per cent of diphenylamine gives unstable smokeless powder. Powder containing 40% of diphenylamine inflames spontaneously when heated in an open vessel at 110° for an hour and a half. Diphenylamine attacks nitrocellulose, but it does not attack it as rapidly as do the products themselves of the decomposition of nitrocellulose in air; and 1 or 2% of the substance, or even less, in smokeless powder is as good a stabilizer as has yet been found.

Transformations of Diphenylamine During Aging of Powder

Desmaroux,[13] Marqueyrol and Muraour,[14] and Marqueyrol and Loriette [15] have studied the diphenylamine derivatives which give a dark color to old powder, and have concluded that they are produced by impurities in the ether which is used in the manufacture or by the oxidizing action of the air during drying and storage. Their presence is not evidence that the powder has decomposed, but indicates that a certain amount of the diphenylamine has been consumed and that correspondingly less of it remains available for use as a stabilizer.

The transformations of diphenylamine [16] in consequence of its reaction with the products of the decomposition of nitrocellulose are indicated by the following formulas. None of these substances imparts any very deep color to the powder.

$$\langle\bigcirc\rangle-NH-\langle\bigcirc\rangle$$

$$\downarrow$$

$$\langle\bigcirc\rangle-N(NO)-\langle\bigcirc\rangle$$

$$\overset{NO_2}{\underset{\langle\bigcirc\rangle}{|}}-NH-\langle\bigcirc\rangle-NO_2 \qquad NO_2-\langle\bigcirc\rangle-NH-\langle\bigcirc\rangle-NO_2$$

$$NO_2-\langle\bigcirc\rangle-NH-\langle\bigcirc\rangle-NO_2 \quad (\text{with } NO_2)$$

13 *Mém. poudres*, **21**, 238 (1924).
14 *Ibid.*, **21**, 259, 272 (1924).
15 *Ibid.*, **21**, 277 (1924).
16 Davis and Ashdown, *Ind. Eng. Chem.*, **17**, 674 (1925).

The diphenylamine is converted first into diphenylnitrosamine which is as good a stabilizer as diphenylamine itself. Since both of these substances may be detected by simple tests upon an alcoholic extract of a sample of the powder, the fitness of the powder for continued storage and use may be easily demonstrated. A strip of filter paper on which the alcoholic extract has been allowed to evaporate is colored blue by a drop of ammonium persulfate solution if unchanged diphenylamine is present. Likewise the extract, if it contains diphenylamine, is colored blue by the addition of a few drops of a saturated aqueous solution of ammonium persulfate. Since the alcoholic extract is often colored, the test is best carried out by comparing the colors of two equal portions of the extract, one with and one without the addition of ammonium persulfate. Diphenylnitrosamine gives no color with ammonium persulfate. One-tenth of a milligram of diphenylnitrosamine imparts an intense blue color to a few cubic centimeters of cold concentrated sulfuric acid. It gives no color with a cold 1% alcoholic solution of α-naphthylamine, but an orange color if the solution is heated.[17] None of the other diphenylamine derivatives which occur in smokeless powder give these tests.

Diphenylnitrosamine rearranges under the influence of mineral acids to form p-nitrosodiphenylamine. The latter substance is evidently formed in smokeless powder and is oxidized and nitrated by the products of the decomposition to form 2,4'- and 4,4'-dinitrodiphenylamine. Davis and Ashdown[16] have isolated both of these substances from old powder, and have also prepared[17] them by the nitration of diphenylnitrosamine in glacial acetic acid solution. Both substances on further nitration yield 2,4,4'-trinitrodiphenylamine, which represents the last stage in the nitration of diphenylamine by the products of the decomposition of smokeless powder. This material has been isolated from a sample of U. S. pyrocellulose powder which was kept at 65° in a glass-stoppered bottle for 240 days after the first appearance of red fumes. The several nitro derivatives of diphenylamine may be distinguished by color reactions with alcoholic solutions of ammonia, sodium hydroxide, and sodium cyanide, and some insight into the past history of the powder may be gained from tests on the alcohol extract with these reagents, but their pres-

[17] Davis and Ashdown, *J. Am. Chem. Soc.*, **46**, 1051 (1924).

ence is evidence of instability, and no powder in which the diphenylnitrosamine is exhausted is suitable for further storage and use.

Absorption of Moisture [18]

Nitrocellulose itself is hygroscopic, but its tendency to take up moisture is modified greatly by other substances with which it is incorporated. Colloided with nitroglycerin in the absence of solvent, it yields a product which shows no tendency to take up moisture from a damp atmosphere. Colloided with ether-alcohol, as in the case of the *poudre B* and the straight pyrocellulose powders which were used in the first World War, it yields a powder which is hygroscopic both because of the hygroscopicity of the nitrocellulose itself and because of the hygroscopicity of the alcohol and ether which it contains. In water-dried powder the alcohol and ether of the surface layer have been largely removed or replaced with water, the hygroscopicity of the surface layer is reduced, and the interior of the grain is prevented to a considerable extent from attracting to itself the moisture which it would otherwise attract. In certain coated and *progressive burning* powders, the surface layers are made up of material of greatly reduced hygroscopicity and the interiors are rendered inaccessible to atmospheric influences.

The tendency of straight nitrocellulose powder to take up moisture and the effect of the absorbed moisture in reducing the ballistic power of the powder are shown by the table below.

Period of exposure, hrs.	0	24	48	72	96
Total volatiles, %	3.26	3.55	3.71	3.84	3.93
External moisture, %	1.02	1.15	1.40	1.47	1.57
Residual solvent, %	2.24	2.40	2.31	2.37	2.35
Velocity, ft. per sec.	1706.6	1699.0	1685.4	1680.4	1669.0
Pressure, lb. per sq. in.	31,100	31,236	30,671	29,636	28,935

A sample of water-dried powder was exposed to an atmosphere practically saturated with water vapor. Portions were removed each day; one part was fired in the gun, and another part was analyzed for *total volatile* matter (TV) and for volatile matter driven off by an hour's heating at 100° (*external moisture*, EM).

[18] Cf. Davis, *Army Ordnance*, **2**, 9 (1921).

The amount of total volatile matter increased regularly during the period of exposure, as did also the amount of volatile matter resident at or near the surface of the powder grains. The amount of volatile matter in the interior of the powder grains (*residual solvent*, RS) did not alter materially during the experiment.

Total volatiles in powder is determined by dissolving the sample in a solvent, precipitating in a porous and granular condition, evaporating off the volatile matter, and drying the residue to constant weight. External moisture is the amount of volatile matter which is driven off by some convenient method of desiccation. The difference between the two is residual solvent, $TV - EM = RS$, and is supposed to correspond to volatile matter resident within the interior of the grain and not accessible to desiccating influences. Various methods of determining external moisture have been in use among the nations which use straight nitrocellulose powder and in the same nation among the manufacturers who produce it. At the time of the first World War, for example, external moisture was determined in Russia by heating the sample at 100° for 6 hours, in France by heating at 60° for 4 hours, and in the United States by heating at 60° in a vacuum for 6 hours. These several methods, naturally, all give different results for external moisture and consequently different results for residual solvent.

There appears really to be no method by which true external moisture may be determined, that is, no method by which only the surface moisture is removed in such fashion that the residual solvent in the powder is found to be the same both before and after the powder has been allowed to take up moisture. Samples of powder were taken and residual solvent was determined by the several methods indicated in the next table. The samples were exposed 2 weeks to an atmosphere practically saturated with water vapor, and residual solvent was again determined as before. The surprising result was secured in every case, as indicated, that the amount of residual solvent was less after the powder had been exposed to the moist atmosphere than it was before it had been exposed. Yet the powder had taken up large quantities of moisture during the exposure. It is clear that the exposure to the moist atmosphere had made the volatile matter of the interior of the grains more accessible to desiccating influences. Evidently the moisture had opened up the interior of the grains, presumably by precipitating the nitrocellulose and producing minute cracks and pores in the colloid. Verification of this explanation is found in the effect of alcohol on colloided pyrocellulose powder. The powder took up alcohol from an atmosphere saturated with alcohol vapor, but alcohol does not precipitate the colloid, it produces no cracks or pores, and in every case residual solvent was found to be greater after the powder had been exposed to alcohol vapor than it had been before such exposure. The following table shows data for typical samples of powder before and after exposures of 2 weeks to atmospheres saturated respectively with water and with alcohol.

Method of Determining External Moisture and Residual Solvent	Exposure to Water Residual solvent			Exposure to Alcohol Residual solvent		
	Before	After	Difference	Before	After	Difference
1 hr. at 100° in open oven..	3.12	2.82	−0.30	2.41	4.57	+2.16
6 hrs. at 100° in open oven..	2.81	2.36	−0.45	2.22	3.92	+1.70
6 hrs. at 55° in vacuum.....	2.91	2.54	−0.37	2.27	4.10	+1.83
55° to constant weight in open oven.............	3.00	2.72	−0.28	2.58	4.25	+1.67
Over sulfuric acid to constant weight...........	2.95	2.39	−0.56	2.32	4.10	+1.78

Samples of pyrocellulose powder, varying in size from 0.30 caliber single-perforated to large multiperforated grains for the 10-inch gun, were exposed to a moist atmosphere until they no longer gained any weight. They were then desiccated by the rather vigorous method of heating for 6 hours at 100°. All the samples lost more weight than they had gained. As the exposures to moisture and subsequent desiccations were repeated, the differences between the weights gained by taking up moisture and the weights lost by drying became less and less until finally the powders on desiccation lost, within the precision of the experiments, exactly the amounts of volatile matter which they had taken up. At this point it was judged that all residual solvent had been driven out of the powder and that further treatment would produce no additional cracks and pores in the grains. The gain or loss (either one, for the two were equal) calculated as per cent of the weight of the desiccated sample gave the apparent hygroscopicities listed below. Since all the powders were made from the

CALIBER	APPARENT HYGROSCOPICITY, %
0.30	3.00
75 mm.	2.75
4.7 inches	2.42
6 inches	2.41
10 inches	2.11

same material, namely, straight pyrocellulose, the differences in the apparent hygroscopicity are presumed to be caused by the drying treatment not being vigorous enough to drive out all the moisture from the interior of the grains of greater web thickness. The drying, however, was so vigorous that the powders became unstable after a few more repetitions of it. The losses on desiccation became greater because of decomposition, and the gains on exposure to moisture became greater because of the hygroscopicity of the decomposition products.

Although hygroscopicity determined in this way is apparent and not absolute, it supplies nevertheless an important means of estimating the effects both of process of manufacture and of composition upon the attitude of the powder toward moisture. Thus, samples of pyrocellulose powder for the 4.7-inch gun, all of them being from the same batch and pressed through the same die, one air-dried, one water-dried, one dried under benzene at 60°, and one under ligroin at 60°, showed apparent hygroscopicities of 2.69%, 2.64%, 2.54%, and 2.61%, which are the same within the experimental error. Milky grains [19] of 75-mm. powder showed an apparent hygroscopicity of 2.79%, compared with 2.75% for the normal amber-colored grains. The experiment with this powder was continued until considerable decomposition was evident; the successive gains and losses were as follows, calculated as per cent of the original weight of the sample.

[19] Grains which had a milky appearance because of the precipitation of the colloid during the water-dry treatment. This result follows if the grains contain more than 7 or 7.5% of ether-alcohol when they are submitted to water-drying.

Gain, %	Loss, %
2.315	3.030
2.439	2.623
2.259	2.337
2.279	2.319
2.179	2.577
2.448	2.554
2.325	2.630
2.385	3.022

Experiments with 75-mm. powders, made from pyrocellulose with the use of ether-alcohol and with various other substances incorporated in the colloids, gave the following results for hygroscopicity. Hydrocellulose does not dissolve in the nitrocellulose

Pyrocellulose with
5% hydrocellulose	2.79%
10% crystalline DNX	2.09%
10% DNX oil	1.99%
10% crystalline DNT	1.93%
15% " "	1.41%
20% " "	1.23%
25% " "	1.06%

colloid, and does not affect its hygroscopicity. The aromatic nitro compounds dissolve, and they have a marked effect in reducing the absorption of moisture. They are explosive non-volatile solvents and contribute to the energy of the powder.

Other non-volatile solvents which are not explosive are discussed below in the section on gelatinizing agents. These tend to reduce the potential of the powder, but their action in this respect is counteracted in practice by using guncotton in place of part or all the nitrocellulose. The guncotton is colloided by the gelatinizing agent, either in the presence or in the absence of a volatile solvent, and the resulting powder is non-hygroscopic and as strong or stronger than straight pyrocellulose powder.

Control of Rate of Burning

Cordite is *degressive* burning, for its burning surface decreases as the burning advances. Powder in strips, in flakes, and in single-perforated tubes has a burning surface which is very nearly constant if the size of the strips or flakes, or the length of the tubes, is large relative to their thickness. Multiperforated grains are *progressive* burning, for their burning surface actually increases as the burning advances, and, other things in the gun being equal, they produce gas at a rate which accelerates more rapidly and, in consequence, gives a greater velocity to the projectile.

A progressive burning strip ballistite was used to some extent by the French in major caliber guns during the first World War. It consisted of a central thick strip or slab of ballistite, 50%

FIGURE 79. Progressive Burning Colloided Smokeless Powder. 12-Inch powder at different stages of its burning. A grain of 12-inch powder, such as appears at the left, was loaded into a 75-mm. gun along with the usual charge of 75-mm. powder (of the same form as the 12-inch grain but of less web thickness). When the gun was fired, a layer of colloid having a thickness equal to one-half the web of the 75-mm. powder was burned off from every grain in the gun. This consumed the 75-mm. powder completely. The 12-inch grain was extinguished when thrown from the muzzle of the gun; it was picked up from the ground—and is the second grain in the above picture. The next grain was shot twice from a 75-mm. gun, the last grain three times. After three shootings, the perforations are so large that a fourth shooting would cause them to meet one another, and the grain to fall apart, leaving slivers.

nitroglycerin and 50% soluble nitrocellulose, made without volatile solvent, and sandwiched between two thin strips of powder made, without volatile solvent, from 50% soluble nitrocellulose and 50% crystalline dinitrotoluene. The two compositions were rolled to the desired thicknesses separately between warm rolls, and were then combined into the laminated product by pressing between warm rolls. The outer layers burned relatively slowly with a temperature of about 1500°; the inner slab burned rapidly with a temperature of about 3000°.

Progressive burning *coated powders,* usually flakes or single-perforated short cylinders, are made by treating the grains with a gelatinizing agent, or non-volatile, non-explosive solvent for nitrocellulose, dissolved in a volatile liquid, generally benzene or acetone, tumbling them together in a sweetie barrel or similar device, and evaporating off the volatile liquid by warming while the tumbling is continued. The material which is applied as a coating is known in this country as a *deterrent,* in England as a

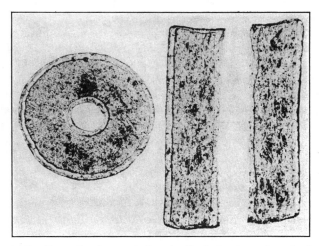

FIGURE 80. Cross Section and Longitudinal Section of a Grain of .50 Caliber Extruded Smokeless Powder, Deterrent Coated (25×). (Courtesy Western Cartridge Company.)

moderant. At the time of the first World War *sym*-dimethyldiphenylurea was already used widely as a deterrent in rifle powder intended for use in shooting matches and in military propellants designed to produce especially high velocities. The substance was called *centralite* because its use had been developed in Germany at the Central War Laboratory at Spandau. The ethyl analog, diethyldiphenylurea, at first known as *ethyl centralite,* is usually called Centralite No. 1 and has generally superseded the methyl compound (or Centralite No. 2) for use in smokeless powder. Although many other substances have been tried and have been patented, this remains the most widely used of any. *Butyl centralite* is a better gelatinizing agent for nitrocellulose than either the methyl or the ethyl compound, and is likely to find more extensive use in the future.

Gelatinizing Agents

Gelatinizing agents, of which the centralites are examples, are often incorporated in colloided straight nitrocellulose and double-base powders where they cause the materials to burn more slowly, where they serve as flash reducing agents, and where they reduce the tendency of the powders to take up moisture. They reduce the amount of volatile solvent which is needed in the manufacture of nitrocellulose powders, and facilitate the manufacture of double-base powders without any volatile solvent at all. The centralites happen also to be effective stabilizers, but this is not a general property of gelatinizing agents.

Marqueyrol and Florentin [20] have published a list of patents for gelatinizing agents, esters, amides, urea derivatives, halogen compounds, ketones, and alcohols, and have reported their study of many of them with respect to their effectiveness on the CP_1 and CP_2, insoluble and soluble nitrocellulose respectively, which were standard in the French service. To a weighed quantity of the dry nitrocellulose a dilute solution of the gelatinizing agent in 95% alcohol was added in portions of 1 cc. at a time, the alcohol was then evaporated at a temperature of 35–40°, more of the alcohol solution was added, and the evaporation was repeated until gelatinization was complete. The results with the best gelatinizing agents are shown in the table below, where the numbers

	CP_1 (Insoluble)	CP_2 (Soluble)
Ethyl sebacate	320	65
Dimethylphenyl-o-tolylurea	260	65
Dimethyldiphenylurea	...	80
Ethyl succinate	400	90
Ethyl phthalate	360	95

represent the amounts by weight of the several substances which were needed for the complete gelatinization of 100 parts of the nitrocellulose. Ninety parts of ethyl citrate or of benzyl benzoate almost completely gelatinized 100 parts of CP_2, 90 of ethyl malonate incompletely, 90 of ethyl oxalate or of ethyl stearate more incompletely, and 90 of ethyl acetoacetate or of ethyl ricinoleate

[20] *Mém. poudres*, **18**, 150, 163 (1921).

but very little. Four hundred parts of triphenyl phosphate almost completely gelatinized 100 parts of CP_1, and 400 of ethyl malonate or of ethyl oxalate produced only incomplete gelatinization.

Marqueyrol and Florentin point out that the lower members of the series of esters, the acetates, butyrates, valerates, etc., are good solvents for nitrocellulose—ethyl and amyl acetate have long been used for the purpose—but the higher members, the stearates and oleates, gelatinize nitrocellulose but very little. To the esters of the dibasic acids the opposite rule appears to apply; the higher members are better than the lower. Acetone is a well-known solvent both for soluble and for "insoluble" nitrocellulose, but acetophenone gelatinizes even soluble nitrocellulose only feebly.

Experiments by the present writer [21] with a variety of other gelatinizing agents have shown that the amounts necessary to produce complete gelatinization of pyrocellulose are different if different solvents are used for applying them. In general they are more effective in benzene than in alcohol, and more in alcohol than in ligroin. Half-gram samples of dry pyrocellulose were treated in 30-cc. beakers with known quantities of the gelatinizing agents dissolved in convenient volumes (15–30 cc.) of alcohol, benzene, or ligroin. The volatile liquids were evaporated off slowly at 60°, the residues were warmed at 60° for 10 minutes longer (during which time a considerable improvement in the gelatinization was generally observed), and were then examined to determine their condition. If complete gelatinization had not occurred, other experiments were carried out with fresh samples. The results, summarized in the table on page 322, are accurate to the nearest 10%. They support several conclusions. *Sym*-dialkyl ureas are excellent gelatinizing agents for nitrocellulose, and the property remains if additional aliphatic or aromatic groups are introduced into the molecule. The heavier the alkyl groups, the greater appears to be the gelatinizing power. Of the aromatic substituted ureas, those in which there are less than three aromatic groups appear to be without action. Among the alkyl esters of sebacic and phthalic acids, those which contain the heavier alkyl groups are generally the better gelatinizing agents. The alkyl esters of aliphatic and of aromatic substituted

[21] Davis, *Ind. Eng. Chem.*, **14**, 1140 (1922).

PARTS BY WEIGHT NECESSARY FOR THE COMPLETE GELATINIZATION OF
100 PARTS OF PYROCELLULOSE

	In Alcohol	In Benzene	In Ligroin
Methylurea	No action with 100 parts		
Ethyleneurea	No action with 100 parts		
Sym-dimethylurea	60	70	...
Sym-diethylurea	50	50	...
Unsym-dimethylurea	No action with 100 parts		
Tetramethylurea	80	...	...
Benzylurea	No action with 100 parts		
Sym-diphenylurea	No action with 100 parts		
Sym-di-p-tolylurea	No action with 100 parts		
Unsym-diphenylurea	No action with 100 parts		
Triphenylurea	...	35	...
α,α-diphenyl-p-tolylurea	...	40	...
Tetraphenylurea	No action with 160 parts	30	...
Ethyltriphenylurea	80	...	...
Sym-dimethyldiphenylurea	70	25	...
Sym-diethyldiphenylurea	70	30	...
Sym-di-n-butyldiphenylurea	60	20	...
Unsym-dimethyldiphenylurea	60	...	...
Carbamic acid ethyl ester	140	80	...
Methylcarbamic acid ethyl ester	90	60	...
Ethylcarbamic acid ethyl ester	90	60	...
Phenylcarbamic acid ethyl ester	20	90	...
Phenylcarbamic acid phenyl ester	No action with 200 parts		
Phenylcarbamic acid benzyl ester	No action with 100 parts		
Diphenylcarbamic acid phenyl ester	80	70	...
Methyl sebacate	80	70	105
Ethyl sebacate	80	50	90
Iso-amyl sebacate	70	95	90
Methyl phthalate	95	70	115
Ethyl phthalate	95	50	100
Iso-amyl phthalate	95	50	80
DNX oil	120	130	330
Trinitrotoluene	...	300	...

carbamic acids are excellent gelatinizing agents, but the aromatic esters appear to be without action unless the total number of aromatic groups is equal to three.

Flashless Charges and Flashless Powder

The discharge of a machine gun shooting ordinary charges of smokeless powder produces bright flashes from the muzzle which at night disclose the position of the gun to the enemy. When a large gun is fired, there is a large and dazzlingly bright muzzle flash, from a 12-inch gun for example a white-hot flame 150 feet or more in length. The light from such a flame reflects from the heavens at night and is visible for a distance of as much as 30 miles, much farther than the sound from the gun may be heard. The enemy by the use of appropriate light-ranging apparatus may determine the position of the flash, and may undertake to bombard and destroy the battery from a great distance.

Smokeless powder, burning in the chamber of the gun at the expense of its own combined oxygen, produces gas which contains hydrogen, carbon monoxide, carbon dioxide, etc., and this gas, being both hot and combustible, takes fire when it emerges from the muzzle and comes into contact with the fresh oxygen supply of the outer air. One part of the brilliancy of the flash is the result of the emergent gas being already preheated, often to a temperature at which it would be visible anyway, generally to a point far above the temperature of its inflammation in air. In thinking about this latter temperature, it is necessary to take account of the fact that, other things being equal, a small cloud of gas from a small gun loses its temperature more quickly and becomes completely mixed with the air more rapidly than a large cloud of gas from a large gun. The gas emerging from a small gun would need to be hotter in the first place if it is to inflame than gas of the same composition emerging from a large gun. It is for this reason perhaps that it is easier to secure flashless discharges with guns of small caliber (not over 6 inch) than with those of major caliber.

There are four ways, distinguishably different in principle, by which flashlessness has been secured, namely,

(1) by the addition to the charge of certain salts, particularly potassium chloride or potassium hydrogen tartarate, or by the use of powdered tin or of some other substance which, dispersed throughout the gas from the powder, acts as an anti-oxidant and prevents its inflammation; [22]

(2) by incorporating carbonaceous material in the smokeless powder, by which means the composition of the gas is altered and

---

[22] Cf. Demougin, *Mém. poudres,* **25,** 130 (1932–33); Fauveau and Le Paire, *ibid.,* **25,** 142 (1932–33); Prettre, *ibid.,* **25,** 160, 169 (1932–33).

the number of mols of gas is actually increased while the temperature is lowered;

(3) by incorporating in the powder a *cool explosive*, such as ammonium nitrate, guanidine nitrate, or nitroguanidine, which explodes or burns with the production of gas notably cooler than the gas from the combustion of ordinary smokeless powder; and

(4) by contriving the ignition of the powder, the acceleration of its burning rate, and the design of the gun itself, any or all these factors, in such fashion that the projectile takes up energy from the powder gas more quickly and more effectively than is ordinarily the case, and thereby lowers the temperature of the gas to a point where the flash is extinguished.

The first of these methods is applicable to all calibers; the second and the third are successful only with calibers of less than 6 inches; the fourth has not yet been sufficiently studied and exploited. It is true, however, that an improved igniter, with the same gun and with the same powder, may determine the difference between a flash and a flashless discharge.

The use of salts to produce flashlessness appears to derive from an early observation of Dautriche that a small amount of black powder, added to the smokeless powder charge of small-arms ammunition, makes the discharge flashless. During the first World War the French regularly loaded a part of their machine gun ammunition with a propellant consisting of 9 parts of smokeless powder and 1 of black powder.

The Germans in their cannon used *anti-flash bags* or *Vorlage*, loaded at the base of the projectiles, between the projectiles and the propelling charges. These consisted of two perforated discs of artificial silk or cotton cloth sewed together in the form of doughnut-shaped bags. The bags were filled with coarsely pulverized potassium chloride. The artillerymen were not informed of the nature of the contents of the bags but were advised against using any whose contents had hardened to a solid cake, and were instructed in their tactical use as follows.

In firing with *Vorlage*, there is produced a red fire [a red glow] at the muzzle and in front of the piece. The smoke is colored red [by the glow]. This light however gives no reflection in the heavens. In fact it is visible and appreciable at a distance only if the piece is placed in such a way that the enemy can see its muzzle. In the daytime, *Vorlage* must be used only when the weather is so dark that the flashes of the shots without them are more visible than the clouds of smoke which they produce. The opaqueness of the background against which the battery stands out or the obscurity of the setting which surrounds it are also at times of a kind to justify the use of *Vorlage* in the daytime.

The anti-flash bags reduced the range by 4.5 to 8%.

Fauveau and Le Paire [23] studied the anti-flash effect of potassium chloride and of other salts and concluded that the lowering of the temperature of the gas which undoubtedly results from their volatilization and dissociation is insufficient to account for the extinction of the flash. Prettre [24] found that the chlorides of sodium and of lithium, and other alkali metal salts which are volatile, had the same effect as potassium chloride. He found that small amounts of potassium chloride, volatilized in mixtures of carbon monoxide and air, had a powerful anti-oxidant action and a correspondingly large effect in raising the temperature of inflammation of the gas. Some of his results are shown in the table below. He found that potassium chloride was without effect

MILLIGRAMS OF KCl PER LITER OF GASEOUS MIXTURE	TEMPERATURE (°C.) OF INFLAMMATION OF AIR CONTAINING		
	24.8% CO	44.1% CO	67.3% CO
0.0	656	657	680
0.4	...	750	800
0.5	730	...	820
0.7	...	810	900
1.0	790	850	1020
1.3	810	...	...
2.0	890	950	...
2.5	...	1000	...
3.0	970	...	...
3.5	1010	...	...

upon the temperature of inflammation of mixtures of hydrogen and air.

The French have used anti-flash bags (*sachets antilueurs*) filled with the crude potassium hydrogen tartarate (about 70%

[23] *Loc. cit.*
[24] *Loc. cit.*

pure) or *argols* which is a by-product of the wine industry. The flat, circular, cotton bags containing the argols were assembled along with the smokeless powder and black powder igniter in silk cartridge bags to make up the complete charge. Since the anti-flash material tended to reduce the ballistic effect of the charge, it was necessary when firing flashless rounds to add an *appoint* or additional quantity of smokeless powder. Thus, for ordinary firing of the 155-mm. gun, the charge consisted of 10 kilograms of *poudre BM7* along with an igniter system containing a total of 115 grams of black powder. For a flashless round, 3 *sachets* containing 500 grams of argols each were used and an additional 305 grams of smokeless powder to restore the ballistics to normal.

Another method of securing flashlessness was by the use of pellets (*pastilles*) of a compressed intimate mixture of 4 parts of potassium nitrate and 1 part of crystalline DNT. Pellets for use in the 155-mm. gun weighed 1 gram each, and were about 2 mm. thick and 15 mm. in diameter. Two or three hundred of these were sewed up in a silk bag which was loaded into the gun along with the bag containing the powder. The pellets burned with the same velocity as *poudre B,* and had but very little effect upon the ballistics. They of course produced a certain amount of smoke and the discharge gave a red glow from the muzzle of the gun.

Oxanilide functions well as an anti-flash agent if it is distributed throughout the powder charge, but not if it is loaded into the gun in separate bags like the materials which have just been mentioned. It is made into a thick paste with glue solution, the paste is extruded in the form of little worms or pellets, and these are dried. Pellets to the amount of 15% of the powder charge produce flashlessness in the 6-inch gun, but the charge is more difficult to light than ordinarily and requires a special igniter.

Oxanilide and many other carbonaceous materials, incorporated in the grains of colloided powder, yield powders which are flashless in guns of the smaller calibers and, in many cases, are as powerful, weight for weight, as powders which contain none of the inert, or at least non-explosive, ingredients. If nitrocellulose burning in the gun produces 1 mol of carbon dioxide and a certain amount of other gaseous products, then nitrocellulose plus 1 mol of carbon under the same conditions will produce 2 mols of carbon monoxide along with substantially the same amount of the other gaseous products. There will be more gas and cooler gas. A colloided powder made from pyrocellulose 85 parts and hydrocellulose 15 parts is flashless in the 75-mm. gun, and gives practically the same ballistic results as a hotter and more expensive powder made from straight nitrocellulose. The strength of the powder may be increased without affecting its flashlessness by substituting part of the pyrocellulose by guncotton of a higher nitrogen content.

Among the materials which have been incorporated into colloided powder for the purpose of reducing or extinguishing the flash are (1) substances, such as starch, hydrocellulose, and anthracene, which are insoluble in the colloid and are non-explosive. They, of course, must exist in a state of fine subdivision to be suitable for this use. Other anti-flash agents are (2) solid or liquid non-explosive substances, such as diethyldiphenylurea and dibutyl phthalate, which are solvents for nitrocellulose and dissolve in the colloid. They reduce the hygroscopicity of the powder and they reduce the amount of volatile solvent which is needed for the manufacture. Still others are (3) the explosive solid and liquid aromatic nitro compounds which are solvents for nitrocellulose and are effective in reducing both the flash and the hygroscopicity. All or any of the substances in these three classes may be used either in a straight nitrocellulose or in a nitrocellulose-nitroglycerin powder. Several flashless powders have been described in the section on the "Classification of Smokeless Powders." Many varieties have been covered by numerous patents. We cite only a single example,[25] for a smokeless, flashless, non-hygroscopic propellent powder made from about 76–79% nitrocellulose (of at least 13% nitrogen content), about 21–24% dinitrotoluene, and about 0.8–1.2% diphenylamine. During the first World War the French and the Italians used a *superattenuated* ballistite, made without volatile solvent, and containing enough aromatic dinitro compound (in place of part of the nitroglycerin) to make it flashless. In a typical case the powder was made from 30 parts CP_1, 30 CP_2, 15 DNT, and 25 nitroglycerin.

[25] U. S. Pat. 2,228,309 (1941), Ellsworth S. Goodyear.

Ball-Grain Powder

The process for the manufacture of ball-grain powder which Olsen and his co-workers have devised [26] combines nicely with Olsen's process for the quick stabilization of nitrocellulose to form a sequence of operations by which a finished powder may be produced more rapidly and more safely than by the usual process. It supplies a convenient means of making up a powder which contains non-volatile solvents throughout the mass of the grains or deterrent or accelerant coatings upon their surface.

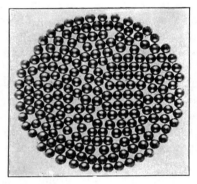

FIGURE 81. Ball Grains (Smokeless Powder) (3×). (Courtesy Western Cartridge Company.)

Nitrocellulose, pulped and given a preliminary or sour boiling, may be used directly without poaching. Deteriorated smokeless powder, containing nitro derivatives of diphenylamine and acidic decomposition products, may be reduced to a coarse powder under water in a hammer mill, and may then be used. Whichever is used, the first necessity is to stabilize it by complete removal of the acid. For this purpose, the material in the presence of water (which may contain a little chalk in suspension or urea in solution) is introduced into a still where it is dissolved with agitation in ethyl acetate to form a heavy syrup or lacquer, and is treated with some substance which is adsorbed by nitrocellulose more readily than acid is adsorbed. It is a curious fact that nitrocellulose is dissolved or dispersed by ethyl acetate much more readily

[26] U. S. Pat. 2,027,114 (1936), Olsen, Tibbitts, and Kerone; also U. S. Pats. 2,111,075 (1938), 2,175,212 (1939), 2,206,916 (1940).

in the presence of water than when water is absent. Diphenylamine is dissolved in the ethyl acetate before the latter is added
to the water and nitrocellulose in the still. At the same time,
centralite or DNT or any other substance which it may be desired
to incorporate in the powder is also dissolved and added. The
water phase and the lacquer are then stirred for 30 minutes by
which operation the nitrocellulose is stabilized.[27] Starch or gum
arabic solution to secure the requisite colloidal behavior is then
introduced into the still, the still is closed, the temperature is
raised so that the lacquer becomes less viscous, and the mixture

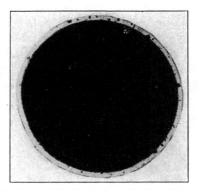

Figure 82. Cross Section of Ball Grain, Double Base, Deterrent Coated
(112×). (Courtesy Western Cartridge Company.)

under pressure is agitated vigorously until the lacquer is broken
up into small globules of the correct size. The pressure is then
reduced, and the ethyl acetate is distilled off and recovered. If
the distillation is carried out too rapidly, the grains are shaped
like kernels of popcorn. If it is carried out at such a rate that the
volatile solvent is evaporated from the surface of the globules no
faster than it moves from the interior to the surface, if the distillation is slow at first and more rapid afterwards, then smooth
ball grains are formed, dense and of homogeneous structure.

[27] Diphenylamine in the presence of water thus has an action beyond
that which it has when it is added to the nitrocellulose gel (in the absence
of a separate water phase) during the manufacture of smokeless powder
by the usual process. Being preferentially adsorbed by the nitrocellulose, it
drives any acid which may be present out of the nitrocellulose and into
the water. After that it fulfils its usual function in the powder.

After the material has cooled, the powder grains are transferred in a slurry to another still and are treated with an emulsion of nitroglycerin dissolved in toluene, or of some other coating agent dissolved in a solvent in which nitrocellulose itself is insoluble, and the volatile solvent is distilled off, leaving the nitroglycerin or other material deposited on the surface of the grains. As much nitroglycerin as 15% of the weight of the powder may be applied in this way. A coating of centralite may, if desired, be put on top of it. The grains are sieved under water and are then dried for use in shotguns. If the powder is to be used in rifles, it is passed in a slurry between warm steel rollers by which all the grains are reduced to the same least dimension or web thickness. Previous to the drying, all the operations in the manufacture of ball-grain powder are carried out under water, and are safe. After the drying, the operations involve the same hazards, by no means insurmountable, as are involved in the ordinary process. The grains are glazed with graphite and blended.

CHAPTER VII

DYNAMITE AND OTHER HIGH EXPLOSIVES

Invention of Dynamite

Dynamite and the fulminate blasting cap both resulted from Alfred Nobel's effort to make nitroglycerin more safe and more convenient to use.[1] Having discovered that nitroglycerin is exploded by the explosion of a small firecracker-like device filled with black powder, he tried the effect of mixing the two materials, and in 1863 was granted a patent[2] which covered the use of a liquid explosive, such as nitroglycerin or methyl or ethyl nitrate, in mixture with gunpowder in order to increase the effectiveness of the latter. The amount of the liquid was limited by the requirement that the mixtures should be dry and granular in character. The explosives were supposed to be actuated by fire, like black powder, but the liquid tended to slow down the rate of burning, and they were not notably successful. The same patent also covered the possibility of substituting a part of the saltpeter by nitroglycerin. Because this substance is insoluble in water and non-hygroscopic, it acts as a protective covering for the salt and makes the use of sodium nitrate possible in these mixtures.

Nobel's next patent,[3] granted in 1864, related to improvements in the manufacture of nitroglycerin and to the exploding of it by heating or by means of a detonating charge. He continued his experiments and in 1867 was granted a patent[4] for an explosive prepared by mixing nitroglycerin with a suitable non-explosive, porous absorbent such as charcoal or siliceous earth. The resulting material was much less sensitive to shock than nitroglycerin. It was known as *dynamite,* and was manufactured and sold also

---

[1] For an account of Nobel and his inventions see de Mosenthal, *Jour. Soc. Chem. Ind.,* 443 (1899).

[2] Brit. Pat. 2359 (1863).

[3] Brit. Pat. 1813 (1864).

[4] Brit. Pat. 1345 (1867).

under the name of Nobel's Safety Powder. The absorbent which was finally chosen as being most satisfactory was diatomaceous earth or kieselguhr (guhr or fuller's earth). Nobel believed that dynamite could be exploded by a spark or by fire if it was confined closely, but preferred to explode it under all conditions by means of a special exploder or cap containing a strong charge of

FIGURE 83. Alfred Nobel (1833-1896). First manufactured and used nitroglycerin commercially, 1863; invented dynamite and the fulminate blasting cap, 1867; straight dynamite, 1869; blasting gelatin and gelatin dynamite, 1875; and ballistite, 1888. He left the major part of his large fortune for the endowment of prizes, now known as the Nobel Prizes, for notable achievements in physics, in chemistry, in physiology and medicine, in literature, and in the promotion of peace.

mercury fulminate, crimped tightly to the end of the fuse in order that it might detonate more strongly. He stated that the form of the cap might be varied greatly but that its action depended upon the sudden development of an intense pressure or shock.

Dynamite with an inactive base (guhr dynamite) is not manufactured commercially in this country. Small quantities are used for experimental purposes where a standard of comparison is needed in studies on the strength of various explosives.

The next important event in the development of these explosives was Nobel's invention of *dynamite with an active base*,[5] an explosive in which the nitroglycerin was absorbed by a mixture of materials which were themselves not explosive separately, such as potassium, sodium, or ammonium nitrate mixed with wood meal, charcoal, rosin, sugar, or starch. The nitroglycerin formed a thin coating upon the particles of the solid materials, and caused them to explode if a fulminate cap was used. The patent suggested a mixture of barium nitrate 70 parts, rosin or charcoal 10, and nitroglycerin 20, with or without the addition of sulfur, as an example of the invention. Nitroglycerin alone was evidently not enough to prevent the deliquescence of sodium and ammonium nitrate in these mixtures, for a later patent[6] of Nobel claimed the addition of small amounts of paraffin, ozokerite, stearine, naphthalene, or of any similar substance which is solid at ordinary temperatures and is of a fatty nature, as a coating for the particles to prevent the absorption of moisture by the explosive and the resulting danger from the exudation of nitroglycerin.

Dynamite with an active base is manufactured and used extensively in this country and in Canada and Mexico. It is known as *straight dynamite*, or simply as dynamite, presumably because its entire substance contributes to the energy of its explosion. The standard 40% straight dynamite which is used in comparative tests at the U. S. Bureau of Mines contains[7] nitroglycerin 40%, sodium nitrate 44%, calcium carbonate (anti-acid) 1%, and wood pulp 15%. Since the time when this standard was adopted, the usage of the term "straight" has altered somewhat in consequence of changes in American manufacturing practice, with the result that this standard material is now better designated as 40% straight nitroglycerin (straight) dynamite. This name distinguishes it from 40% l. f. or 40% low-freezing (straight) dynamite which contains, instead of straight nitroglycerin, a mixture of nitric esters produced by nitrating a mix-

[5] Brit. Pat. 442 (1869).

[6] Brit. Pat. 1570 (1873).

[7] C. A. Taylor and W. H. Rinkenbach, "Explosives, Their Materials, Constitution, and Analysis," *U. S. Bur. Mines Bull.* 219, Washington, 1923, p. 133.

ture of glycerin and glycol or of glycerin and sugar. Practically all active-base dynamites now manufactured in the United States, whether straight or ammonia or gelatin, are of this l. f. variety. American straight dynamites contain from 20 to 60% of mixed nitric esters absorbed on wood pulp and mixed with enough sodium or potassium nitrate to maintain the oxygen balance and to take care of the oxidation of part or occasionally of all the wood pulp.

Judson powder is a special, low-grade dynamite in which 5 to 15% of nitroglycerin is used as a coating on a granular *dope* made by mixing ground coal with sodium nitrate and sulfur, warming the materials together until the sulfur is melted, forming into grains which harden on cooling and are screened for size. It is intermediate in power between black powder and ordinary dynamite and is used principally for moving earth and soft rock in railroad work.

Nobel's inventions of *blasting gelatin* and *gelatin dynamite* are both covered by the same patent.[8] Seven or 8% of collodion cotton dissolved in nitroglycerin converted it to a stiff jelly which was suitable for use as a powerful high explosive. Solvents, such as acetone, ether-alcohol, and nitrobenzene, facilitated the incorporation of the two substances in the cold, but Nobel reported that collodion cotton dissolved readily in nitroglycerin without additional solvent if the nitroglycerin was warmed gently on the water bath. A cheaper explosive of less power could be made by mixing the gelatinized nitroglycerin with black powder or with mixtures composed of an oxidizing agent, such as a nitrate or chlorate, and a combustible material, such as coal dust, sulfur, sawdust, sugar, starch, or rosin. A typical gelatin dynamite consists of nitroglycerin 62.5%, collodion cotton 2.5%, saltpeter 27.0%, and wood meal 8%. A softer jelly is used for making gelatin dynamite than is suitable for use by itself as a blasting gelatin, and somewhat less collodion is used in proportion to the amount of nitroglycerin.

All straight nitroglycerin explosives can be frozen. Straight dynamite when frozen becomes less sensitive to shock and to initiation, but blasting gelatin becomes slightly more sensitive.

[8] Brit. Pat. 4179 (1875).

When the explosives are afterwards thawed, the nitroglycerin shows a tendency to exude.

Invention of Ammonium Nitrate Explosives

In 1867 two Swedish chemists, C. J. Ohlsson and J. H. Norrbin, patented an explosive, called *ammoniakkrut,* which consisted of ammonium nitrate either alone or in mixture with charcoal, sawdust, naphthalene, picric acid, nitroglycerin, or nitrobenzene. Theoretical calculations had shown that large quantities of heat and gas were given off by the explosions of these mixtures. The proportions of the materials were selected in such manner that all the carbon should be converted to carbon dioxide and all the hydrogen to water. Some of these explosives were difficult to ignite and to initiate, but the trouble was remedied by including some nitroglycerin in their compositions and by firing them with fulminate detonators. They were used to some extent in Sweden. Nobel purchased the invention from his fellow-countrymen early in the 1870's, and soon afterwards took out another patent [9] in connection with it, but still found that the hygroscopicity of the ammonium nitrate created real difficulty. He was not able to deal satisfactorily with the trouble until after the invention of gelatin dynamite. In present manufacturing practice in this country the tendency of the ammonium nitrate to take up water is counteracted by coating the particles with water-repelling substances, oils, or metallic soaps.

In 1879 Nobel took out a Swedish patent for *extra-dynamite* (ammon-gelatin-dynamit), one example of which was a fortified gelatin dynamite consisting of nitroglycerin 71%, collodion 4%, charcoal 2%, and ammonium nitrate 23%. Another contained much less nitroglycerin, namely, 25%, along with collodion 1%, charcoal 12%, and ammonium nitrate 62%, and was crumbly and plastic between the fingers rather than clearly gelatinous.

In these explosives, and in the ammonium nitrate *permissible* explosives which contain still less nitroglycerin, it is supposed that the nitroglycerin or the nitroglycerin jelly, which coats the particles of ammonium nitrate, carries the explosive impulse originating in the detonator, that this causes the ammonium nitrate to decompose explosively to produce nitrogen and water

[9] The above-cited Brit. Pat. 1570 (1873).

and oxygen, the last named of which enters into a further explosive reaction with the charcoal or other combustible material. Other explosive liquids or solids, such as liquid or solid DNT, TNT, or TNX, nitroglycol, nitrostarch, or nitrocellulose, may be used to sensitize the ammonium nitrate and to make the mixture more easily detonated by a blasting cap. Non-explosive combustible materials, such as rosin, coal, sulfur, cereal meal, and paraffin, also work as sensitizers for ammonium nitrate, and a different hypothesis is required to explain their action.

Guhr Dynamite

Guhr dynamite is used rather widely in Europe. It is not hygroscopic. Liquid water however, brought into contact with it, is absorbed by the kieselguhr and displaces the nitroglycerin which separates in the form of an oily liquid. The nitroglycerin thus set free in a wet bore hole might easily seep away into a fissure in the rock where it would later be exploded accidentally by a drill or by the blow of a pick. Water does not cause the separation of nitroglycerin from blasting gelatin or gelatin dynamite. It tends to dissolve the soluble salts which are present in straight dynamite and to liberate in the liquid state any nitroglycerin with which they may be coated.

Guhr dynamite, made from 1 part of kieselguhr and 3 parts of nitroglycerin, is not exploded by a blow of wood upon wood, but is exploded by a blow of iron or other metal upon iron. In the drop test it is exploded by the fall of a 1-kilogram weight through 12 to 15 cm., or by the fall of a 2-kilogram weight through 7 cm. The frozen material is less sensitive: a drop of more than 1 meter of the kilogram weight or of at least 20 cm. of the 2-kilogram weight is necessary to explode it. Frozen or unfrozen it is exploded in a paper cartridge by the impact of a bullet from a military rifle. A small sample will burn quietly in the open, but will explode if it is lighted within a confined space. A cartridge explodes if heated on a metal plate.

The velocity of detonation of guhr dynamite varies with the density of loading and with the diameter of the charge, but does not reach values equal to the maxima under best conditions for nitroglycerin and blasting gelatin. Velocities of 6650 to 6800 meters per second, at a density of loading of 1.50 (the highest which is practical) have been reported. Naoúm,[10] working with charges in an iron pipe 34 mm. in internal diameter and at a density of loading of 1.30, found for nitroglycerin guhr dynamite a velocity of detonation of 5650 meters per second, and, under the same conditions, for nitroglycol guhr dynamite one of 6000 meters per second.

FIGURE 84. Determination of the Velocity of Detonation of Dynamite by the Dautriche Method. (Courtesy Hercules Powder Company.) Compare Figure 9, page 17.

Dynamites, like guhr dynamite and straight dynamite, which contain nitroglycerin in the subdivided but liquid state communicate explosion from cartridge to cartridge more readily, and in general are more easy to initiate, than blasting gelatin and gelatin dynamite in which no liquid nitroglycerin is present. A cartridge of guhr dynamite 30 mm. in diameter will propagate its explosion through a distance of 30 cm. to a similar cartridge.

[10] Phokion Naoúm, "Nitroglycerine and Nitroglycerine Explosives," trans E. M. Symmes, Baltimore, The Williams and Wilkins Company, 1928 p. 277.

Straight Dynamite

Straight dynamite containing 60% or less of mixed nitric esters—but not more because of the danger of exudation—is used extensively in the United States, but has found little favor in

FIGURE 85. Dynamite Manufacture. (Courtesy Hercules Powder Company.) Rubbing the dry ingredients of dynamite through a screen into the bowl of a mixing machine.

Europe. It is made simply by mixing the explosive oil with the absorbent materials; the resulting loose, moist-appearing or greasy mass, from which oil ought not to exude under gentle pressure, is put up in cartridges or cylinders wrapped in paraffined paper and dipped into melted paraffin wax to seal them against moisture.

The strength of straight nitroglycerin dynamite is expressed by the per cent of nitroglycerin which it contains. Thus, "40% straight nitroglycerin dynamite" contains 40% of nitroglycerin, but "40% ammonia dynamite," "40% gelatin dynamite," etc., whatever their compositions may be, are supposed to have the same strength or explosive force as 40% straight nitroglycerin dynamite. Munroe and Hall [11] in 1915 reported for typical straight nitroglycerin dynamites the compositions which are

FIGURE 86. Dynamite Manufacture. (Courtesy Hercules Powder Company.) Hoppers underneath the mixing machine, showing the buggies which carry the mixed dynamite to the packing machines.

shown in the following table. Although these dynamites are not now manufactured commercially in the United States, their explosive properties, studied intensively at the U. S. Bureau of

| | STRENGTH | | | | | | | | | |
	15%	20%	25%	30%	35%	40%	45%	50%	55%	60%
Nitroglycerin	15	20	25	30	35	40	45	50	55	60
Combustible material	20	19	18	17	16	15	14	14	15	16
Sodium nitrate......	64	60	56	52	48	44	40	35	29	23
Calcium or magnesium carbonate...	1	1	1	1	1	1	1	1	1	1

Mines and reported as a matter of interest, do not differ greatly from those of the l. f. dynamites by which they have been superseded in common use. The combustible material stated to be used in these compositions consists of a mixture of wood pulp,

[11] Charles E. Munroe and Clarence Hall, "A Primer on Explosives for Metal Miners and Quarrymen," *U. S. Bur. Mines Bull.* 80, Washington, 1915, p. 22.

flour, and brimstone for the grades below 40% strength, wood
pulp alone for the 40% and stronger. In commercial practice the
dope sometimes contains coarse combustible material, like rice
hulls, sawdust, or bran, which makes the explosive more bulky
and has the effect of reducing the velocity of detonation. Tests
at the U. S. Bureau of Mines on standard straight dynamites in
cartridges 1¼ inches in diameter showed for the 30% grade a
velocity of detonation of 4548 meters per second, for the 40%

FIGURE 87. Dynamite Manufacture. (Courtesy Hercules Powder Company.) Dumping the mixed dynamite onto the conveyor belt which raises
it to the hopper of the semi-automatic packing machine.

grade 4688 meters per second, and for the 60% grade 6246 meters
per second. The 40% dynamite was exploded in one case out of
three by an 11-cm. drop of a 2-kilogram weight, in no case out
of five by a 10-cm. drop. Cartridges 1¼ inches in diameter and
8 inches long transmitted explosion from one to another through
a distance of 16 inches once in two trials, but not through a
distance of 17 inches in three trials. The 40% dynamite gave a
small lead block compression of 16.0 mm., and an expansion
(average of three) in the Trauzl test of 278 cc.[12]

[12] Clarence Hall, W. O. Snelling, and S. P. Howell, "Investigations of
Explosives Used in Coal Mines," *U. S. Bur. Mines Bull.* 15, Washington,
1912, pp. 171, 173.

Munroe and Hall [13] also reported the following compositions for typical ordinary and low-freezing ammonia dynamites, the combustible material in each case being a mixture of wood pulp, flour, and brimstone. Low-freezing dynamites at present in use in this country contain nitroglycol or nitrosugar instead of the above-mentioned nitrosubstitution compounds. In Europe dinitrochlorohydrin, tetranitrodiglycerin, and other nitric esters are used.

Strength	Ordinary					Low-Freezing				
	30%	35%	40%	50%	60%	30%	35%	40%	50%	60%
Nitroglycerin............	15	20	22	27	35	13	17	17	21	27
Nitrosubstitution compounds..............	..	..	..	..	..	3	4	4	5	6
Ammonium nitrate.......	15	15	20	25	30	15	15	20	25	30
Sodium nitrate..........	51	48	42	36	24	53	49	45	36	27
Combustible material....	18	16	15	11	10	15	14	13	12	9
Calcium carbonate or zinc oxide................	1	1	1	1	1	1	1	1	1	1

Three of the standard French ammonia dynamites, according to Naoúm,[14] have the compositions and explosive properties listed below.

Nitroglycerin	40	20	22
Ammonium nitrate	45	75	75
Sodium nitrate	5	...	...
Wood or cereal meal	10	5	...
Charcoal	...	...	3
Lead block expansion	400.0 cc.	335.0 cc.	330.0 cc.
Lead block crushing	22.0 mm.	15.5 mm.	16.0 mm.
Density	1.38	1.20	1.33

Taylor and Rinkenbach [15] report typical analyses of American ammonium nitrate dynamite (I below) and ammonium nitrate sodium nitrate dynamite (II below). These formulas really represent ammonium nitrate permissible explosives, very close in their

[13] Op. cit., p. 23.
[14] Op. cit., p. 285.

[15] Op. cit., pp. 136, 138.

compositions to Monobel (III below) which is permissible in
this country for use in coal mines. Naoúm [16] reports that this

	I	II	III
Nitroglycerin	9.50	9.50	10.0
Ammonium nitrate	79.45	69.25	80.0
Sodium nitrate	...	10.20	...
Carbonaceous combustible material [17]	9.75	9.65	...
Wood meal	...	...	10.0
Anti-acid	0.40	0.50	...
Moisture	0.90	0.90	...

Monobel (density about 1.15) gives a lead block expansion of
about 350 cc. and a lead block crushing of 12 mm. He states that

Figure 88. Dynamite Manufacture. (Courtesy Hercules Powder Company.) Cartridges of dynamite as they come from the semi-automatic
packing machine.

Monobel belongs to the class of typical ammonium nitrate explosives rather than to the dynamites, and points out that no specific effect can be ascribed to the 10% nitroglycerin which it contains, for an explosive containing only a small quantity, say 4%,

[16] *Op. cit.*, p. 286.
[17] The carbonaceous combustible material contains 0.40% grease or oil
which was added to the ammonium nitrate to counteract its hygroscopicity.
Note that the figures in the first two columns of the table represent results
of analyses; those in the third column represent the formula according to
which the explosive is mixed.

of nitroglycerin, or none at all, will give essentially the same performance. But the ammonium nitrate explosive with no nitroglycerin in it is safer to handle and more difficult to detonate.

Blasting Gelatin

Blasting gelatin exists as a yellowish, translucent, elastic mass of density about 1.63. Strong pressure does not cause nitroglycerin to exude from it. Its surface is rendered milky by long contact with water, but its explosive strength is unaffected. It is less sensitive to shock, blows, and friction than nitroglycerin, guhr dynamite, and straight dynamite, for its elasticity enables it more readily to absorb the force of a blow, and a thin layer explodes under a hammer more easily than a thick one. Blasting gelatin freezes with difficulty. When frozen, it loses its elasticity and flexibility, and becomes a hard, white mass. Unlike guhr dynamite and straight dynamite, it is more sensitive to shock when frozen than when in the soft and unfrozen state.

Unlike nitroglycerin, blasting gelatin takes fire easily from a flame or from the spark of a fuse. Its combustion is rapid and violent, and is accompanied by a hissing sound. If a large quantity is burning, the combustion is likely to become an explosion, and the same result is likely to follow if even a small quantity of the frozen material is set on fire.

Pulverulent explosives or explosive mixtures are easier to initiate and propagate detonation for a greater distance than liquid explosives, especially viscous ones, and these are easier to detonate and propagate more readily than colloids. The stiffer the colloid the more difficult it becomes to initiate, until, with increasingly large proportions of nitrocellulose in the nitroglycerin gel, tough, horny colloids are formed, like ballistite and cordite, which in sizable aggregates can be detonated only with difficulty. Blasting gelatin is more difficult to detonate than any of the forms of dynamite in which the nitroglycerin exists in the liquid state. Naoúm [18] reports that a freshly prepared blasting gelatin made from 93 parts of nitroglycerin and 7 parts of collodion cotton is exploded by a No. 1 (the weakest) blasting cap and propagates detonation even in 25-mm. cartridges across a gap of about 10 mm. A blasting gelatin containing 9% of collodion cotton requires a No. 4 blasting cap to make it explode and propagates

[18] *Op. cit.,* p. 316.

its explosion to an adjacent cartridge only when initiated by a No. 6 blasting cap.

Blasting gelatin and gelatin dynamite on keeping become less sensitive to detonation, and, after long storage in a warm climate, may even become incapable of being detonated. The effect has been thought to be due to the small air bubbles which make newly prepared blasting gelatin appear practically white but which disappear when the material is kept in storage and becomes translucent and yellowish. But this cannot be the whole cause of the effect, for the colloid becomes stiffer after keeping. The loss of sensitivity is accompanied by a rapid dropping off in the velocity of detonation and in the brisance. According to Naoúm,[19] blasting gelatin containing 7% collodion cotton when newly prepared gave a lead block expansion of 600 cc., after 2 days 580 cc., and one containing 9% collodion gave when freshly made an expansion of 580 cc., after 2 days 545 cc.

Blasting gelatin under the most favorable conditions has a velocity of detonation of about 8000 meters per second. In iron pipes it attains this velocity only if its cross section exceeds 30 mm. in diameter, and it attains it only at a certain distance away from the point of initiation, so that in the Dautriche method where short lengths are used lower values are generally obtained. In tubes of 20–25 mm. diameter, and with samples of a sensitivity reduced either by storage or by an increased toughness of the colloid, values as low as 2000–2500 meters per second have been observed.

Gelatin Dynamite

Blasting gelatin is not used very widely in the United States; the somewhat less powerful gelatin dynamite, or simply gelatin as it is called, is much more popular. Gelatin dynamite is essentially a straight dynamite in which a gel is used instead of the liquid nitroglycerin or l. f. mixture of nitric esters. It is a plastic mass which can be kneaded and shaped. The gel contains between 2 and 5.4% collodion cotton, and is not tough and really elastic like blasting gelatin. Correspondingly it is initiated more easily and has a higher velocity of detonation and better propagation. The gel is prepared by mixing the nitroglycerin and collodion cotton, allowing to stand at 40–45°C. for some hours or over

19 Op. cit., p. 322.

night, and then incorporating mechanically with the dope materials which have been previously mixed together. Munroe and Hall [20] in 1915 gave the compositions listed below as typical of gelatin dynamites offered for sale at that time in this country. Instead of straight nitroglycerin, l. f. mixtures of nitric esters are now used.

	STRENGTH						
	30%	35%	40%	50%	55%	60%	70%
Nitroglycerin	23.0	28.0	33.0	42.0	46.0	50.0	60.0
Nitrocellulose	0.7	0.9	1.0	1.5	1.7	1.9	2.4
Sodium nitrate	62.3	58.1	52.0	45.5	42.3	38.1	29.6
Combustible material [21]	13.0	12.0	13.0	10.0	9.0	9.0	7.0
Calcium carbonate	1.0	1.0	1.0	1.0	1.0	1.0	1.0

The three standard explosives which are used in Great Britain are called respectively blasting gelatin, gelatin dynamite, and *Gelignite*. Gelignite, let us note, is a variety of gelatin dynamite as the latter term is used in this country. It is the most widely used of the three and may indeed be regarded as the standard explosive.

	BLASTING GELATIN	GELATIN DYNAMITE	GELIGNITE
Nitroglycerin	92	75	60
Collodion cotton	8	5	4
Wood meal	..	5	8
Potassium nitrate	..	15	28

The gelatin dynamites most widely used in Germany contain about 65 parts of gelatinized nitroglycerin and about 35 parts of dope or absorbent material. The dope for an explosive for domestic use consists of 76.9% sodium nitrate, 22.6% wood meal, and 0.5% chalk, and for one for export of 80% potassium nitrate, 19.5% wood meal, and 0.5% chalk. A weaker *Gelignite II* and certain high-strength gelatin dynamites, as tabulated below, are also manufactured for export.

	GELIGNITE II	HIGH-STRENGTH GELATIN DYNAMITE		
		80%	81%	75%
Nitroglycerin	47.5	75	75.8	70.4
Collodion cotton	2.5	5	5.2	4.6
Potassium nitrate	37.5	15	15.2	19.3
Wood meal with chalk	3.5	5	3.8	5.7
Rye meal	9.0	..	...	...

[20] *Op. cit.*, p. 23.

[21] Wood pulp was used in the 60% and 70% grades. Flour, wood pulp, and, in some examples, rosin and brimstone were used in the other grades.

The gelatin dynamites manufactured in Belgium are called *Forcites*. The reported compositions of several of them are tabulated below. *Forcite extra* is an ammonia gelatin dynamite.

	For- cite Extra	For- cite Su- per- ieure	Su- per For- cite No. 1	For- cite No. 1	For- cite No. 1P	For- cite No. 2	For- cite No. 2P
Nitroglycerin	64	64	64	49	49	36	36
Collodion cotton	3.5	3	3	2	2	3	2
Sodium nitrate	..	24	..	36	..	35	..
Potassium nitrate	..	..	23	..	37	..	46
Ammonium nitrate	25	..	..	..	..	..	..
Wood meal	6.5	8	9	13	11	11	..
Bran	..	..	..	..	..	14	15
Magnesium carbonate	1	1	1	1	1	1	1

In France gelatin dynamites are known by the names indicated in the following table where the reported compositions of several of them are tabulated.

	Dynamite-gomme- extra-forte	Dynamite-gomme- potasse	Dynamite-gomme- soude	Gélatine A	Gélatine B-potasse	Gélatine B-soude	Gomme F	Gélignite
Nitroglycerin	92–93	82–83	82–83	64	57.5	57	49	58
Collodion cotton	8–7	6–5	6–5	3	2.5	3	2	2
Potassium nitrate	...	9–10	...	..	32.0	..	36	28
Sodium nitrate	...	...	9–10	24	..	34	..	..
Wood meal	...	2–3	2–3	·8	8.0	6	10	9
Flour	...	...	...	..	..	..	3	3
Magnesium carbonate	...	...	...	1	..	..	..	..

Permissible Explosives

The atmosphere of coal mines frequently contains enough methane (fire damp) to make it explode from the flame of a black powder or dynamite blast. Dust also produces an explosive atmosphere, and it may happen, if dust is not already present, that one blast will stir up clouds of dust which the next blast will cause to explode. Accidents from this cause became more and more frequent as the industrial importance of coal increased during the nineteenth century and as the mines were dug deeper and contained more fire damp, until finally the various nations which were producers of coal appointed commissions to study and develop means of preventing them. The first of these was appointed in France in 1877, the British commission in 1879, the Prussian commission in 1881, and the Belgian and Austrian commissions at later dates. The Pittsburgh testing station of the U. S. Geological Survey was officially opened and regular work was commenced there on December 3, 1908, with the result that the first American list of explosives permissible for use in gaseous and dusty coal mines was issued May 15, 1909. On July 1, 1909, the station was taken over by the U. S. Bureau of Mines,[22] which, since January 1, 1918, has conducted its tests at the Explosives Experiment Station at Bruceton, not far from Pittsburgh, in Pennsylvania.

Explosives which are approved for use in gaseous and dusty coal mines are known in this country as *permissible* explosives, in England as *permitted* explosives, and are to be distinguished from *authorized* explosives which conform to certain conditions with respect to safety in handling, in transport, etc. Explosives which are safe for use in coal mines are known in France as *explosifs antigrisouteux*, in Belgium as *explosifs S. G. P. (sécurité, grisou, poussière)*, in Germany as *schlagwettersichere Sprengstoffe* while the adjective *handhabungssichere* is applied to those which are safe in handling. Both kinds, permissible and authorized, are *safety explosives, explosifs de sûreté, Sicherheitssprengstoffe*.

A mixture of air and methane is explosive if the methane content lies between 5 and 14%. A mixture which contains 9.5% of methane, in which the oxygen exactly suffices for complete combustion, is the one which explodes most violently, propagates the explosion most easily, and produces the highest temperature. This mixture ignites at about 650° to 700°. Since explosives in general produce temperatures which are considerably above 1000°, explo-

[22] A few of the interesting and important publications of the U. S. Bureau of Mines are listed in the footnote, Vol. I, pp. 22–23.

sive mixtures of methane and air would always be exploded by them if it were not for the circumstance, discovered by Mallard and Le Chatelier,[23] that there is a certain delay or period of induction before the gaseous mixture actually explodes. At 650° this amounts to about 10 seconds, at 1000° to about 1 second, and at 2200° there is no appreciable delay and the explosion is presumed to follow instantaneously after the application of this temperature however momentary. Mallard and Le Chatelier concluded that an explosive having a temperature of explosion of 2200° or higher would invariably ignite fire damp. The French commission which was studying these questions at first decided that the essential characteristic of a permissible explosive should be that its calculated temperature of explosion should be not greater than 2200°, and later designated a temperature of 1500° as the maximum for explosives permissible in coal seams and 1900° for those intended to be used in the accompanying rock.

The flame which is produced by the explosion of a brisant explosive is of extremely short duration, and its high temperature continues only for a small fraction of a second, for the hot gases by expanding and by doing work immediately commence to cool themselves. If they are produced in the first place at a temperature below that of the instantaneous inflammation of fire damp, they may be cooled to such an extent that they are not sufficiently warm for a sufficiently long time to ignite fire damp at all. Black powder, burning slowly, always ignites explosive gas mixtures. But any high explosive may be made safe for use in gaseous mines by the addition to it of materials which reduce the initial temperature of the products of its explosion. Or, in cases where this initial temperature is not too high, the same safety may be secured by limiting the size of the charge and by firing the shot in a well-tamped bore hole under such conditions that the gases are obliged to do more mechanical work and are cooled the more in consequence.

Permissible explosives may be divided into two principal classes: (1) those which are and (2) those which are not based upon a high explosive which is cool in itself, such as ammonium nitrate, or guanidine nitrate, or nitroguanidine. The second class may be subdivided further, according to composition, into as

[23] *Ann. Min.,* [8] 11, 274 (1887).

many classes as there are varieties in the compositions of high explosives, or it may be subdivided, irrespective of composition, according to the means which are used to reduce the explosion temperature. Thus, an explosive containing nitroglycerin, nitrostarch, chlorate or perchlorate, or tetranitronaphthalene, or an explosive which is essentially black powder, may have its temperature of explosion reduced by reason of the fact that (a) it contains an excess of carbonaceous material, (b) it contains water physically or chemically held in the mixture, or (c) it contains volatile salts or substances which are decomposed by heat. Ammonium nitrate may also be used as a means of lowering the temperature of explosion, and thus defines another subdivision (d) which corresponds to an overlapping of the two principal classes, (a) and (b).

Ammonium nitrate, although it is often not regarded as an explosive, may nevertheless be exploded by a suitable initiator. On complete detonation it decomposes in accordance with the equation

$$2NH_4NO_3 \longrightarrow 4H_2O + 2N_2 + O_2$$

but the effect of feeble initiation is to cause decomposition in another manner with the production of oxides of nitrogen. By using a booster of 20–30 grams of Bellite (an explosive consisting of a mixture of ammonium nitrate and dinitrobenzene) and a detonator containing 1 gram of mercury fulminate, Lobry de Bruyn [24] succeeded in detonating 180 grams of ammonium nitrate compressed in a 8-cm. shell. The shell was broken into many fragments. A detonator containing 3 grams of mercury fulminate, used without the booster of Bellite, produced only incomplete detonation. Lheure [25] secured complete detonation of cartridges of ammonium nitrate [26] loaded in bore holes in rock by means of a trinitrotoluene detonating fuse which passed completely through them.

The sensitiveness of ammonium nitrate to initiation is increased by the addition to it of explosive substances, such as nitroglycerin, nitrocellulose, or aromatic nitro compounds, or of

[24] Rec. trav. chim., 10, 127 (1891).

[25] Ann. Min., [10] 12, 169 (1907).

[26] On the explosibility of ammonium nitrate, see also Munroe, Chem. Met. Eng., 26, 535 (1922); Cook, ibid., 31, 231 (1924); Sherrick, Army Ordnance, 4, 237, 329 (1924).

non-explosive combustible materials, such as rosin, sulfur, charcoal, flour, sugar, oil, or paraffin. Substances of the latter class
react with the oxygen which the ammonium nitrate would otherwise liberate; they produce additional gas and heat, and increase
both the power of the explosive and the temperature of its explosion. Pure ammonium nitrate has a temperature of explosion of
about 1120° to 1130°. Ammonium nitrate explosives permissible
in the United States generally produce instantaneous temperatures between 1500° and 2000°.

Among the first permissible explosives developed in France
were certain ones of the Belgian Favier type which contained
no nitroglycerin and consisted essentially of ammonium nitrate,
sometimes with other nitrates, along with a combustible material such as naphthalene or nitrated naphthalene or other aromatic nitro compounds. These explosives have remained the
favorites in France for use in coal mines. The method of manufacture is simple. The materials are ground together in a wheel
mill, and the mass is broken up, sifted, and packed in paraffined
paper cartridges. The compositions of the mixtures are those
which calculations show to give the desired temperatures of explosion. *Grisourites roches,* permissible for use in rock, have
temperatures of explosion between 1500° and 1900°; *Grisounites
couches,* for use in coal, below 1500°. Several typical compositions are listed below.

	Grisou-naphtalite-roche	Grisou-naphtalite-roche salpêtrée	Grisou-naphtalite-couche	Grisou-naphtalite-couche salpêtrée	Grisou-tétrylite-couche
Ammonium nitrate......	91.5	86.5	95	90	88
Potassium nitrate......	..	5.0	..	5	5
Dinitro-naphthalene.	8.5	8.5	..	..	..
Trinitro-naphthalene.	..	..	5	5	..
Tetryl........	..	..	..	..	7

The French also have permissible explosives containing both ammonium nitrate and nitroglycerin (gelatinized), with and without saltpeter. These are called *Grisou-dynamites* or *Grisoutines*.

	Grisou-dynamite-roche	Grisou-dynamite-roche salpêtrée	Grisou-dynamite-couche	Grisou-dynamite-couche salpêtrée
Nitroglycerin......	29.0	29.0	12.0	12.0
Collodion cotton....	1.0	1.0	0.5	0.5
Ammonium nitrate.	70.0	65.0	87.5	82.5
Potassium nitrate...	..	5.0	..	5.0

The effect of ammonium nitrate in lowering the temperature of explosion of nitroglycerin mixtures is nicely illustrated by the data of Naoúm [27] who reports that guhr dynamite (75% actual nitroglycerin) gives a temperature of 2940°, a mixture of equal amounts of guhr dynamite and ammonium nitrate 2090°, and a mixture of 1 part of guhr dynamite and 4 of ammonium nitrate 1468°.

In ammonium nitrate explosives in which the ingredients are not intimately incorporated as they are in the Favier explosives, but in which the granular particles retain their individual form, the velocity of detonation may be regulated by the size of the nitrate grains. A relatively slow explosive for producing lump coal is made with coarse-grained ammonium nitrate, and a faster explosive for the procurement of coking coal is made with fine-grained material.

The first explosives to be listed as permissible by the U. S. Bureau of Mines were certain *Monobels* and *Carbonites,* and Monobels are still among the most important of American permissibles. Monobels contain about 10% nitroglycerin, about 10% carbonaceous material, wood pulp, flour, sawdust, etc., by the physical properties of which the characteristics of the explosive are somewhat modified, and about 80% ammonium nitrate of which, however, a portion, say 10%, may be substituted by a volatile salt such as sodium chloride.

[27] *Op. cit.,* p. 403.

In Europe the tendency is to use a smaller amount of nitroglycerin, say 4 to 6%, or, as in the Favier explosives, to omit it altogether. Ammonium nitrate permissible explosives which contain nitroglycerin may be divided broadly into two principal classes, those of low ammonium nitrate content in which the oxygen is balanced rather accurately against the carbonaceous material and which are cooled by the inclusion of salts, and those which have a high ammonium nitrate content but whose temperature of explosion is low because of an incomplete utilization of the oxygen by a relatively small amount of carbonaceous material. Explosives of the latter class are more popular in England and in Germany. Several examples of commercial explosives of each sort are listed in the following table.

	I	II	III	IV	V	VI	VII	VIII
Ammonium nitrate	52.0	53.0	60.0	61.0	66.0	73.0	78.0	83.0
Potassium nitrate	21.0	...	...	...	...	2.8	5.0	7.0
Sodium nitrate	...	12.0	5.0	3.0	...	...	...	...
Barium nitrate	...	...	...	...	...	...	...	2.0
Na or K chloride	...	...	21.0	20.5	22.0	15.0	8.0	...
Hydrated ammonium oxalate	16.0	19.0	...	...	...	...	...	...
Ammonium chloride	6.0	...	...	...	...	...	...	...
Cereal or wood meal	...	4.0	4.0	7.5	2.0	1.0	5.0	2.0
Glycerin	...	...	...	3.0	...	...	...	...
Powdered coal	...	...	...	...	4.0	...	...	...
Nitrotoluene	...	...	6.0	1.0	...	...	...	...
Dinitrotoluene	...	...	...	...	...	5.0	...	...
Trinitrotoluene	...	6.0	...	...	...	...	...	2.0
Nitroglycerin	5.0	5.0	4.0	4.0	4.0	3.2	4.0	4.0

The *Carbonites* which are permissible are straight dynamites whose temperatures of explosion are lowered by the excess of carbon which they contain. As a class they merge, through the *Ammon-Carbonites*, with the class of ammonium nitrate explosives. The Carbonites, have the disadvantage that they produce gases which contain carbon monoxide, and for that reason have largely given way for use in coal mines to ammonium nitrate permissibles which contain an excess of oxygen. Naoúm [28] reports the compositions and explosive characteristics of four German Carbonites as follows.

[28] *Op. cit.*, p. 401.

	I	II	III	IV
Nitroglycerin	25.0	25.0	25.0	30.0
Potassium nitrate	30.5	34.0	. . .	. . .
Sodium nitrate	. . .	. . .	30.5	24.5
Barium nitrate	4.0	1.0	. . .	. . .
Spent tan bark meal	40.0	1.0	. . .	. . .
Meal	. . .	38.5	39.5	40.5
Potassium dichromate	. . .	. . .	5.0	5.0
Sodium carbonate	0.5	0.5	. . .	. . .
Heat of explosion, Cal./kg.	576	506	536	602
Temperature of explosion	1874°	1561°	1666°	1639°
Velocity of detonation, meters/sec. ...	2443	2700	3042	2472
Lead block expansion	235 cc.	213 cc.	240 cc.	258 cc.

The salts which are most frequently used in permissible explosives are sodium chloride and potassium chloride, both of which are volatile (the potassium chloride more readily so), ammonium chloride and ammonium sulfate, which decompose to form gases, and the hydrated salts, alum $Al_2(SO_4)_3 \cdot K_2SO_4 \cdot 24H_2O$; ammonium alum $Al_2(SO_4)_3 \cdot (NH_4)_2SO_4 \cdot 24H_2O$; chrome alum $Cr_2(SO_4)_3 \cdot K_2SO_4 \cdot 24H_2O$; aluminum sulfate $Al_2(SO_4)_3 \cdot 18H_2O$; ammonium oxalate $(NH_4)_2C_2O_4 \cdot H_2O$; blue vitriol $CuSO_4 \cdot 5H_2O$; borax $Na_2B_4O_7 \cdot 10H_2O$; Epsom salt $MgSO_4 \cdot 7H_2O$; Glauber's salt $Na_2SO_4 \cdot 10H_2O$; and gypsum $CaSO_4 \cdot 2H_2O$, all of which give off water, while the ammonium salts among them yield other volatile products in addition. Hydrated sodium carbonate is not suitable for use because it attacks both ammonium nitrate and nitroglycerin.[29]

Sprengel Explosives

Explosives of a new type were introduced in 1871 by Hermann Sprengel, the inventor of the mercury high-vacuum pump, who patented [30] a whole series of mining explosives which were prepared by mixing an oxidizing substance with a combustible one "in such proportions that their mutual oxidation and de-oxidation should be theoretically complete." The essential novelty of his invention lay in the fact that the materials were mixed just before the explosive was used, and the resultant explosive mixture was

[29] C. G. Storm, "The Analysis of Permissible Explosives," *U. S. Bur. Mines Bull.* **96**, Washington, 1916.

[30] Brit. Pats. **921, 2642** (1871).

fired by means of a blasting cap. Among the oxidizing agents
which he mentioned were potassium chlorate, strong nitric acid,
and liquid nitrogen dioxide; among the combustible materials
nitrobenzene, nitronaphthalene, carbon disulfide, petroleum, and
picric acid.[31] Strong nitric acid is an inconvenient and unpleasant
material to handle. It can eat through the copper capsule of a
blasting cap and cause the fulminate to explode. Yet several
explosives containing it have been patented, *Oxonite*, for example,
consisting of 58 parts of picric acid and 42 of fuming nitric acid,
and *Hellhoffite*, 28 parts of nitrobenzene and 72 of nitric acid.
These explosives are about as powerful as 70% dynamite, but
are distinctly more sensitive to shock and to blows. Hellhoffite
was sometimes absorbed on kieselguhr to form a plastic mass, but
it still had the disadvantage that it was intensely corrosive and
attacked paper, wood, and the common metals.

The peculiarities of the explosives recommended by Sprengel
so set them apart from all others that they define a class; explosives which contain a large proportion of a liquid ingredient and
which are mixed *in situ* immediately before use are now known
as Sprengel explosives. They have had no success in England, for
the reason that the mixing of the ingredients has been held to
constitute manufacture within the meaning of the Explosives Act
of 1875 and as such could be carried out lawfully only on licensed
premises. Sprengel explosives have been used in the United States,
in France, and in Italy, and were introduced into Siberia and
China by American engineers when the first railroads were built
in those countries. *Rack-a-rock*, patented by S. R. Divine,[32] is
particularly well known because it was used for blasting out Hell
Gate Channel in New York Harbor. On October 10, 1885, 240,399
pounds of it, along with 42,331 pounds of dynamite, was exploded
for that purpose in a single blast. It was prepared for use by
adding 21 parts of nitrobenzene to 79 parts of potassium chlorate
contained in water-tight copper cartridges.

[31] Sprengel was aware in 1871 that picric acid alone could be detonated
by means of fulminate but realized also that more explosive force could
be had from it if it were mixed with an oxidizing agent. Picric acid alone
was evidently not used practically as an explosive until after Turpin in
1886 had proposed it as a bursting charge for shells.

[32] Brit. Pats. 5584, 5596 (1881); 1461 (1882); 5624, 5625 (1883).

The *Prométhées,* authorized in France under the name of *explosifs O No. 3,* are prepared by dipping cartridges of a compressed oxidizing mixture of potassium chlorate 80 to 95% and manganese dioxide 5 to 20% into a liquid prepared by mixing nitrobenzene, turpentine, and naphtha in the proportions 50/20/30 or 60/15/25. The most serious disadvantage of these explosives was an irregularity of behavior resulting from the circumstance that different cartridges absorbed different quantities of the combustible oil, generally between 8 and 13%, and that the absorption was uneven and sometimes caused incomplete detonation. Similar explosives are those of Kirsanov, a mixture of 90 parts of turpentine and 10 of phenol absorbed by a mixture of 80 parts of potassium chlorate and 20 of manganese dioxide, and of Fielder, a liquid containing 80 parts of nitrobenzene and 20 of turpentine absorbed by a mixture of 70 parts of potassium chlorate and 30 of potassium permanganate.

The *Panclastites,* proposed by Turpin in 1881, are made by mixing liquid nitrogen dioxide with such combustible liquids as carbon disulfide, nitrobenzene, nitrotoluene, or gasoline. They are very sensitive to shock and must be handled with the greatest caution after they have once been mixed. In the first World War the French used certain ones of them, under the name of *Anilites,* in small bombs which were dropped from airplanes for the purpose of destroying personnel. The two liquids were enclosed in separate compartments of the bomb, which therefore contained no explosive and was safe while the airplane was carrying it. When the bomb was released, a little propeller on its nose, actuated by the passage through the air, opened a valve which permitted the two liquids to mix in such fashion that the bomb was then filled with a powerful high explosive which was so sensitive that it needed no fuze but exploded immediately upon impact with the target.

Liquid Oxygen Explosives

Liquid oxygen explosives were invented in 1895 by Linde who had developed a successful machine for the liquefaction of gases. The *Oxyliquits,* as he called them, prepared by impregnating cartridges of porous combustible material with liquid oxygen or liquid air are members of the general class of Sprengel explosives, and have the unusual advantage from the point of view of safety that they rapidly lose their explosiveness as they lose their liquid oxygen by evaporation. If they have failed to fire in a bore hole, the workmen need have no fear of going into the place with a pick or a drill after an hour or so has elapsed.

Liquid oxygen explosives often explode from flame or from the spurt of sparks from a miner's fuse, and frequently need no detonator, or, putting the matter otherwise, some of them are themselves satisfactory detonators. Like other detonating explosives, they may explode from shock. Liquid oxygen explosives made from carbonized cork and from kieselguhr mixed with petroleum were used in the blasting of the Simplon tunnel in 1899. The explosive which results when a cartridge of spongy metallic aluminum absorbs liquid oxygen is of theoretical interest because its explosion yields no gas; it yields only solid aluminum oxide and heat, much heat, which causes the extremely rapid gasification of the excess of liquid oxygen and it is this which produces the explosive effect. Lampblack is the absorbent most commonly used in this country.

Liquid oxygen explosives were at first made up from liquid air more or less self-enriched by standing, the nitrogen (b.p. $-195°$) evaporating faster than the oxygen (b.p. $-183°$), but it was later shown that much better results followed from the use of pure liquid oxygen. Rice reports [33] that explosives made from liquid oxygen and an absorbent of crude oil on kieselguhr mixed with lampblack or wood pulp and enclosed in a cheesecloth bag within a corrugated pasteboard insulator were 4 to 12% stronger than 40% straight nitroglycerin dynamite in the standard Bureau of Mines test with the ballistic pendulum. They had a velocity of detonation of about 3000 meters per second. They caused the ignition of fire damp and produced a flame which lasted for 7.125 milliseconds as compared with 0.342 for an average permissible explosive (no permissible producing a flame of more than 1 millisecond duration). The length of the flame was $2\frac{1}{2}$ times that of the flame of the average permissible. In the Trauzl lead block an explosive made up from a liquid air (i.e., a mixture of liquid

[33] George S. Rice, "Development of Liquid Oxygen Explosives during the War," *U. S. Bur. Mines Tech. Paper* 243, Washington, 1920, pp. 14–16. Also, S. P. Howell, J. W. Paul, and J. L. Sherrick, "Progress of Investigations on Liquid Oxygen Explosives," *U. S. Bur. Mines Tech. Paper* 294, Washington, 1923, pp. 33, 35, 51.

oxygen and liquid nitrogen) containing 33% of oxygen gave no explosion; with 40% oxygen an enlargement of 9 cc.; with 50% 80 cc., with 55% 147 cc.; and with 98% oxygen an enlargement of 384 cc., about 20% greater than the enlargement produced by 60% straight dynamite. The higher temperatures of explosion of the liquid oxygen explosives cause them to give higher results in the Trauzl test than correspond to their actual explosive power.

Liquid oxygen explosives are used in this country for open-cut mining or strip mining, not underground, and are generally prepared near the place where they are to be used. The cartridges are commonly left in the "soaking box" for 30 minutes, and on occasions have been transported in this box for several miles.

One of the most serious faults of liquid oxygen explosives is the ease with which they inflame and the rapidity with which they burn, amounting practically and in the majority of cases to their exploding from fire. Denues [34] has found that treatment of the granular carbonaceous absorbent with an aqueous solution of phosphoric acid results in an explosive which is non-inflammable by cigarettes, matches, and other igniting agents. Mono- and diammonium phosphate, ammonium chloride, and phosphoric acid were found to be suitable for fireproofing the canvas wrappers. Liquid oxygen explosives made up from the fireproofed absorbent are still capable of being detonated by a blasting cap. Their strength, velocity of detonation, and length of life after impregnation are slightly but not significantly shorter than those of explosives made up from ordinary non-fireproofed absorbents containing the same amount of moisture.

Chlorate and Perchlorate Explosives

The history of chlorate explosives goes back as far as 1788 when Berthollet attempted to make a new and more powerful gunpowder by incorporating in a stamp mill a mixture of potassium chlorate with sulfur and charcoal. He used the materials in the proportion 6/1/1. A party. had been organized to witness the manufacture, M. and Mme. Lavoisier, Berthollet, the Commissaire M. de Chevraud and his daughter, the engineer M. Lefort, and others. The mill was started, and the party went away for

[34] A. R. T. Denues, "Fire Retardant Treatments of Liquid Oxygen Explosives," *U. S. Bur. Mines Bull.* 429, Washington, 1940.

breakfast. Lefort and Mlle. de Chevraud were the first to return. The material exploded, throwing them to a considerable distance and causing such injuries that they both died within a few minutes. In 1849 the problem of chlorate gunpowder was again attacked by Augendre vho invented a *white powder* made from potassium chlorate 4 parts, cane sugar 1 part, and potassium ferrocyanide 1 part. However, no satisfactory propellent powder for use in guns has yet been made from chlorate. Chlorate powders are used in toy salutes, maroons, etc., where a sharp explosion accompanied by noise is desired, and chlorate is used in primer compositions and in practical high explosives of the Sprengel type (described above) and in the Cheddites and Silesia explosives.

Many chlorate mixtures, particularly those which contain sulfur, sulfides, and picric acid, are extremely sensitive to blows and to friction. In the *Street explosives,* later called Cheddites because they were manufactured at Chedde in France, the chlorate is phlegmatized by means of castor oil, a substance which appears to have remarkable powers in this respect. The French *Commission des Substances Explosives* in 1897 commenced its first investigation of these explosives by a study of those which are listed below, and concluded [35] that their sensitivity to shock is

	I	II	III
Potassium chlorate	75.0	74.6	80.0
Picronitronaphthalene	20.0	...	...
Nitronaphthalene	...	5.5	12.0
Starch	...	14.9	...
Castor oil	5.0	5.0	8.0

less than that of No. 1 dynamite (75% guhr dynamite) and that when exploded by a fulminate cap they show a considerable brisance which however is less than that of dynamite. Later studies showed that the Cheddites had slightly more force than No. 1 dynamite, although they were markedly less brisant because of their lower velocity of detonation. After further experimentation four Cheddites were approved for manufacture in France, but the output of the Poudrerie de Vonges where they were made consisted principally of Cheddites No. 1 and No. 4.

[35] *Mém. Poudres,* **9,** 144 (1897–1898); **11,** 22 (1901); **12,** 117, 122 (1903–1904); **13,** 144, 282 (1905–1906); **15,** 135 (1909–1910); **16,** 66 (1911–1912).

	O No. 1 Formula 41	O No. 1 Formula 60 bis	O No. 2 Formula 60 bis M Cheddite No. 4	O No. 5 Cheddite No. 1
Potassium chlorate.........	80	80	79	..
Sodium chlorate............	..	..	..	79
Nitronaphthalene...........	12	13	1	..
Dinitrotoluene.............	..	2	15	16
Castor oil.................	8	5	5	5

The Cheddites are manufactured by melting the nitro compounds in the castor oil at 80°, adding little by little the pulverized chlorate dried and still warm, and mixing thoroughly. The mixture is emptied out onto a table, and rolled to a thin layer which hardens on cooling and breaks up under the roller and is then sifted and screened.

Sodium chlorate contains more oxygen than potassium chlorate, but has the disadvantage of being hygroscopic. Neither salt ought to be used in mixtures which contain ammonium nitrate or ammonium perchlorate, for double decomposition might occur with the formation of dangerous ammonium chlorate. Potassium chlorate is one of the chlorates least soluble in water, potassium perchlorate one of the least soluble of the perchlorates. The latter salt is practically insoluble in alcohol. The perchlorates are intrinsically more stable and less reactive than the chlorates, and are much safer in contact with combustible substances. Unlike the chlorates they are not decomposed by hydrochloric acid, and they do not yield an explosive gas when warmed with concentrated sulfuric acid. The perchlorates require a higher temperature for their decomposition than do the corresponding chlorates.

SOLUBILITY: PARTS PER 100 PARTS OF WATER

	$KClO_3$	$NaClO_3$	$KClO_4$	NH_4ClO_4
At 0°	3.3	82.	0.7	12.4
At 100°	56.	204.	18.7	88.2

Mixtures of aromatic nitro compounds with chlorate are dangerously sensitive unless they are phlegmatized with castor oil or a similar material, but there are other substances, such as rosin, animal and vegetable oils, and petroleum products, which give mixtures which are not unduly sensitive to shock and friction and may be handled with reasonable safety. Some of these, such as *Pyrodialyte* [36] and the *Steelites*,[37] were studied by the *Commission des Substances Explosives*. The former consisted of 85 parts of potassium chlorate and 15 of rosin, 2 parts of alcohol being used during the incorporation. The latter, invented by Everard Steele of Chester, England, contained an oxidized rosin (*résidée* in French) which was made by treating a mixture of 90 parts of colophony and 10 of starch with 42 Bé nitric acid. After washing, drying, and powdering, the *résidée* was mixed with powdered potassium chlorate, moistened with methyl alcohol, warmed, and stirred gently while the alcohol was evaporated. *Colliery Steelite*

	STEELITE No. 3	STEELITE No. 5	STEELITE No. 7	COLLIERY STEELITE
Potassium chlorate	75	83.33	87.50	72.5–75.5
Résidée	25	16.67	12.50	23.5–26.5
Aluminum		5.00	...	
Castor oil		....	....	0.5–1.0
Moisture		....	....	0–1

passed the Woolwich test for safety explosives and was formerly on the British permitted list but failed in the Rotherham test. In Germany the *Silesia* explosives have been used to some extent. *Silesia No. 4* consists of 80 parts of potassium chlorate and 20 of rosin, and *Silesia IV 22*, 70 parts of potassium chlorate, 8 of rosin, and 22 of sodium chloride, is cooled by the addition of the volatile salt and is on the permissible list.

The *Sebomites*,[38] invented by Eugène Louis, contained animal fat which was solid at ordinary temperature, and were inferior to the Cheddites in their ability to transmit detonation. *Explosifs P* (*potasse*) and *S* (*soude*)[39] and the *Minélites*,[40] containing petroleum hydrocarbons, were studied in considerable detail by Dautriche, some of whose results for velocities of detonation are reported in the table on pages 362–363 where they are compared with

[36] *Mém. Poudres*, **11**, 53 (1901).
[37] *Ibid.*, **15**, 181 (1909–1910).
[38] *Ibid.*, **13**, 280 (1905–1906); **15**, 137 (1909–1910).
[39] *Ibid.*, **15**, 212 (1909–1910).
[40] *Ibid.*, **16**, 224 (1911–1912).

his results for Cheddite 60, fourth formula.[41] His experimental results [42] illustrate very clearly the principle that there is an optimum density of loading at which the velocity of detonation is greatest and that at higher densities the velocity drops and the detonation is incomplete and poorly propagated. The Cheddite 60,

	EXPLOSIFS		MINÉLITES		
	P	S	A	B	C
Potassium chlorate	90	..	90	90	89
Sodium chlorate	..	89	..	..	..
Heavy petroleum oil	..	..	3	..	..
Vaseline	..	..	..	3	4
Paraffin	10	11	7	7	5
Pitch	..	..	..	..	2

fourth formula, when ignited burns slowly with a smoky flame. *Explosifs P* and *S* and the *Minélites* burn while the flame of a Bunsen burner is played upon them but, in general, go out when the flame is removed. *Minélite B,* under the designation *O No. 6 B,* was used by the French during the first World War in grenades and mines. A similar explosive containing 90 parts of sodium chlorate instead of 90 of potassium chlorate was used in grenades and in trench mortar bombs.

Chlorate explosives which contain aromatic nitro compounds have higher velocities of detonation and are more brisant than those whose carbonaceous material is merely combustible. The addition of a small amount of nitroglycerin increases the velocity of detonation still farther. Brisant chlorate explosives of this sort were developed in Germany during the first World War and were known as *Koronit* and *Albit* (*Gesteinskoronit, Kohlen-koronit, Wetteralbit,* etc.). They found considerable use for a time but have now been largely superseded by low-percentage dynamites and by perchlorate explosives. Two of them, manufactured by the Dynamit A.-G., had according to Naoúm [43] the compositions and explosive characteristics which are indicated

[41] The composition of this explosive was the same as that which is given in the table on page 359 as that of *O* No. 2, formula 60 *bis* M, or Cheddite No. 4.

[42] In several cases Dautriche reported temperatures, but the velocity of detonation appears to be unaffected by such temperature variations as those between summer and winter.

[43] *Op. cit.,* p. 428.

Explosive	In Tubes of	Diameter	Density of Loading	Velocity of Detonation, M./Sec.
Explosif P	copper	20–22 mm.	0.62	2137
			1.00	3044
			1.05	3185
			1.36	3621
			1.48	3475
			1.54	Incomplete
			0.99	2940
			1.24	3457
			1.45	3565
			1.59	Incomplete
Explosif P	paper	29 mm.	0.95	2752
			1.30	3406
			1.35	3340
			0.90	2688
			1.21	3308
			1.36	3259
			1.41	Incomplete
Explosif S	copper	20–22 mm.	0.88	2480
			1.25	2915
			0.81	2191
			0.92	2457
			1.33	2966
			1.45	2940
			1.54	2688
			1.56	Incomplete
			1.58	Incomplete
Explosif S	paper	29 mm.	1.05	2335
			1.16	2443
			1.29	2443
			1.39	Incomplete
			1.47	Incomplete
Cheddite 60 4th formula	copper	20–22 mm.	1.51	3099
			1.62	2820
			0.84	2457
			1.39	3045
			1.48	3156
Cheddite 60 4th formula	paper	29 mm.	1.25	2774
			1.31	2915
			1.40	2843
			1.50	Incomplete

Minélite A in powder	copper	20–22 mm.	0.87	2800
			0.99	2930
			1.17	3125
			1.24	3235
			1.38	Incomplete
			1.52	Incomplete
			0.89	2435
			0.95	2835
			1.20	3235
			1.39	3125
			1.45	Incomplete
			0.87	2395
			1.27	3355
			1.39	Incomplete
Minélite A in powder	paper	29 mm.	1.08	2670
			1.19	2835
			1.25	Incomplete
			1.28	Incomplete
			1.19	2895
			1.24	Incomplete
Minélite A in grains	copper	20–22 mm.	0.87	2150
			1.12	2415
			1.20	2550
			1.29	3025
			1.33	2480
			1.35	Incomplete
			1.30	2895
			0.85	2100
			1.17	2415
			1.27	2750
Minélite B in powder	copper	20–22 mm.	0.97	2350
			1.07	2895
			1.24	3235
			1.33	3090
			1.45	Incomplete
			1.57	Inçomplete
			1.00	2925
			1.12	2925
			1.26	3165
			1.02	2585
			1.14	2910
			1.30	3180
			1.41	Complete
			1.38	3160
Minélite C in powder	copper	20–22 mm.	1.28	3125
			1.37	Incomplete
			1.48	Incomplete

below. It is interesting that the explosive which contained a small amount of nitroglycerin was more brisant, as well as softer and more plastic, and less sensitive to shock, to friction, and to initiation than the drier explosive which contained no nitroglycerin. It required a No. 3 blasting cap to explode it, but the material which contained no nitroglycerin was exploded by a weak No. 1.

	GESTEINS-KORONIT T1	GESTEINS-KORONIT T2
Sodium chlorate	72.0	75.0
Vegetable meal	1.0–2.0	1.0–2.0
Di- and trinitrotoluene	20.0	20.0
Paraffin	3.0–4.0	3.0–4.0
Nitroglycerin	3.0–4.0	...
Heat of explosion, Cal./kg.	1219.0	1241.0
Temperature of explosion	3265.0°	3300.0°
Velocity of detonation, m./sec.	5000.0	4300.0
Density of cartridge	1.57	1.46
Lead block expansion	290.0 cc.	280.0 cc.
Lead block crushing	20.0 mm.	19.5 mm.

During the first World War when Germany needed to conserve as much as possible its material for military explosives, blasting explosives made from perchlorate came into extensive use. The Germans had used in their trench mortar bombs an explosive, called *Perdit,* which consisted of a mixture of potassium perchlorate 56%, with dinitrobenzene 32% and dinitronaphthalene 12%. After the War, the perchlorate recovered from these bombs and that from the reserve stock came onto the market, and perchlorate explosives, *Perchlorit, Perchloratit, Persalit, Perkoronit,* etc., were used more widely than ever. The sale of these explosives later ceased because the old supply of perchlorate became exhausted and the new perchlorate was too high in price. Each of these explosives required a No. 3 cap for its initiation. Perchlorate explosives in general are somewhat less sensitive to initiation than chlorate explosives. A small amount of nitroglycerin in perchlorate explosives plays a significant part in propagating the explosive wave and is more important in these compositions than it is in ammonium nitrate explosives. Naoúm [44] reports the following particulars concerning two of the Perkoronites.

[44] *Op. cit.,* p. 430.

	PERKORONIT A	PERKORONIT B
Potassium perchlorate	58	59
Ammonium nitrate	8	10
Di- and trinitrotoluene, vegetable meal....	30	31
Nitroglycerin	4	..
Heat of explosion, Cal./kg.	1170.0	1160.0
Temperature of explosion	3145.0°	3115.0°
Velocity of detonation, m./sec.	5000.0	4400.0
Density of cartridge	1.58	1.52
Lead block expansion	340.0 cc.	330.0 cc.
Lead block crushing	20.0 mm.	18.0 mm.

Potassium perchlorate and ammonium perchlorate permissible explosives, cooled by means of common salt, ammonium oxalate, etc., and containing either ammonium nitrate or alkali metal nitrate with or without nitroglycerin, are used in England, Belgium, and elsewhere. They possess no novel features beyond the explosives already described. Explosives containing ammonium perchlorate yield fumes which contain hydrogen chloride. Potassium perchlorate produces potassium chloride.

Early in the history of these explosives the French *Commission des Substances Explosives* published a report on two ammonium perchlorate Cheddites.[45] The manufacture of these explosives,

	I	II
Ammonium perchlorate	82	50
Sodium nitrate	..	30
Dinitrotoluene	13	15
Castor oil	5	5

however, was not approved for the reason that the use of castor oil for phlegmatizing was found to be unnecessary. Number I took fire easily and burned in an 18-mm. copper gutter at a rate of 4.5 mm. per second, and produced a choking white smoke. Cheddite 60, for comparison, burned irregularly in the copper gutter, with a smoke which was generally. black, at a rate of 0.4–0.5 mm. per second. Number II took fire only with the greatest difficulty, and did not maintain its own combustion. The maximum velocities of detonation in zinc tubes 20 mm. in diameter were about 4020 meters per second for No. I and about 3360 for No. II.

[45] *Mém. poudres,* **14,** 206 (1907–1908).

The *Commission* published in the same report a number of interesting observations on ammonium perchlorate. Pieces of cotton cloth dipped into a solution of ammonium perchlorate and dried were found to burn more rapidly than when similarly treated with potassium chlorate and less rapidly than when similarly treated with sodium chlorate. Ammonium perchlorate inflamed in contact with a hot wire and burned vigorously with the production of choking white fumes, but the combustion ceased as soon as the hot wire was removed. Its sensitivity to shock, as determined by the drop test, was about the same as that of picric acid, but its sensitivity to initiation was distinctly less. A 50-cm. drop of a 5-kilogram weight caused explosions in about 50% of the trials. A cartridge, 16 cm. long and 26 mm. in diameter, was filled with ammonium perchlorate gently tamped into place (density of loading about 1.10) and was primed with a cartridge of the same diameter containing 25 grams of powdered picric acid (density of loading about 0.95) and placed in contact with one end of it. When the picric acid booster was exploded, the cartridge of perchlorate detonated only for about 20 mm. of its length and produced merely a slight and decreasing furrow in the lead plate on which it was resting. When a booster of 75 grams of picric acid was used, the detonation was propagated in the perchlorate for 35 mm. The temperature of explosion of ammonium perchlorate was calculated to be 1084°.

The French used two ammonium perchlorate explosives during the first World War.

	I	II
Ammonium perchlorate	86	61.5
Sodium nitrate	..	30.0
Paraffin	14	8.5

The first of these was used in 75-mm. shells, the second in 58-mm. trench mortar bombs.

Hydrazine perchlorate melts at 131–132°, burns tranquilly, and explodes violently from shock.

Guanidine perchlorate is relatively stable to heat and to mechanical shock but possesses extraordinary explosive power and sensitivity to initiation. Naoúm [46] states that it gives a lead block expansion of about 400 cc. and has a velocity of detonation of about 6000 meters per second at a density of loading of 1.15.

[46] Naoúm, "Schiess- und Sprengstoffe," Dresden and Leipzig, 1927, p. 137.

Ammonium Nitrate Military Explosives

The *Schneiderite* (*Explosif S* or *Sc*) which the French used during the first World War in small and medium-size high-explosive shells, especially in the 75 mm., was made by incorporating 7 parts of ammonium nitrate and 1 of dinitronaphthalene in a wheel mill, and was loaded by compression. Other mixtures, made in the same way, were used in place of Schneiderite or as a substitute for it.

	NX	NT	NTN	NDNT	N2TN
Ammonium nitrate	77	70	80	85	50
Sodium nitrate	..	..	..	..	30
Trinitrotoluene	..	30	..	5	..
Trinitroxylene	23	..	..	..	..
Dinitronaphthalene	..	..	..	10	..
Trinitronaphthalene	..	..	20	..	20

Amatol, developed by the British during the first World War, is made by mixing granulated ammonium nitrate with melted trinitrotoluene, and pouring or extruding the mixture into the shells where it solidifies. The booster cavity is afterwards drilled out from the casting. The explosive can be cut with a hand saw. It is insensitive to friction and is less sensitive to initiation and more sensitive to impact than trinitrotoluene. It is hygroscopic, and in the presence of moisture attacks copper, brass, and bronze.

Amatol is made up in various proportions of ammonium nitrate to trinitrotoluene, such as 50/50, 60/40, and 80/20. The granulated, dried, and sifted ammonium nitrate, warmed to about 90°, is added to melted trinitrotoluene at about 90°, and the warm mixture, if 50/50 or 60/40, is ladled into the shells which have been previously warmed somewhat in order that solidification may not be too rapid, or, if 80/20, is *stemmed* or extruded into the shells by means of a screw operating within a steel tube. Synthetic ammonium nitrate is preferred for the preparation of amatol. The pyridine which is generally present in gas liquor and tar liquor ammonia remains in the ammonium nitrate which is made from these liquors and causes frothing and the formation of bubbles in the warm amatol—with the consequent probability of cavitation in the charge. Thiocyanates which are often present in ammonia from the same sources likewise cause frothing, and phenols if present tend to promote exudation.

The velocity of detonation of TNT-ammonium nitrate mixtures decreases regularly with increasing amounts of ammonium nitrate, varying from about 6700 meters per second for TNT to about 4500 meters per second for 80/20 amatol. The greater the proportion of ammonium nitrate the less the brisance and the greater the heaving power of the amatol. 50/50 Amatol does not contain oxygen enough for the complete combustion of its trinitrotoluene, and gives a smoke which is dark colored but less black than the smoke from straight TNT. 80/20 Amatol is less brisant than TNT. It gives an insignificant white smoke. Smoke boxes are usually loaded with 80/20 amatol in order that the artilleryman may observe the bursting of his shells. The best smoke compositions for this purpose contain a large proportion of aluminum and provide smoke by day and a brilliant flash of light by night.

The name of *ammonal* is applied both to certain blasting explosives which contain aluminum and to military explosives, based upon ammonium nitrate, which contain this metal. Military ammonals are brisant and powerful explosives which explode with a bright flash. They are hygroscopic, but the flake aluminum which they contain behaves somewhat in the manner of the shingles on a roof and helps materially to exclude moisture. At the beginning of the first World War the Germans were using in major caliber shells an ammonal having the first of the compositions listed below. After the War had advanced and TNT

	GERMAN I	AMMONAL II	FRENCH AMMONAL
Ammonium nitrate	54	72	86
Trinitrotoluene	30	12	..
Aluminum flakes	16	16	8
Stearic acid	..	..	6

had become more scarce, ammonal of the second formula was adopted. The French also used ammonal in major caliber shells during the first World War. All three of the above-listed explosives were loaded by compression. Experiments have been tried with an ammonal containing ammonium thiocyanate; the mixture was melted, and loaded by pouring but was found to be unsatisfactory because of its rapid decomposition. Ammonal yields a flame which is particularly hot, and consequently gives an unduly high result in the Trauzl lead block test.

CHAPTER VIII

NITROAMINES AND RELATED SUBSTANCES

The nitroamines are substituted ammonias, substances in which a nitro group is attached directly to a trivalent nitrogen atom. They are prepared in general either by the nitration of a nitrogen base or of one of its salts, or they are prepared by the splitting off of water from the nitrate of the base by the action of concentrated sulfuric acid upon it. At present two nitroamines are of particular interest to the explosives worker, namely, nitroguanidine and cyclotrimethylenetrinitramine (cyclonite). Both are produced from synthetic materials which have become available in large commercial quantities only since the first World War, the first from cyanamide, the second from formaldehyde from the oxidation of synthetic methyl alcohol.

Nitroamide (Nitroamine)

Nitroamide, the simplest of the nitroamines, is formed by the action of dilute acid on potassium nitrocarbamate, which itself results from the nitration of urethane and the subsequent hydrolysis of the nitro ester by means of alcoholic potassium hydroxide.

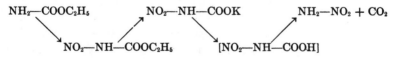

$$NH_2—COOC_2H_5 \qquad NO_2—NH—COOK \qquad NH_2—NO_2 + CO_2$$
$$NO_2—NH—COOC_2H_5 \qquad [NO_2—NH—COOH]$$

Nitroamide is strongly acidic, a white crystalline substance, melting at 72–73° with decomposition, readily soluble in water, alcohol, and ether, and insoluble in petroleum ether. It explodes on contact with concentrated sulfuric acid. The pure material decomposes slowly on standing, forming nitrous oxide and water; it cannot be preserved for more than a few days. When an aqueous solution of nitroamide is warmed, gas bubbles begin to come off at about 60–65°, and decomposition is complete after boiling for a short time.

The solution which results when ammonium nit ate is dissolved in a large excess of concentrated sulfuric acid evidently contains nitroamide.[1] If the solution is warmed directly, no nitric acid distils from it but at about 150° it gives off nitrous oxide which corresponds to the dehydration of the nitroamide by the action of the strong acid. The nitroamide moreover, by the action of the same acid, may be hydrated to yield nitric acid, slowly if the solution is digested at 90° to 120°, under which conditions the nitric acid distils out, and rapidly at ordinary temperature in the nitrometer where mercury is present which reacts with the nitric acid as fast as it is formed.

$$NH_3 \cdot HONO_2 \text{ minus } H_2O \longrightarrow NH_2-NO_2 \begin{cases} \text{minus } H_2O \longrightarrow N_2O \\ \text{plus} \quad H_2O \longrightarrow NH_3 + HONO_2 \end{cases}$$

The two reactions, hydration and dehydration, or, more exactly, the formation of nitrous oxide and of nitric acid, are more or less general reactions of the substituted nitroamines. The extent to which one or the other occurs depends largely upon the groups which are present in the molecule. Thus, tetryl on treatment with concentrated sulfuric acid forms nitric acid, and it gives up one and only one of its nitro groups in the nitrometer, but the reaction is not known by which nitrous oxide is eliminated from it. Methylnitramine, on the other hand, gives nitrous oxide readily enough but shows very little tendency to produce nitric acid.

Solutions of nitrourea and nitroguanidine in concentrated sulfuric acid contain actual nitroamide, and these substances give up their nitro group nitrogen in the nitrometer. Nitroamide has been isolated [2] both from an aqueous solution of nitrourea and from a solution of the same substance in concentrated sulfuric acid.

$$NH_2-CO-NH-NO_2 \rightleftharpoons HNCO + NH_2-NO_2$$

The reaction is reversible, for nitroamide in aqueous solution combines with cyanic acid to form nitrourea.

[1] Davis and Abrams, *J. Am. Chem. Soc.*, **47**, 1043 (1925).
[2] Davis and Blanchard, *J. Am. Chem. Soc.*, **51**, 1790 (1929).

Methylnitramine

Methylnitramine is produced when aniline reacts with tetryl in benzene solution, and when ammonia water or barium hydroxide solution acts upon dinitrodimethyloxamide. The structure of tetryl was first proved by its synthesis from picryl chloride and the potassium salt of methylnitramine.

Methylnitramine is a strong monobasic acid, very readily soluble in water, alcohol, chloroform, and benzene, less soluble in ether, and sparingly soluble in petroleum ether. It crystallizes from ether in flat needles which melt at 38°. It is not decomposed by boiling in aqueous solution even in the presence of an excess of alkali. On distillation it yields dimethylnitramine, m.p. 57°, methyl alcohol, nitrous oxide and other products. Methylnitramine owes its acidity to the fact that it is tautomeric.

$$CH_3-N\begin{matrix} H \\ NO_2 \end{matrix} \rightleftharpoons CH_3-N=N\begin{matrix} O \\ OH \end{matrix}$$

Dimethylnitramine, in which there is no hydrogen atom attached to the atom which carries the nitro group, cannot tautomerize, and is not acidic.

Methylnitramine decomposes explosively in contact with concentrated sulfuric acid. If the substance is dissolved in water, and if concentrated sulfuric acid is added little by little until a considerable concentration is built up, then the decomposition proceeds more moderately, nitrous oxide is given off, and dimethyl ether (from the methyl alcohol first formed) remains dissolved in the sulfuric acid. The same production of nitrous oxide occurs even in the nitrometer in the presence of mercury. If methylnitramine and a small amount of phenol are dissolved together in water, and if concentrated sulfuric acid is then added little by little, a distinct yellow color shows that a trace of nitric acid has been formed. The fact that methylnitramine gives a blue color with the diphenylamine reagent shows the same thing.

Methylnitramine is conveniently prepared [3] by nitrating methylurethane with absolute nitric acid, drowning in water, neutralizing with sodium carbonate, extracting with ether, and

[3] Franchimont and Klobbie, *Rec. trav. chim.*, **7**, 354 (1887).

then passing ammonia gas into the ether solution of methyl-nitrourethane,

$$CH_3-NH-COOC_2H_5$$

$$\begin{array}{c} H\!\!\shortmid\!\!-NH_2 \\ | \\ CH_3-N\!\!\shortmid\!\!-COOC_2H_5 + NH_3 \\ | \\ NO_2 \end{array}$$

$$CH_3-N-COOC_2H_5 \qquad CH_3-NH-NO_2\cdot NH_3$$
$$\quad | \qquad\qquad\qquad\qquad\qquad$$
$$\quad NO_2 \qquad\qquad\qquad CH_3-N\!\!<\!\!\begin{array}{c}H\\NO_2\end{array}$$

A white crystalline precipitate of the ammonium salt of methyl-nitramine is deposited. This is dissolved in alcohol, and the solution is boiled—whereby ammonia is driven off—and concentrated to a small volume. The product is procured by completing the evaporation in a vacuum desiccator over sulfuric acid.

The heavy metal salts of methylnitramine are primary explosives, but have not been investigated extensively.

Urea Nitrate

Although urea has the properties of an amide (carbamide) rather than those of an amine, it nevertheless acts as a monoacid base in forming salts among which the nitrate and the oxalate are noteworthy because they are sparingly soluble in cold water, particularly in the presence of an excess of the corresponding acid. The nitrate, white monoclinic prisms which melt at 152° with decomposition, is procured by adding an excess of nitric acid (1.42) to a strong aqueous solution of urea. The yield is increased if the mixture is chilled and allowed to stand for a time. Urea nitrate is stable and not deliquescent. It has interest as a powerful and cool explosive, but suffers from the disadvantage that it is corrosively acidic in the presence of moisture.

Pure urea is manufactured commercially by pumping ammonia and carbon dioxide into an autoclave where they are heated together under pressure while more of each gas is pumped in. Ammonium carbamate is formed at first, this loses water from its molecule to form urea, and the autoclave finally becomes filled with a strong solution of urea which is drawn off and crystallized.

$$2NH_3 + CO_2 \longrightarrow NH_3 \cdot HO—CO—NH_2 \longrightarrow H_2O + NH_2—CO—NH_2$$

Urea is sometimes incorporated in blasting explosives for the purpose of lowering the temperature of explosion. Its use as a stabilizer has already been mentioned.

Nitrourea

Nitrourea is a cool but powerful explosive, and would be useful if it were not for the fact that it tends to decompose spontaneously in the presence of moisture. The mechanism of its reactions is the same as that of the reactions of nitroguanidine, which differs from it in containing an $>NH$ group where nitrourea contains a $>CO$, but the reactions of nitrourea are very much more rapid. The nitro group promotes the *urea dearrangement,* so that nitrourea when dissolved in water or when warmed breaks down into cyanic acid and nitroamide much more readily than urea breaks down under like conditions into cyanic acid and ammonia. The imido group in place of the carbonyl hinders it; guanidine dearranges less readily than urea, and nitroguanidine is substantially as stable as urea itself.

Nitrourea is prepared by adding dry urea nitrate (200 grams) in small portions at a time with gentle stirring to concentrated sulfuric acid (1.84) (300 cc.) while the temperature of the mixture is kept below 0°. The milky liquid is poured without delay into a mixture of ice and water (1 liter), the finely divided white precipitate is collected on a filter, sucked as dry as may be, and, without washing, is immediately dissolved while still wet in boiling alcohol.[4] The liquid on cooling deposits pearly leaflets of nitrourea. It is chilled and filtered, and the crystals are rinsed with cold alcohol and dried in the air. The product, which melts at 146° to 153° with decomposition, is sufficiently pure for use in synthesis, and may be preserved for several years unchanged in hard glass bottles. If slightly moist nitrourea is allowed to stand in contact with soft glass, that is, in contact with a trace

[4] The product at this point contains acid enough to prevent it from decomposing in boiling alcohol. For a second recrystallization it is unsafe to heat the alcohol above 60°.

of alkali, it decomposes completely within a short time forming water, ammonia, nitrous oxide, urea, biuret, cyanuric acid, etc. Pure nitrourea, recrystallized from benzene, ether, or chloroform, in which solvents it is sparingly soluble, melts with decomposition at 158.4–158.8°.

In water and in hydrophilic solvents nitrourea dearranges rapidly into cyanic acid and nitroamide. Alkalis promote the reaction. If an aqueous solution of nitrourea is warmed, bubbles of nitrous oxide begin to come off at about 60°. If it is allowed to stand over night at room temperature, the nitrourea disappears completely and the liquid is found to be a solution of cyanic acid. Indeed, nitrourea is equivalent to cyanic acid for purposes of synthesis. It reacts with alcohols to form carbamic esters (urethanes) and with primary and second amines to form mono- and *unsym*-di-substituted ureas.

Guanidine Nitrate

Guanidine nitrate is of interest to us both as an explosive itself and a component of explosive mixtures, and as an intermediate in the preparation of nitroguanidine. All other salts of guanidine require strong mixed acid to convert them to nitroguanidine, but the nitrate is converted by dissolving it in concentrated sulfuric acid and pouring the solution into water.

Guanidine is a strong monoacid base, indistinguishable from potassium hydroxide in an electrometric titration. There is considerable evidence [5] which indicates that the charge of the guanidonium ion resides upon its carbon atom.

$$\underset{\text{Guanidine}}{\overset{\displaystyle NH_2}{\underset{\displaystyle NH}{NH_2\!-\!\overset{|}{\underset{\|}{C}}}}} \; + \; H^+ \; \rightleftharpoons \; \underset{\text{Guanidonium ion}}{\overset{\displaystyle NH_2}{\underset{\displaystyle NH_2}{NH_2\!-\!\overset{|}{\underset{|}{C^+}}}}}$$

Guanidine itself is crystalline, deliquescent, and strongly caustic, and takes up carbon dioxide from the air.

Guanidine was first obtained by Strecker in 1861 by the oxidation with hydrochloric acid and potassium chlorate of guanine (a substance found in guano and closely related to uric acid).

[5] Davis, Yelland, and Ma, *J. Am. Chem. Soc.*, **59**, 1993 (1937).

Guanidine or its salts may be prepared, among other ways, by the interaction (1) of orthocarbonic ester or (2) of chloropicrin

1. $CCl_3 \cdot NO_2 + 3NH_3 \longrightarrow NH_2-C(NH)-NH_2 + HNO_2 + 3HCl$

2. $C(OC_2H_5)_4 + 3NH_3 \longrightarrow NH_2-C(NH)-NH_2 + 4C_2H_5-OH$

with aqueous ammonia at 150°, by the interaction (3) of carbon tetrabromide with alcoholic ammonia in a sealed tube at 100°,

3. $CBr_4 + 3NH_3 \longrightarrow NH_2-C(NH)-NH_2 + 4HBr$

by the interaction (4) of cyanogen iodide with alcoholic ammonia in a sealed tube at 100°, whereby cyanamide and ammonium iodide are formed first and then combine with one another to

4. $I-C\equiv N + 2NH_3 \longrightarrow NH_2-C\equiv N + NH_3 \cdot HI \longrightarrow$
$$NH_2-C(NH)-NH_2 \cdot HI$$

form guanidine iodide, by the combination (5) of cyanamide, already prepared, with an ammonium salt by heating the materials with alcohol in a sealed tube at 100°, and (6) by heating

6. $NH_4NCS \rightleftharpoons NH_3 + HNCS \rightleftharpoons NH_2-CS-NH_2 \rightleftharpoons NH_2-C\equiv N + H_2S$

$NH_4NCS + NH_2-C\equiv N \longrightarrow NH_2-C(NH)-NH_2 \cdot HNCS$

ammonium thiocyanate at 170–190° for 20 hours, or until hydrogen sulfide no longer comes off, whereby the material is converted into guanidine thiocyanate. The reaction depends upon the fact that the ammonium thiocyanate is in part converted into thiourea, and that this breaks down into hydrogen sulfide, which escapes, and cyanamide which combines with the unchanged ammonium thiocyanate to form the guanidine salt. The yield from this process is excellent.

For many years guanidine thiocyanate was the most easily prepared and the most commonly used of the salts of guanidine. Other salts were made from it by metathetical reactions. Nitroguanidine, prepared from the thiocyanate by direct nitration with mixed acids, was found to contain traces of sulfur compounds which attacked nitrocellulose and affected the stability of smokeless powder, and this is one of the reasons why nitroguanidine powders did not come into early use. Guanidine thiocyanate is deliquescent. Strong solutions of it dissolve filter paper.

Cyanamide itself is not a suitable raw material for the preparation of guanidine salts, for it is difficult to prepare and to purify, and it polymerizes on keeping. The evaporation of an aqueous solution of cyanamide yields the dimer, dicyandiamide, and the heating, or even the long keeping, of the dry substance produces the trimer, melamine.

$$NH_2—C(NH)—NH—C≡N$$
Dicyandiamide

$$NH_2—C≡N$$
Cyanamide

$$HN=C \quad C=NH$$

NH
‖
C
⟨ NH   NH
HN=C   C=NH
⟩ NH
Melamine

Cyanamide, colorless crystals, m.p. 40°, is readily soluble in water, alcohol, and ether. An aqueous solution of cyanamide gives a black precipitate of copper cyanamide with ammoniacal copper sulfate solution, and a yellow precipitate of silver cyanamide with ammoniacal silver nitrate. The precipitates are almost unique among the compounds of copper and silver in the respect that they are insoluble in ammonia water.

Before the development of the cyanamide process for the fixation of nitrogen, cyanamide was prepared by the interaction of cyanogen chloride or bromide (from the action of the halogen on potassium cyanide) with ammonia in water or ether solution.

$$KCN + Cl_2 \longrightarrow KCl + Cl—CN$$
$$2NH_3 + Cl—CN \longrightarrow NH_4Cl + NH_2—CN$$

If the reaction, say, with cyanogen chloride, is carried out in ether solution, ammonium chloride precipitates and is filtered off, and the cyanamide is procured as a syrup by allowing the ether solution to evaporate spontaneously and later as crystals by allowing the syrup to stand over sulfuric acid in a desiccator. Cyanamide may also be prepared by removing the component atoms of hydrogen sulfide from thiourea by means of mercuric oxide. Thionyl chloride effects the corresponding removal of water from urea.

$$NH_2—CS—NH_2 \text{ minus } H_2S \text{ (HgO)} \longrightarrow NH_2—CN + HgS + H_2O$$
$$NH_2—CO—NH_2 \text{ minus } H_2O \text{ (Cl}_2SO) \longrightarrow NH_2—CN + SO_2 + 2HCl$$

The cyanamide process has made cyanamide and its derivatives more easily available for commercial synthesis. Coke and limestone are heated together in the electric furnace for the production of calcium carbide. This substance, along with a small amount of calcium chloride which acts as a catalyst, is then heated at 800–1000° in a stream of nitrogen gas.

$$2CaCO_3 + 5C \longrightarrow 2Ca{\Large\langle}^C_C{\scriptstyle|||} + 3CO_2$$

$$Ca{\Large\langle}^C_C{\scriptstyle|||} + N_2 \longrightarrow CaNCN + C$$

The resulting dark-colored mixture of calcium cyanamide and carbon is known as *lime nitrogen (Kalkstickstoff)* and is used in fertilizers. If steam is passed through it, it yields ammonia.

$$CaNCN + 3H_2O \text{ (steam)} \longrightarrow CaCO_3 + 2NH_3$$

Water, whether cool or warm, produces some cyanamide, which is readily soluble, and some calcium hydrogen cyanamide, white, microcrystalline, and sparingly soluble, but water plus acid for the removal of the calcium (sulfuric acid, oxalic acid, or carbon dioxide) yields a solution of cyanamide which is directly applicable for use in certain reactions.

$$2CaNCN + 2H_2O \longrightarrow Ca(OH)_2 + Ca{\Large\langle}^{NH-CN}_{NH-CN}$$

$$Ca{\Large\langle}^{NH-CN}_{NH-CN} + CO_2 + H_2O \longrightarrow CaCO_3 + 2NH_2-CN$$

On hydrolysis with dilute sulfuric acid it yields urea. On treatment with ammonium sulfide it prefers to react with the hydrogen sulfide part of the molecule to form thiourea, not with the ammonia part to form guanidine, and the reaction is the commercial source of many tons of thiourea for the rubber industry. On evaporation for crystals, the solution yields dicyandiamide which constitutes a convenient source for the preparation of guanidine nitrate.

Dicyandiamide crystallizes from water in handsome flat needles or plates which melt at 208.0–208.1° and decompose if heated slightly above the melting point. A saturated aqueous solution contains—

TEMPERATURE, °C.	GRAMS PER 100 CC. OF SOLUTION
0	1.3
10	2.0
20	3.4
30	5.0
40	7.6
50	11.4
60	16.1
70	22.5
80	30.0
90	37.9
100	46.7

The preparation of guanidine nitrate from dicyandiamide by the action of aqua regia has been patented,[6] but the reaction evidently depends solely upon the hydrolysis of the cyan group and does not require the use of a vigorous oxidizing agent. Marqueyrol and Loriette in a French patent of September 26, 1917, described a process for the preparation of nitroguanidine direct from dicyandiamide without the isolation of any intermediate products. The process depends upon the hydrolysis of the dicyandiamide by means of 61% sulfuric acid to form guanylurea or dicyandiamidine (sulfate) which is then further hydrolyzed to form carbon dioxide, which escapes, and guanidine and ammonia, which remain in the reaction mixture in the form of sulfates.

$$NH_2—C(NH)—NH—CN + H_2O \longrightarrow$$
Dicyandiamide

$$NH_2—C(NH)—NH—CO—NH_2 + H_2O \longrightarrow$$
Guanylurea

$$NH_2—C(NH)—NH_2 + CO_2 + NH_3$$
Guanidine

The guanidine sulfate, without removal from the mixture, is then nitrated to nitroguanidine.[7] The process yields a nitroguanidine which is suitable for use in nitrocellulose powder, but it suffers from the disadvantages that the dicyandiamide, which corresponds after all to two molecules of cyanamide, yields in theory

[6] Ulpiani, Ger. Pat. 209,431 (1909).

[7] The procedure, under conditions somewhat different from those described in the patent, is illustrated by our process for the preparation of β-nitroguanidine; see page 383.

only one molecular equivalent of guanidine, that the actual yield is considerably less than the theory because of the loss of guanidine by hydrolysis to carbon dioxide and ammonia, and that the final nitration of the guanidine sulfate, which is carried out in the presence of water and of ammonium sulfate, requires strong and expensive mixed acid.

Werner and Bell [8] reported in 1920 that dicyandiamide heated for 2 hours at 160° with 2 mols of ammonium thiocyanate gives 2 mols of guanidine thiocyanate in practically theoretical yield. Ammonium thiocyanate commends itself for the reaction because it is readily fusible. The facts suggest that another fusible ammonium salt might work as well, ammonium nitrate melts at about 170°, and, of all the salts of guanidine, the nitrate is the one which is most desired for the preparation of nitroguanidine. When dicyandiamide and 2 mols of ammonium nitrate are mixed and warmed together at 160°, the mixture first melts to a colorless liquid which contains biguanide (or guanylguanidine) nitrate, which presently begins to deposit crystals of guanidine nitrate, and which after 2 hours at 160° solidifies completely to a mass of that substance.[9] The yield is practically theoretical. The reaction consists, first, in the addition of ammonia to the cyan group of the dicyandiamide, then in the ammoniolytic splitting of the biguanide to form two molecules of guanidine.

$$NH_2—C(NH)—NH—CN + NH_3 \cdot HNO_3 \longrightarrow$$
Dicyandiamide

$$NH_2—C(NH)—NH—C(NH)—NH_2 \cdot HNO_3 + NH_3 \cdot HNO_3 \longrightarrow$$
Biguanide nitrate

$$2NH_2—C(NH)—NH_2 \cdot HNO_3$$
Guanidine nitrate

The nitric acid of the original 2 mols of ammonium nitrate is exactly sufficient for the formation of 2 mols of guanidine nitrate. But the intermediate biguanide is a strong diacid base; the ammonium nitrate involved in its formation supplies only one equivalent of nitric acid; and there is a point during the early part of the process when the biguanide mononitrate tends to attack the unchanged ammonium nitrate and to liberate ammonia from it. For this reason the process works best if a small excess of

[8] J. Chem. Soc., 118, 1133 (1920).

[9] Davis, J. Am. Chem. Soc., 43, 2234 (1921); Davis, U. S. Pat. 1,440,063 (1922), French Pat. 539,125 (1922).

ammonium nitrate is used. The preparation may be carried out by heating the materials together either in the dry state or in an autoclave in the presence of water or of alcohol.

Guanidine nitrate is not deliquescent. It is readily soluble in alcohol, very readily in water, and may be recrystallized from either solvent. The pure material melts at 215–216°.

Preparation of Guanidine Nitrate. An intimate mixture of 210 grams of dicyandiamide and 440 grams of ammonium nitrate is placed in a 1 liter round-bottom flask, and the flask is arranged for heating in an oil bath which has a thermometer in the oil. The oil bath is warmed until the thermometer indicates 160°, and the temperature is held at this point for 2 hours. At the end of that time the flask is removed and allowed to cool, and its contents is extracted on the steam bath by warming with successive portions of water. The combined solution is filtered while hot for the removal of white insoluble material (ammeline and ammelide), concentrated to a volume of about a liter, and allowed to crystallize. The mother liquors are concentrated to a volume of about 250 cc. for a second crop, after the removal of which the residual liquors are discarded. The crude guanidine nitrate may be recrystallized by dissolving it in the least possible amount of boiling water and allowing to cool, etc., or it may be dried thoroughly and used directly for the preparation of nitroguanidine. A small amount of ammonium nitrate in it does not interfere with its conversion to nitroguanidine by the action of concentrated sulfuric acid.

Nitroguanidine

Nitroguanidine exists in two forms.[10] The α-form invariably results when guanidine nitrate is dissolved in concentrated sulfuric and the solution is poured into water. It is the form which is commonly used in the explosives industry. It crystallizes from water in long, thin, flat, flexible, lustrous needles which are tough and extremely difficult to pulverize; $N_\alpha = 1.518$, $N_\beta = $ a little greater than 1.668, $N_\gamma = $ greater than 1.768, double refraction 0.250. When α-nitroguanidine is decomposed by heat, a certain amount of β-nitroguanidine is found among the products.

β-Nitroguanidine is produced in variable amount, usually along with some of the α-compound, by the nitration of the mixture of guanidine sulfate and ammonium sulfate which results from the hydrolysis of dicyandiamide by sulfuric acid. Conditions have

[10] Davis, Ashdown, and Couch, *J. Am. Chem. Soc.*, **47**, 1063 (1925).

been found, as described later, which have yielded exclusively the β-compound in more than thirty trials. It crystallizes from water in fernlike clusters of small, thin, elongated plates; $N_\alpha = 1.525$, N_β not determined, $N_\gamma = 1.710$, double refraction 0.185. It is converted into the α-compound by dissolving in concentrated sulfuric acid and pouring the solution into water.

Both α- and β-nitroguanidine, if dissolved in hot concentrated nitric acid and allowed to crystallize, yield the same nitrate, thick, rhomb-shaped prisms which melt at 147° with decomposition. The nitrate loses nitric acid slowly in the air, and gives α-nitroguanidine when recrystallized from water. Similarly, both forms recrystallized from strong hydrochloric acid yield a hydrochloride which crystallizes in needles. These lose hydrogen chloride rapidly in the air, and give α-nitroguanidine when recrystallized from water. The two forms are alike in all their chemical reactions, in their derivatives and color reactions.

Both forms of nitroguanidine melt at 232° if the temperature is raised with moderate slowness, but by varying the rate of heating melting points varying between 220° and 250° may be obtained.

Neither form can be converted into the other by solution in water, and the two forms can be separated by fractional crystallization from this solvent. They appear to differ slightly in their solubility in water, the two solubility curves lying close together but apparently crossing each other at about 25°, where the solubility is about 4.4 grams per liter, and again at about 100°, where the solubility is about 82.5 grams per liter. Between these temperatures the β-form appears to be the more soluble.

Preparation of α-*Nitroguanidine.* Five hundred cc. of concentrated sulfuric acid in a 1-liter beaker is cooled by immersing the beaker in cracked ice, and 400 grams of well-dried guanidine nitrate is added in small portions at a time, while the mixture is stirred with a thermometer and the temperature is not allowed to rise above 10°. The guanidine nitrate dissolves rapidly, with very little production of heat, to form a milky solution. As soon as all crystals have disappeared, the milky liquid is poured into 3 liters of cracked ice and water, and the mixture is allowed to stand with chilling until precipitation and crystallization are complete. The product is collected on a filter, rinsed with water for the removal of sulfuric acid, dissolved in boiling water (about 4 liters), and allowed to crystallize by standing over night. Yield 300–310 grams, about 90% of the theory.

The rapid cooling of a solution of α-nitroguanidine produces small needles, which dry out to a fluffy mass but which are still too coarse to be incorporated properly in colloided powder. An

FIGURE 89. α-Nitroguanidine (25×). Small crystals from the rapid cooling of a hot aqueous solution.

extremely fine powder may be procured by the rapid cooling of a mist or spray of hot nitroguanidine solution, either by spraying it against a cooled surface from which the material is removed continuously, or by allowing the spray to drop through a tower up which a counter current of cold air is passing.

Preparation of β-Nitroguanidine. Twenty-five cc. of 61% aqueous sulfuric acid is poured upon 20 grams of dicyandiamide contained in a 300-cc. round-bottom flask equipped with a reflux condenser. The mixture warms up and froths considerably. After the first vigorous reaction has subsided, the material is heated for 2 hours in an oil bath at 140° (thermometer in the oil). The reaction mass, chilled in a freezing mixture, is treated with ice-cold nitrating acid prepared by mixing 20 cc. of fuming nitric acid (1.50) with 10 cc. of concentrated sulfuric acid (1.84). After the evolution of red fumes has stopped, the mixture is heated for 1 hour in the boiling-water bath, cooled, and drowned in 300 cc. of cracked ice and water. The precipitate, collected on a filter, rinsed with water for the removal of acid, and recrystallized from water, yields about 6 grams of β-nitroguanidine, about 25% of the theory.

Saturated solutions of nitroguanidine in sulfuric acid of various concentrations contain [11] the amounts indicated below.

CONCENTRATION OF SOLVENT SULFURIC ACID, %	NITROGUANIDINE (GRAMS) PER 100 CC.	
	at 0°	at 25°
45	5.8	10.9
40	3.4	8.0
35	2.0	5.2
30	1.3	2.9
25	0.75	1.8
20	0.45	1.05
15	0.30	0.55
0	0.12	0.42

Nitroguanidine on reduction is converted first into nitrosoguanidine and then into aminoguanidine (or guanylhydrazine). The latter substance is used in the explosives industry for the preparation of tetracene. In organic chemical research it finds use because of the fact that it reacts readily with aldehydes and ketones to form products which yield crystalline and easily characterized nitrates.

$NO_2-NH-C(NH)-NH_2$
Nitroguanidine

$N-NH-C(NH)-NH_2$
Aminoguanidine

$CH=NH-NH-C(NH)-NH_2$
Benzalaminoguanidine

Preparation of Benzalaminoguanidine Nitrate (Benzaldehyde Guanylhydrazone Nitrate). Twenty-six grams of zinc dust and 10.4 grams of nitroguanidine are introduced into a 300-cc. Erlenmeyer flask, 150 cc. of water is added, then 42 cc. of glacial acetic acid at such a rate that the temperature of the mixture does not rise above 40°. The liquid at first turns yellow because of the formation of nitrosoguanidine but

[11] Davis, *J. Am. Chem. Soc.*, **44**, 868 (1922).

becomes colorless again when the reduction is complete. After all the zinc has disappeared, 1 mol of concentrated nitric acid is added, then 1 mol of benzaldehyde, and the mixture is shaken and scratched to facilitate the separation of the heavy granular precipitate of benzalaminoguanidine nitrate. The product, recrystallized from water or from alcohol, melts when pure at 160.5°.

Nitroguanidine and nitrosoguanidine both give a blue color with the diphenylamine reagent, and both give the tests described below, but the difference in the physical properties of the substances is such that there is no likelihood of confusing them.

Tests for Nitroguanidine. To 0.01 gram of nitroguanidine in 4 cc. of cold water 2 drops of saturated ferrous ammonium sulfate solution is added, then 1 cc. of 6 N sodium hydroxide solution. The mixture is allowed to stand for 2 minutes, and is filtered. The filtrate shows a fuchsine color but fades to colorless on standing for half an hour. Larger quantities of nitroguanidine give a stronger and more lasting color.

One-tenth gram of nitroguanidine is treated in a test tube with 5 cc. of water and 1 cc. of 50% acetic acid, and the mixture is warmed at 40–50° until everything is dissolved. One gram of zinc dust is added, and the mixture is set aside in a beaker of cold water for 15 minutes. After filtering, 1 cc. of 6% copper sulfate solution is added. The solution becomes intensely blue, and, on boiling, gives off gas, becomes turbid, and presently deposits a precipitate of metallic copper. If, instead of the copper sulfate solution, 1 cc. of a saturated solution of silver acetate [12] is added, and the solution is boiled, then a precipitate of metallic silver is formed.

Many of the reactions of nitroguanidine, particularly its decomposition by heat and the reactions which occur in aqueous and in sulfuric acid solutions, follow directly from its dearrangement.[13] Nitroguanidine dearranges in two modes, as follows.

$$\begin{matrix} H \\ \\ H \end{matrix} \!\! N-C(NH)-N\!\! \begin{matrix} NO_2 \\ \\ H \end{matrix} \rightleftharpoons NH_2-NO_2 + HNCNH \rightleftharpoons NH_2CN$$
$$\text{Cyanamide}$$

$$\begin{matrix} H \\ \\ H \end{matrix} \!\! N\!-\!C(NH)-N\!\! \begin{matrix} NO_2 \\ \\ H \end{matrix} \rightleftharpoons NH_3 + HNCN-NO_2 \rightleftharpoons NC-N\!\! \begin{matrix} NO_2 \\ \\ H \end{matrix}$$
$$\text{Nitrocyanamide}$$

[12] Two grams of silver acetate, 2 cc. of glacial acetic acid, diluted to 100 cc., warmed, filtered, and allowed to cool.

[13] Davis and Abrams, *Proc. Am. Acad. Arts and Sciences,* **61,** 437 (1926).

A solution of nitroguanidine in concentrated sulfuric acid comports itself as if the nitroguanidine had dearranged into nitroamide and cyanamide. When it is warmed, nitrous oxide containing a small amount of nitrogen comes off first (from the dehydration of the nitroamide) and carbon dioxide (from the hydrolysis of the cyanamide) comes off later and more slowly. Long-continued heating at an elevated temperature produces ammonia and carbon dioxide quantitatively according to the equation,

$$NH_2—C(NH)—NH—NO_2 + H_2O \longrightarrow N_2O + 2NH_3 + CO_2$$

The production of nitrous oxide is not exactly quantitative because of secondary reactions. A solution of nitroguanidine in concentrated sulfuric acid, after standing for some time, no longer gives a precipitate of nitroguanidine when it is diluted with water.

A freshly prepared solution of nitroguanidine in concentrated sulfuric acid contains no nitric acid, for none can be distilled out of it, but it is ready to produce nitric acid (by the hydration of the nitroamide) if some material is present which will react with it. Thus, it gives up its nitro group quantitatively in the nitrometer, and it is a reagent for the nitration of such substances as aniline, phenol, acet-p-toluide, and cinnamic acid which are conveniently nitrated in sulfuric acid solution.

In aqueous solution nitroguanidine dearranges in both of the above-indicated modes, but the tendency toward dearrangement is small unless an acceptor for the product of the dearrangement is present. It results that nitroguanidine is relatively stable in aqueous solution; after many boilings and recrystallizations the same solution finally becomes ammoniacal. Ammonia, being alkaline, tends to promote the decomposition of nitroamide in aqueous solution. Also, because of its mass action effect, it tends to inhibit dearrangement in the second mode which produces ammonia. If nitroguanidine is warmed with aqueous ammonia, the reaction is slow. But, if it is warmed with water and a large excess of ammonium carbonate, nitrous oxide comes off rapidly, the ammonia combines with the cyanamide from the dearrangement, and guanidine carbonate is formed in practically quantitative amount.

Preparation of Guanidine Carbonate from Nitroguanidine. Two hundred and eight grams of nitroguanidine, 300 grams of ammonium carbonate, and 1 liter of water are heated together in a 2-liter flask in the water bath. The flask is equipped with a reflux condenser and with a thermometer dipping into the mixture. When the thermometer indicates 65–70°, nitrous oxide escapes rapidly, and it is necessary to shake the flask occasionally to prevent the undissolved nitroguanidine from being carried up into the neck. The temperature is raised as rapidly as may be done without the reaction becoming too violent. After all the material has gone into solution, the flask is removed from the water bath and the contents boiled under reflux for 2 hours by the application of a free flame. The liquid is then transferred to an evaporating dish and evaporated to dryness on the steam or water bath. During this process all the remaining ammonium carbonate ought to be driven off. The residue is taken up in the smallest possible amount of cold water, filtered for the removal of a small amount of melamine, and the filtrate is stirred up with twice its volume of 95% alcohol which causes the precipitation of guanidine carbonate (while the traces of urea which will have been formed remain in solution along with any ammonium carbonate which may have survived the earlier treatment). The guanidine carbonate is filtered off, rinsed with alcohol, and dried. The filtrate is evaporated to dryness, taken up in water, and precipitated with alcohol for a second crop—total yield about 162 grams or 90% of the theory. The product gives no color with the diphenylamine reagent; it is free from nitrate and of a quality which would be extremely difficult to procure by any process involving the double decomposition of guanidine nitrate.

In the absence of ammonia and in the presence of a primary aliphatic amine, nitroguanidine in aqueous solution dearranges in the second of the above-indicated modes, ammonia is liberated, and the nitrocyanamide combines with the amine to form an alkylnitroguanidine.

$$HNCN—NO_2 + CH_3—NH_2 \longrightarrow CH_3—NH—C(NH)—NH—NO_2$$
Nitrocyanamide   Methylnitroguanidine

The structure of the N-alkyl,N'-nitroguanidine is demonstrated by the fact that it yields the amine and nitrous oxide on hydrolysis, indicating that the alkyl group and the nitro group are attached to different nitrogen atoms.

$$CH_3—NH—C(NH)—NH—NO_2 + H_2O \longrightarrow$$
$$CH_3—NH_2 + NH_3 + N_2O + CO_2$$

The same N-alkyl,N'-nitroguanidines are produced by the nitration of the alkyl guanidines.[14]

[14] Davis and Elderfield, *J. Am. Chem. Soc.*, **55**, 731 (1933).

Nitroguanidine, warmed with an aqueous solution of hydrazine, yields N-amino,N'-nitroguanidine,[15] white crystals from water, m. p. 182°. This substance explodes on an iron anvil if struck with a heavy sledge hammer allowed to drop through a distance of about 8 inches. It may perhaps have some interest as an explosive.

Flashless colloided powder containing nitroguanidine produces a considerable amount of gray smoke made up of solid materials from the decomposition of the substance. The gases smell of ammonia. The powder produces more smoke than the other flashless powders which are used in this country.

Nitroguanidine decomposes immediately upon melting and cannot be obtained in the form of a liquid, as can urea, dicyandiamide, and other substances which commence to decompose when heated a few degrees above their melting points. A small quantity heated in a test tube yields ammonia, water vapor, a white sublimate in the upper part of the tube, and a yellow residue of mellon which is but little affected if warmed to a bright red heat. The products which are formed are precisely those which would be predicted from the dearrangements,[16] namely, water and nitrous oxide (from nitroamide), cyanamide, melamine (from the polymerization of cyanamide), ammonia, nitrous oxide again and cyanic acid (from nitrocyanamide), cyanuric acid (from the polymerization of cyanic acid), ammeline and ammelide (from the co-polymerization of cyanic acid and cyanamide) and, from the interaction and decomposition of these substances, carbon dioxide, urea, melam, melem, mellon, nitrogen, prussic acid, cyanogen, and paracyanogen. All these substances have been detected in, or isolated from, the products of the decomposition of nitroguanidine by heat.

There is no doubt whatever that nitroguanidine is a cool explosive, but there appears to be a disagreement as to the temperature which it produces. A package of nitroguanidine, exploded at night by means of a blasting cap, produces no visible flash. If 10 or 15% of the substance is incorporated in nitrocellulose powder, it makes the powder flashless. Vieille [17] found that the gases from the explosion of nitroguanidine were much less erosive

[15] Phillips and Williams, *J. Am. Chem. Soc.*, **50**, 2465 (1928).
[16] Davis and Abrams, *Proc. Am. Acad. Arts and Sciences*, **61**, 443 (1926).
[17] *Mém. poudres*, **11**, 195 (1901).

than those from other explosives of comparable force, and considered the fact to be in harmony with his general conclusion that the hotter explosives are the more erosive. In his experiments the explosions were made to take place in a steel bomb equipped with a crusher gauge and with a removable, perforated, steel plug through the perforation in which the hot gases from the explosion were allowed to escape. They swept away, or eroded off, a certain amount of the metal. The plug was weighed before and after the experiment, its density had been determined, and the number of cubic millimeters of metal lost was reported as a measure of the erosion. Some of Vieille's results are indicated in the following table.

EXPLOSIVE	CHARGE (Grams)	PRESSURE (Kg./sq. cm.)	EROSION	EROSION PER GRAM	FORCE
Poudre BF	3.45	2403	20.3	5.88	
	3.50	2361	22.7	6.58	
	3.55	2224	24.7	6.96 6.4	9,600
	3.55	2253	25.5	7.19	
	3.55	2143	20.1	5.66	
Cordite	3.55	2500	64.2	18.1	10,000
Ballistite VF	3.47	2509	84.5	24.3	
	3.51	2370	83.2	23.7	
	3.55	2542	90.2	25.4 24.3	10,000
	3.55	2360	85.9	24.2	
	3.55	2416	84.5	23.8	
Black military	10.00	2167	22.3	2.2	3,000
Black sporting	8.88	1958	40.0	4.5	3,000
Blasting gelatin	3.35	2458	105.0	31.4	10,000
Nitromannite	3.54	2361	83.5	23.6	10,000
Nitroguanidine	3.90	2019	8.8	2.3	9,000

These experiments [18] were carried out in a bomb of 17.8 cc. capacity, which corresponds, for the example cited, to a density of loading of 0.219 for the nitroguanidine which was pulverulent

[18] The cordite used in these experiments was made from 57% nitroglycerin, 5% vaseline, and 38% high nitration guncotton colloided with acetone; the ballistite VF of equal amounts by weight of nitroglycerin and high nitration guncotton colloided with ethyl acetate. The black military powder was made from saltpeter 75, sulfur 10, and charcoal 15; the black sporting powder from saltpeter 78, sulfur 10, and charcoal 12. The blasting gelatin contained 94% nitroglycerin and 6% soluble nitrocotton.

material "firmly agglomerated in a manner to facilitate the naturally slow combustion of that substance."

An experiment with 18.11 grams nitroguanidine in a bomb of 75.0 cc. capacity (density of loading 0.241) showed an erosion of 2.29 per gram of explosive.

The temperature (907°) which Vieille accepted as the temperature produced by the explosion of nitroguanidine had been determined earlier by Patart [19] who published in 1904 an account of manometric bomb experiments with guanidine nitrate and with nitroguanidine. The explosives were agglomerated under a pressure of 3600 kilograms per square centimeter, broken up into grains 2 or 3 mm. in diameter, and fired in a bomb of 22 cc. capacity. Some of Patart's experimental results are tabulated below. Calculated from these data, Patart reported for guanidine

	PRESSURE, KILOGRAMS PER SQUARE CENTIMETER	
DENSITY OF LOADING	Guanidine Nitrate	Nitroguanidine
0.15..............	1128 ⎫ 1038 ⎬1083 ... ⎭	1304 ⎫ 1584 ⎬1435 1416 ⎭
0.20..............	1556 ⎫1486 1416 ⎭	2060 ⎫2091 2122 ⎭
0.25..............	2168 ⎫2098 2028 ⎭	3092 ⎫3080 3068 ⎭
0.30..............	3068 ⎫2941 2814 ⎭	4118 ⎫4078 4038 ⎭
0.35..............	3668 ⎫3699 3730 ⎭	... ...

nitrate, covolume 1.28, force 5834, and temperature of explosion 929°; for nitroguanidine, covolume 1.60, force 7140, and temperature of explosion 907°. He appears to have felt that these calculated temperatures of explosion were low, for he terminated his article by calling attention to the extraordinary values of the covolume deduced from the pressures in the closed vessel, and subpended a footnote:

It may be questioned whether the rapid increase of the pressure with the density of loading, rather than being the consequence of a constant reaction giving place to a considerable covolume, is not due simply to the mode of de-

[19] *Mém. poudres*, **13**, 153 (1905–1906).

composition being variable with the density of loading and involving a more and more complete decomposition of the explosive. Only an analysis of the gases produced by the reaction can determine this point, as it also can determine the actual temperature of the deflagration.

The later studies of Muraour and Aunis [20] have shown that the temperature of explosion of nitroguanidine may be much higher than Patart calculated, and have given probability to his hypothesis that the density of loading has an effect upon the mode of the explosive decomposition. These investigators found that a platinum wire 0.20 mm. in diameter, introduced into the bomb along with the nitroguanidine, was melted by the heat of the explosion—a result which indicates a temperature of at least 1773°C. They pointed out that nitroguanidine, if compressed too strongly, may take fire with difficulty and may undergo an incomplete decomposition, and hence at low densities of loading may produce unduly low pressures corresponding to a covolume which is too large and to a temperature of explosion which is too low. The pressure of 3600 kilograms per square centimeter, under which Patart compressed his nitroguanidine, is much too high. Nitroguanidine compressed under 650 kilograms per square centimeter, and fired in a manometric bomb of 22 cc. capacity, at a density of loading of 0.2, and with a primer of 1 gram of black powder, gave a pressure of 1737 kilograms per square centimeter; compressed under 100 kilograms per square centimeter and fired in the same way nitroguanidine gave a pressure of 1975 kilograms per square centimeter, or a difference of 238 kilograms. In an experiment with a bomb of 139 cc. capacity, density of loading 0.2, Muraour and Aunis observed a pressure which, correction being made for various heat losses, corresponded to a temperature of 1990°.

Assuming that nitroguanidine explodes to produce carbon dioxide, water, carbon monoxide, hydrogen, and nitrogen,[21] assuming that the equilibrium constant for the reaction, $CO + H_2O \rightleftharpoons CO_2 + H_2$, is 6, and that the molecular heat of formation at con-

[20] Annales des Mines, **9**, 178, 180 (1920); Comp. rend., **190**, 1389, 1547 (1930); Mém. poudres, **25**, 91 (1932–1933).

[21] This assumption however is not true, for powder which contains nitroguanidine produces a gray smoke consisting of solid decomposition products and yields gases which smell of ammonia.

stant volume of nitroguanidine is 17.9 Calories, and taking the values of Nernst and Wohl for the specific heats of the various gases, Muraour and Aunis calculated the following values for the explosion of nitroguanidine, temperature 2098°, covolume 1.077, force 9660, and pressure (at density of loading 0.20) 2463 kilograms per square centimeter. They have also calculated the temperature of explosion of ammonium nitrate 1125°,[22] of "explosive NO" (ammonium nitrate 78.7, trinitrotoluene 21.3) 2970°, and of explosive N4 (ammonium nitrate 90, potassium nitrate 5, trinitronaphthalene 5) 1725°, and have found by experiment that the last named of these explosives, fired at a density of loading of 0.30, did not fuse a platinum wire (0.06-mm. diameter) which had been introduced along with it into the bomb.

Nitroguanidine detonates completely under the influence of a detonator containing 1.5 gram of fulminate. According to Patart [23] 40 grams exploded on a lead block 67 mm. in diameter produced a shortening of 7 mm. Picric acid under the same conditions produced a shortening of 10.5 mm., and Favier explosive (12% dinitronaphthalene, 88% ammonium nitrate) one of 8 mm. Muraour and Aunis [24] experimented with nitroguanidine compressed under 100 kilograms per square centimeter and with trinitrotoluene compressed under 1000 kilograms per square centimeter, in a manometric bomb of 22-cc. capacity and at densities of loading of 0.13, 0.20, 0.25, and 0.30, and reported that the two explosives gave the same pressures.

During the first World War the Germans used in trench mortar bombs an explosive consisting of nitroguanidine 50%, ammonium nitrate, 30%, and paraffin 20%.

Nitrosoguanidine

Nitrosoguanidine is a cool and flashless primary explosive, very much more gentle in its behavior than mercury fulminate and lead azide. It is a pale yellow crystalline powder which explodes on contact with concentrated sulfuric acid or on being heated in a melting point tube at 165°. It explodes from the blow of a car-

---

[22] The temperature of 1121° was calculated by Hall, Snelling, and Howell, "Investigations of Explosives Used in Coal Mines," *U. S. Bur. Mines Bull.* 15, Washington, 1912, p. 32.

[23] *Mém. poudres,* **13,** 159 (1905–1906).

[24] *Ibid.,* **25,** 92–93, footnote (1932–1933).

penter's hammer on a concrete block. Its sensitivity to shock, to friction, and to temperature, and the fact that it decomposes slowly in contact with water at ordinary temperatures, militate against its use as a practical explosive. It may be kept indefinitely in a stoppered bottle if it is dry.

The reactions of nitrosoguanidine in aqueous solution are similar to those of nitroguanidine except that nitrogen and nitrous acid respectively are formed under conditions which correspond to the formation of nitrous oxide and nitric acid from nitroguanidine. It dearranges principally as follows.

$$NH_2—C(NH)—NH—NO \rightleftharpoons NH_2—NO + HNCNH \rightleftharpoons NH_2—CN$$

If it is warmed in aqueous solution, the nitrosoamide breaks down into water and nitrogen, and the cyanamide polymerizes to dicyandiamide. The evaporation of the solution yields crystals of the latter substance. A cold aqueous solution of nitrosoguanidine acidified with hydrochloric acid yields nitrous acid, and may be used for the introduction of a nitroso group into dimethylaniline or some similar substance which is soluble in the acidified aqueous liquid.

Preparation of Nitrosoguanidine.[25] Twenty-one grams of nitroguanidine, 11 grams of ammonium chloride, 18 grams of zinc dust, and 250 cc. of water in an 800-cc. beaker are stirred together mechanically while external cooling is applied to prevent the temperature from rising above 20–25°. After 2 hours or so the gray color of the zinc disappears, the mixture is yellow, and on settling shows no crystals of nitroguanidine. The mixture is then cooled to 0° or below by surrounding the beaker with a mixture of cracked ice and salt; it is filtered, and the filtrate is discarded. The yellow residue, consisting of nitrosoguanidine mixed with zinc oxide or hydroxide and basic zinc chloride, is extracted with 4 successive portions of 250 cc. each of water at 65°. The combined extracts, allowed to stand over night at 0°, deposit nitrosoguanidine which is collected, rinsed with water, and dried at 40°. Yield 8.0–9.2 grams, 45–52% of the theory.

The flashlessness of nitrosoguanidine may be demonstrated safely by igniting about 0.5 gram of it on the back of the hand. The experiment is most striking if carried out in a darkened room. The sample being poured out in a conical heap on the back of the left hand, a match held in the right hand is scratched and allowed to burn until the mate-

[25] Davis and Rosenquist, *J. Am. Chem. Soc.*, **59**, 2114 (1937).

rial which composes the burnt head of the match has become thoroughly heated, it is extinguished by shaking, and the burnt head is then touched to the heap of nitrosoguanidine. The nitrosoguanidine explodes with a zishing sound and with a cloud of gray smoke, but with no visible flash whatsoever. The place on the hand where the nitrosoguanidine was fired will perhaps itch slightly, and the next day will perhaps show a slight rash and peeling of the skin. There is no sensation of being burned, and the explosion is so rapid that the hand remains steady and makes no reflex movement.

Ethylenedinitramine

Ethylenedinitramine, m.p. 174–176° with decomposition, is produced when dinitroethyleneurea is refluxed with water,[26] or it

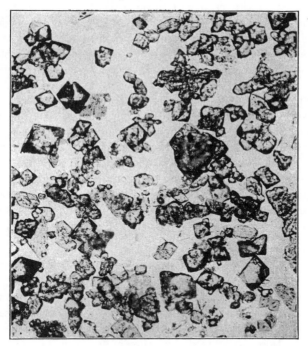

Figure 90. Ethylenedinitramine Crystals (60×).

may be prepared directly, without isolating this intermediate, by the nitration of ethyleneurea with mixed acid.

[26] Franchimont and Klobbie, *Rec. trav. chim.*, **7**, 17, 244 (1887).

$$\text{Ethyleneurea} \quad \begin{array}{l} CH_2-NH \\ | \\ CH_2-NH \end{array}\!\!>\!\!CO \longrightarrow \begin{array}{l} CH_2-N(NO_2) \\ | \\ CH_2-N(NO_2) \end{array}\!\!>\!\!CO \longrightarrow \begin{array}{l} CH_2-NH-NO_2 \\ | \\ CH_2-NH-NO_2 \end{array} \text{Ethylenedinitramine}$$

Dinitroethyleneurea

It is a dibasic acid and forms neutral salts, the silver salt a pulverulent precipitate, the potassium salt needles from alcohol. It is sparingly soluble in water, about 1 part in 200 at 25°, and is not affected by refluxing with this solvent. On refluxing with dilute sulfuric acid it gives nitrous oxide, acetaldehyde, and glycol. Hale [27] has reported that it explodes spontaneously when heated to 180°, in which respect it resembles mercury fulminate and nitroglycerin, but that it corresponds in resistance to shock more nearly to the relatively insensitive high explosives, like TNT and picric acid, which are used as the bursting charges of shells. He found that it is exploded by a 10-inch drop of a 2-kilogram weight, the same as picric acid, and reported that it withstands the standard 120° stability test as well as tetryl.

Dinitrodimethyloxamide

This substance was prepared by Franchimont [28] by dissolving dimethyloxamide in very strong nitric acid (specific gravity 1.523) without cooling, allowing to stand, and pouring into water, and by Thiele and Meyer [29] by dissolving dimethyloxamide in crude nitric acid, adding fuming sulfuric acid to the chilled solution, and pouring onto ice. Dimethyloxamide is prepared readily by the interaction of methylamine with an ester of oxalic acid.

$$\begin{array}{l} COOR \\ | \\ COOR \end{array} + 2NH_2-CH_3 \longrightarrow \begin{array}{l} CO-NH-CH_3 \\ | \\ CO-NH-CH_3 \end{array} \longrightarrow \begin{array}{l} CO-N(NO_2)-CH_3 \\ | \\ CO-N(NO_2)-CH_3 \end{array}$$

Dimethyloxamide   Dinitrodimethyloxamide

Dinitrodimethyloxamide is very slightly soluble in water, sparingly in ether and chloroform, and soluble in alcohol from which it crystallizes in needles which melt at 124° and decompose at a higher temperature. By reduction with zinc and acetic acid in alcohol solution it yields dimethyloxamide. It is not destroyed by refluxing with concentrated hydrochloric acid. Concentrated sulfuric acid splits off nitric acid, and the substance accordingly

[27] U. S. Pat. 2,011,578 (1935).
[28] Rec. trav. chim., 2, 96 (1882); 4, 197 (1884); 13, 311 (1893).
[29] Ber., 29, 961 (1896).

gives up its nitro group in the nitrometer. On treatment with an excess of aqueous ammonia or on refluxing with a slight excess of barium hydroxide solution, it yields the corresponding salt of methylnitramine. Haid, Becker, and Dittmar [30] have reported that dinitrodimethyloxamide, like PETN, tetryl, TNT, and picric acid, gives no red fumes after 30 days at 100° while nitrocellulose in their experiments gave red fumes after 36 hours and dipentaerythrite hexanitrate after 8 days.

Dinitrodimethyloxamide has interesting explosive properties, but it is limited in its use because it develops an acidity when wet with water. It has been reported [31] that 30 parts of dinitrodimethyloxamide and 70 parts of PETN yield a eutectic which melts at 100° and can be poured as a homogeneous liquid. The cast explosive has a velocity of detonation of 8500 meters per second which is equal to that of PETN under the best conditions. The further addition of dimethyl oxalate or of camphor [32] lowers the melting point still more and affects the brisance only slightly but has a significant phlegmatizing action. A mixture of PETN 60%, dinitrodimethyloxamide 30%, and dimethyl oxalate 10% melts at 82°, and has, when cast, a velocity of detonation of 7900 meters per second which is higher than the velocity of detonation of cast picric acid.

Dinitrodimethylsulfamide

This substance was first prepared by Franchimont [33] by dissolving 1 part of dimethylsulfamide in 10 parts of the strongest nitric acid, and drowning in water. Dimethylsulfamide is prepared by the interaction of methylamine and sulfuryl chloride in chilled absolute ether solution.

Dimethylsulfamide   Dinitrodimethylsulfamide

Dinitrodimethylsulfamide is very slightly soluble in water, very readily in hot alcohol, and moderately in chloroform and benzene. Crystals from benzene, m.p. 90°. The vapor of the substance

[30] *Z. ges. Schiess- u. Sprengstoffw.*, **30**, 68 (1935).

[31] Ger. Pat. 499,403, cited by Foulon, *Z. ges. Schiess- u. Sprengstoffw.*, **27,** 191 (1932).

[32] Ger. Pat. 505,852.

[33] *Rec. trav. chim.*, **3**, 419 (1883).

explodes if heated to about 160°. Dinitrodimethylsulfamide has been suggested as an addition to PETN for the preparation of a fusible explosive which can be loaded by pouring.

Cyclotrimethylenetrinitramine (Cyclonite, Hexogen, T4).

The name of *cyclonite*, given to this explosive by Clarence J. Bain because of its cyclic structure and cyclonic nature, is the one by which it is generally known in the United States. The Germans call it *Hexogen*, the Italians *T4*.

FIGURE 91. George C. Hale. Has studied cyclonite, ethylenedinitramine, and many other explosives. Author of numerous inventions and publications in the field of military powder and explosives. Chief Chemist, Picatinny Arsenal, 1921-1929; Chief of the Chemical Department, Picatinny Arsenal, 1929—.

Cyclonite, prepared by the nitration of hexamethylenetetramine, is derived ultimately from no other raw materials than coke, air, and water. It has about the same power and brisance as PETN, and a velocity of detonation under the most favorable conditions of about 8500 meters per second.

Hexamethylenetetramine, $C_6H_{12}N_4$, is obtained in the form of colorless, odorless, and practically tasteless crystals by the evaporation of an aqueous solution of formaldehyde and ammonia. It is used in medicine under the names of *Methenamine, Hexamine, Cystamine, Cystogen,* and *Urotropine,* administered orally as an antiseptic for the urinary tract, and in industry in the manufacture of plastics and as an accelerator for the vulcanization of rubber. It has feebly basic properties and forms a nitrate, $C_6H_{12}N_4 \cdot 2HNO_3$, m.p. 165°, soluble in water, insoluble in alcohol, ether, chloroform, and acetone. The product, $C_3H_6O_6N_6$, prepared by nitrating this nitrate and patented by Henning[34] for possible use in medicine, was actually cyclonite. Herz later patented[35] the same substance as an explosive compound, cyclotrimethylenetrinitramine, which he found could be prepared by treating hexamethylenetetramine directly with strong nitric acid. In his process the tetramine was added slowly in small portions at a time to nitric acid (1.52) at a temperature of 20–30°. When all was in solution, the liquid was warmed to 55°, allowed to stand for a few minutes, cooled to 20°, and the product precipitated by the addition of water. The nitration has been studied further by Hale[36] who secured his best yield, 68%, in an experiment in which 50 grams of hexamethylenetetramine was added during 15 minutes to 550 grams of 100% nitric acid while the temperature was not allowed to rise above 30°. The mixture was then cooled to 0°, held there for 20 minutes, and drowned.

Hexamethylenetetramine   Cyclotrimethylenetrinitramine

The formaldehyde which is liberated by the reaction tends to be oxidized by the nitric acid if the mixture is allowed to stand or is warmed. It remains in the spent acid after drowning and interferes with the recovery of nitric acid from it.

[34] Ger. Pat. 104,280 (1899).
[35] Brit. Pat. 145,791 (1920); U. S. Pat. 1,402,693 (1922).
[36] *J. Am. Chem. Soc.,* 47, 2754 (1925).

Cyclonite is a white crystalline solid, m.p. 202°. It is insoluble in water, alcohol, ether, ethyl acetate, petroleum ether, and carbon tetrachloride, very slightly soluble in hot benzene, and soluble 1 part in about 135 parts of boiling xylene. It is readily soluble in hot aniline, phenol, ethyl benzoate, and nitrobenzene, from all of which it crystallizes in needles. It is moderately soluble in hot acetone, about 1 part in 8, and is conveniently recrystallized from this solvent from which it is deposited in beautiful, transparent, sparkling prisms. It dissolves very slowly in cold concentrated sulfuric acid, and the solution decomposes on standing. It dissolves readily in warm nitric acid (1.42 or stronger) and separates only partially again when the liquid is cooled. The chemical reactions of cyclonite indicate that the cyclotrimethylenetrinitramine formula which Herz suggested for it is probably correct.

Cyclonite is hydrolyzed slowly when the finely powdered material is boiled with dilute sulfuric acid or with dilute caustic soda solution.

$$C_3H_6O_6N_6 + 6H_2O \longrightarrow 3NH_3 + 3CH_2O + 3HNO_3$$

Quantitative experiments have shown that half of its nitrogen appears as ammonia. If the hydrolysis is carried out in dilute sulfuric acid solution, the formaldehyde is oxidized by the nitric acid and nitrous acid is formed.

If cyclonite is dissolved in phenol at 100° and reduced by means of sodium, it yields methylamine, nitrous acid, and prussic acid. Finely powdered cyclonite, suspended in 80% alcohol and treated with sodium amalgam, yields methylamine, ammonia, nitrous acid, and formaldehyde, a result which probably indicates that both hydrolysis and reduction occur under these conditions.

When a large crystal of cyclonite is added to the diphenylamine reagent, a blue color appears slowly on the surface of the crystal. Powdered cyclonite gives within a few seconds a blue color which rapidly becomes more intense. If cinnamic acid is dissolved in concentrated sulfuric acid, and if finely powdered cyclonite is added while the mixture is stirred, gas comes off at a moderate rate, and the mixture, after standing over night and drowning, gives a precipitate which contains a certain amount of p-nitrocinnamic acid.

In the drop test cyclonite is exploded by a 9-inch drop of a 2-kilogram weight. For the detonation of 0.4 gram, the explosive requires 0.17 gram of mercury fulminate. It fails to detonate when struck with a fiber shoe, and detonates when struck with a steel shoe, in the standard frictional impact test of the U. S. Bureau of Mines. In 5 seconds it fumes off at 290°, but at higher temperatures, even as high as 360°, it does not detonate.

CHAPTER IX

PRIMARY EXPLOSIVES, DETONATORS, AND PRIMERS

Primary explosives explode from shock, from friction, and from heat. They are used in primers where it is desired by means of shock or friction to produce fire for the ignition of powder, and they are used in detonators where it is desired to produce shock for the initiation of the explosion of high explosives. They are also used in toy caps, toy torpedoes, and similar devices for the making of noise. Indeed, certain primary explosives were used for this latter purpose long before the history of modern high explosives had yet commenced.

Discovery of Fulminating Compounds

Fulminating gold, silver, and platinum (Latin, *fulmen*, lightning flash, thunderbolt) are formed by precipitating solutions of these metals with ammonia. They are perhaps nitrides or hydrated nitrides, or perhaps they contain hydrogen as well as nitrogen and water of composition, but they contain no carbon and must not be confused with the fulminates which are salts of fulminic acid, HONC. They are dangerously sensitive, and are not suited to practical use.

Fulminating gold is described in the writings of the pseudonymous Basil Valentine,[1] probably written by Johann Thölde (or Thölden) of Hesse and actually published by him during the years 1602–1604. The author called it *Goldkalck,* and prepared it by dissolving gold in an *aqua regia* made by dissolving sal ammoniac in nitric acid, and then precipitating by the addition of potassium carbonate solution. The powder was washed by decantation 8 to 12 times, drained from water, and dried in the air where no sunlight fell on it, "and not by any means over the

[1] We find the description on page 289 of the second part of the third German edition of the collected writings of Basil Valentine, Hamburg, 1700.

fire, for, as soon as this powder takes up a very little heat or warmth, it kindles forthwith, and does remarkably great damage, when it explodes with such vehemence and might that no man would be able to restrain it." The author also reported that warm distilled vinegar converted the powder into a material which was no longer explosive. The name of *aurum fulminans* was given to the explosive by Beguinus who described its preparation in his *Tyrocinium Chymicum*, printed in 1608.

Fulminating gold precipitates when a solution of pure gold chloride is treated with ammonia water. The method of preparation described by Basil Valentine succeeds because the sal ammoniac used for the preparation of the *aqua regia* supplies the necessary ammonia. If gold is dissolved in an *aqua regia* prepared from nitric acid and common salt, and if the solution is then treated with potassium carbonate, the resulting precipitate has no explosive properties. Fulminating gold loses its explosive properties rapidly if it is allowed to stand in contact with sulfur.

Fulminating gold was early used both for war and for entertainment. The Dutch inventor and chemist, Cornelis Drebbel, being in the service of the British Navy, devoted considerable time to the preparation of fulminating gold and used his material as a detonator in petards and torpedoes in the English expedition against La Rochelle in 1628. Pepys, in his diary for November 11, 1663, reports a conversation with a Dr. Allen concerning *aurum fulminans* "of which a grain . . . put in a silver spoon and fired, will give a blow like a musquett and strike a hole through the silver spoon downward, without the least force upward."

Fulminating silver was prepared in 1788 by Berthollet who precipitated a solution of nitrate of silver by means of lime water, dried the precipitated silver oxide, treated it with strong ammonia water which converted it into a black powder, decanted the liquid, and left the powder to dry in the open air. Fulminating silver is more sensitive to shock and friction than fulminating gold. It explodes when touched; it must not be enclosed in a bottle or transferred from place to place, but must be left in the vessel, or better upon the paper, where it was allowed to dry.

The black material which deposits in a reagent bottle of ammoniacal silver nitrate, and sometimes collects on the rim and around the stopper, contains fulminating silver. Explosions are reported to have been caused by the careless turning of the glass stopper of a bottle containing this reagent. After a test (for aldehyde, for example) has been made with ammoniacal silver nitrate solution, the liquid ought promptly to be washed down the sink, and all insoluble matter left in the vessel ought to be dissolved out with dilute nitric acid.

Fulminating platinum was first prepared by E. Davy, about 1825, by adding ammonia water to a solution of platinum sulfate, boiling the precipitate with a solution of potash, washing, and allowing to dry. It was exploded by heat, but not easily by percussion or friction.

Fourcroy prepared a *fulminating mercury* by digesting red oxide of mercury in ammonia water for 8 or 10 days. The material became white and finally assumed the form of crystalline scales. The dried product exploded loudly from fire, but underwent a spontaneous decomposition when left to itself. At slightly elevated temperatures it gave off ammonia and left a residue of mercury oxide.

In the *Journal de physique* for 1779 the apothecary, Bayen, described a fulminating mercurial preparation of another kind. Thirty parts of precipitated, yellow oxide of mercury, washed and dried, was mixed with 4 or 5 parts of sulfur; the mixture exploded with violence when struck with a heavy hammer or when heated on an iron plate. Other mixtures which react explosively when initiated by percussion have been studied more recently,[2] metallic sodium or potassium in contact with the oxide or the chloride of silver or of mercury or in contact with chloroform or carbon tetrachloride.

The explosion of chloroform in contact with an alkali metal may be demonstrated by means of the apparatus illustrated in Figure 92. About 0.3 gram of sodium or of potassium or of the liquid alloy of the two is introduced into a thin-wall glass tube, or, better yet, is sealed up in a small glass bulb, 6 to 8 mm. in diameter, which has a capillary 15 to 20 mm. in length. The tube or bulb containing the alkali metal is placed in the bottom of a narrow test tube into which 1 or 2 cc. of chloroform has already been introduced, and the apparatus is then

² Staudinger, *Z. Elektrochem.*, **31**, 549 (1925); Davis and McLean, *J. Am. Chem. Soc.*, **60**, 720 (1938).

ready for the experiment. Or, if it is desired to prepare in advance an explosive capsule which can safely be kept as long as desired, then the bulb is held in place at the bottom of the test tube by a collar of glass (a section of glass tubing) sintered to the inner wall of the test tube, and the top of the test tube is drawn down and sealed. When the prepared test tube or capsule is dropped onto a concrete pavement from

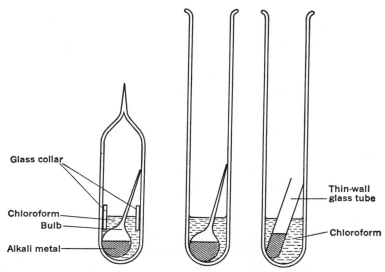

Glass collar

Chloroform
Bulb

Alkali metal

Thin-wall
glass tube

Chloroform

FIGURE 92. Apparatus for Demonstrating the Explosion of Chloroform with an Alkali Metal.

a height of 6 feet, a loud explosion is produced accompanied by a bright flash which is visible even in the direct sunlight. The chemical reaction is as follows, each one of the three chlorine atoms of the chloroform reacting in a different manner.

$$6CHCl_3 + 6Na \longrightarrow 6NaCl + 6HCl +$$

Mercury fulminate appears to have been prepared for the first time by Johann Kunckel von Löwenstern (1630–1703), the same chemist who discovered phosphorus and applied the purple of

Cassius practically to the manufacture of ruby glass. In his post-humous *Laboratorium Chymicum* he says: [3]

> Further evidence that mercury is cold is to be seen when you dissolve it in *aqua fortis* (nitric acid), evaporate the solution to dryness, pour highly rectified *spiritum vini* (alcohol) over the residue, and then warm it slightly so that it begins to dissolve. It commences to boil with amazing vigor. If the glass is somewhat stopped up, it bursts into a thousand pieces, and, in consequence, it must by no means be stopped up. I once dissolved silver and mercury together in *aqua fortis* and poured over it an excess of *spiritum vini,* and set the mixture to putrify *in fimum equinum* (horse manure) after having stopped up the glass with mere sealing wax only. When it happened a few days later that the manure became a little warm, it made such a thunder-crack, with the shattering of the glass, that the stable-servant imagined, since I had put it in a box, either that someone had shot at him through the window or that the Devil himself was active in the stable. As soon as I heard this news, I was able easily to see that the blame was mine, that it must have been my glass. Now this was with silver and mercury, 2 *loth* of each. Mercury does the same thing [4] alone, but silver not at all.

The preparation and properties of mercury fulminate were described in much detail by Edward Howard [5] in 1800 in a paper presented to the Royal Society of London. The method of preparation which he found to be most satisfactory was as follows: 100 grains of mercury was dissolved by heating in 1½ drams of nitric acid (specific gravity 1.3), the solution was cooled and added to 2 ounces of alcohol (specific gravity 0.849) in a glass vessel, the mixture was warmed until effervescence commenced, the reaction was allowed to proceed to completion, and the precipitate which formed was collected on a filter, washed with distilled water, and dried at a temperature not exceeding that of the water bath. Howard found that the fulminate was exploded by means of an electric spark or by concentrated sulfuric acid brought into contact with it. When a few grains were placed on a

---

[3] Kunckel, "Collegium Physico-Chymicum Experimentale, oder Laboratorium Chymicum," ed. Engelleder, Hamburg, 1716, p. 213. Cf. Davis, *Army Ordnance,* **7,** 62 (1926).

[4] Kunckel's meaning in the last sentence is evidently that mercury nitrate reacts with alcohol on warming, and that silver nitrate does not react with alcohol under the same conditions.

[5] *Phil. Trans. Roy. Soc.,* **204** (1800).

cold anvil and struck with a cold hammer, a very stunning disagreeable noise was produced and the faces of the hammer and anvil were indented. A few grains floated in a tinfoil capsule on hot oil exploded at 368°F. (186.7°C.). When a mixture of fine- and coarse-grain black powder was placed on top of a quantity of fulminate and the fulminate was fired, the black powder was blown about but it was not ignited and was recovered unchanged. Howard also attempted by means of alcohol to produce fulminating compounds from gold, platinum, antimony, tin, copper, iron, lead, nickel, bismuth, cobalt, arsenic, and manganese, but silver was the only one of these metals with which he had any success.

Brugnatelli in 1802 worked out a satisfactory method for the preparation of silver fulminate by pouring onto 100 grains of powdered silver nitrate first an ounce of alcohol and then an ounce of nitric acid. After the fulminate had precipitated, the mixture was diluted with water to prevent it from dissolving again and immediately filtered. Silver fulminate explodes more easily from heat and from friction than mercury fulminate and is more spectacular in its behavior. It quickly became an object of amateur interest and public wonderment, one of the standard exhibits of street fakirs and of mountebanks at fairs. Liebig, who was born in 1803, saw a demonstration of silver fulminate in the market place at Darmstadt when he was a boy. He watched the process closely, recognized by its odor the alcohol which was used, went home, and succeeded in preparing the substance for himself. He retained his interest in it, and in 1823 carried out studies on the fulminates in the laboratory of Gay-Lussac at Paris.

Mercury Fulminate

The commercial preparation of mercury fulminate is carried out by a process which is essentially the same as that which Howard originally recommended. Five hundred or 600 grams of mercury is used for each batch, the operation is practically on the laboratory scale, and several batches are run at the same time. Since the reaction produces considerable frothing, capacious glass balloons are used. The fumes, which are poisonous and inflammable, are passed through condensers, and the condensate, which contains alcohol, acetaldehyde, ethyl nitrate, and ethyl nitrite, is utilized by mixing it with the alcohol for the next batch.

Pure fulminate is white, but the commercial material is often grayish in color. The color is improved if a small amount of cupric chloride is added to the nitric acid solution of mercury before it is poured into the alcohol in the balloon, but the resulting white fulminate is actually less pure than the unbleached material.

Preparation of Mercury Fulminate. Five grams of mercury is added to 35 cc. of nitric acid (specific gravity 1.42) in a 100-cc. Erlenmeyer

FIGURE 93. Fulminate Manufacture. (Courtesy Atlas Powder Company.) At left, flasks in which mercury is dissolved in nitric acid. At right, balloons in which the reaction with alcohol occurs.

flask, and the mixture is allowed to stand without shaking until the mercury has gone into solution. The acid liquid is then poured into 50 cc. of 90% alcohol in a 500-cc. beaker in the hood. The temperature of the mixture rises, a vigorous reaction commences, white fumes come off, and crystals of fulminate soon begin to precipitate. Red fumes appear and the precipitation of the fulminate becomes more rapid, then white fumes again as the reaction moderates. After about 20 minutes the reaction is over; water is added, and the crystals are washed with water repeatedly by decantation until the washings are no longer acid to litmus. The product consists of grayish-yellow crystals, and corresponds to a good grade of commercial fulminate. It may be obtained white and entirely pure by dissolving in strong ammonia water, filtering, and reprecipitating by the addition of 30% acetic acid. The pure fulminate is filtered off, washed several times with cold water, and stored under water, or, if a very small amount is desired for experimental purposes, it is dried in a desiccator.

The chemical reactions in the preparation appear to be as follows. (1) The alcohol is oxidized to acetaldehyde, and (2) the nitrous acid which is formed attacks the acetaldehyde to form a nitroso derivative which goes over to the more stable, tautomeric, isonitroso form.

$$CH_3\text{—}CH_2\text{—}OH \rightarrow CH_3\text{—}CHO \longrightarrow \begin{array}{c} CH_2\text{—}CHO \\ | \\ NO \end{array} \rightleftharpoons \begin{array}{c} CH\text{—}CHO \\ \| \\ N\text{—}OH \end{array}$$

$$\qquad\qquad\qquad\qquad\qquad\qquad \text{Nitrosoacetaldehyde} \qquad \text{Isonitrosoacetaldehyde}$$

(3) The isonitrosoacetaldehyde is oxidized to isonitrosoacetic acid, and (4) this is nitrated by the nitrogen dioxide which is present to form nitroisonitrosoacetic acid.

$$\begin{array}{c} CH\text{—}CHO \\ \| \\ N\text{—}OH \end{array} \longrightarrow \begin{array}{c} CH\text{—}COOH \\ \| \\ N\text{—}OH \end{array} \longrightarrow \begin{array}{c} NO_2 \\ | \\ C\text{—}COOH \\ \| \\ N\text{—}OH \end{array}$$

$$\qquad\qquad\qquad \text{Isonitrosoacetic acid} \qquad \text{Nitroisonitrosoacetic acid}$$

(5) The nitroisonitrosoacetic acid loses carbon dioxide to form formonitrolic acid which (6) decomposes further into nitrous acid and fulminic acid, and (7) the fulminic acid reacts with the mercury nitrate to form the sparingly soluble mercury fulminate which precipitates.

$$\begin{array}{c} NO_2 \\ | \\ C\text{—}[COO]H \\ \| \\ N\text{—}OH \end{array} \longrightarrow \begin{array}{c} [NO_2] \\ | \\ C\,H \\ \| \\ N\text{—}OH \end{array} \nearrow \begin{array}{c} HONC \\ \downarrow \\ Hg(ONC)_2 \end{array}$$

$$\qquad\qquad\qquad \text{Formonitrolic acid} \qquad\qquad \text{Mercury fulminate}$$

Fulminate can be prepared from acetaldehyde instead of from alcohol, and from substances which are convertible into acetaldehyde, such as paraldehyde, metaldehyde, dimethyl- and diethylacetal. Methyl alcohol, formaldehyde, propyl alcohol, butyraldehyde, glycol, and glyoxal do not yield fulminate.[6]

Fulminate can, however, be prepared from a compound which contains only one carbon atom. The sodium salt of nitromethane gives with an aqueous solution of mercuric chloride at 0° a white

[6] Wöhler and Theodorovits, *Ber.*, **38**, 1345 (1905).

precipitate of the mercuric salt of nitromethane which gradually becomes yellow and which, digested with warm dilute hydrochloric acid, yields mercury fulminate.[7]

$$CH_2=N\diagdown{}^O_{ONa} \longrightarrow C[H_2=N]\diagup{}^O_{OHgO}\diagdown{}^O_{}N=C[H_2] \longrightarrow Hg\diagdown{}^{ONC}_{ONC}$$

Sodium fulminate, soluble in water, has a molecular weight which corresponds [8] to the simple monomolecular formula, NaONC. These facts, taken together with the fact that mercury fulminate warmed with concentrated aqueous hydrochloric acid yields hydroxylamine and formic acid,[9] prove that fulminic acid is the oxime of carbon monoxide.

$$HO—N=C\diagup + 2H_2O \longrightarrow HO—NH_2 + H—COOH$$

Mercury fulminate dissolves readily in an aqueous solution of potassium cyanide to form a complex compound from which it is reprecipitated by the addition of strong acid. It dissolves in pyridine and precipitates again if the solution is poured into water. A sodium thiosulfate solution dissolves mercury fulminate with the formation of mercury tetrathionate and other inert compounds, and this reagent is used both for the destruction of fulminate and for its analysis.[10] The first reaction appears to be as follows.

$$Hg(ONC)_2 + 2Na_2S_2O_3 + H_2O \longrightarrow HgS_4O_6 + 2NaOH + NaCN + NaNCO$$

The cyanide and cyanate are salts of weak acids and are largely hydrolyzed, and the solution, if it is titrated immediately, appears to have developed four molecules of sodium hydroxide for every molecule of mercury in the sample which was taken. If the solution is allowed to stand, the alkalinity gradually decreases because of a secondary reaction whereby sulfate and thiocyanate are formed.

[7] Nef, *Ann.*, **280**, 275 (1894); Jones, *Am. Chem. J.*, **20**, 33 (1898).

[8] Wöhler, *Ber.*, **43**, 754 (1910).

[9] Carstenjen and Ehrenberg, *J. prak. Chem.*, [2] **25**, 232 (1883); Steiner, *Ber.*, **16**, 1484, 2419 (1883); Divers and Kawita, *J. Chem. Soc.*, **45**, 17 (1884).

[10] Brownsdon, *Chem. News*, **89**, 303 (1904); Philip, *Z. ges. Schiess- u. Sprengstoffw.*, **7**, 109, 156, 180, 198, 221 (1912); Taylor and Rinkenbach, "Explosives, Their Materials, Constitution, and Analysis," *U. S. Bureau of Mines Bulletin* 219, Washington, 1923, p. 62.

$HgS_4O_6 + NaCN + NaNCO + 2NaOH \longrightarrow$
$$HgSO_4 + Na_2SO_4 + 2NaNCS + H_2O$$

This reaction is restrained by a large excess of thiosulfate, and even more effectively by potassium iodide. A moderate excess of thiosulfate is commonly used, and an amount of potassium iodide

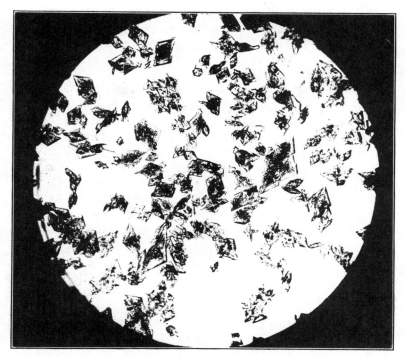

FIGURE 94. Mercury Fulminate Crystals for Use in Primer Composition (30×).

equal to 10 times the weight of the fulminate, and the titration for acidity (methyl orange indicator) is made as rapidly as possible. After that, the same solution is titrated with iodine (starch indicator) to determine the amount of unused thiosulfate and hence, by another method, the amount of actual fulminate in the sample. Speed is not essential in the second titration, for the iodine value does not change greatly with time as does the alkalinity. Blank determinations ought to be made because of the possibility that the iodide may contain iodate, and the apparent analytical results ought to be corrected accordingly.

Mercury fulminate has a specific gravity of 4.45, but a mass of the crystals when merely shaken down has an apparent density (gravimetric density) of about 1.75. In detonators the material is usually compressed to a density of about 2.5, but densities as high as 4.0 have been obtained by vigorous compression. Mercury fulminate crystallizes from water in crystals which contain $\frac{1}{2}H_2O$, from alcohol in crystals which are anhydrous. One liter of water at 12° dissolves 0.71 gram, at 49° 1.74 grams, and at 100° 7.7 grams.

Mercury fulminate is usually stored under water, or, where there is danger of freezing, under a mixture of water and alcohol. When wet it is not exploded by a spark or by ordinary shock, but care must be taken that no part of the individual sample is allowed to dry out, for wet fulminate is exploded by the explosion of dry fulminate. It is not appreciably affected by long storage, either wet or dry, at moderate temperatures. At the temperature of the tropics it slowly deteriorates and loses its ability to explode. At 35°C. (95°F.) it becomes completely inert after about 3 years, at 50°C. (122°F.) after about 10 months. The heavy, dark-colored product of the deterioration of fulminate is insoluble in sodium thiosulfate solution.

When loaded in commercial detonators mercury fulminate is usually compressed under a pressure of about 3000 pounds per square inch, and in that condition has a velocity of detonation of about 4000 meters per second, explodes from a spark, and, in general, has about the same sensitivity to fire and to shock as the loosely compressed material. When compressed under greater and greater pressures, it gradually loses its property of detonating from fire. After being pressed at 25,000–30,000 pounds per square inch, mercury fulminate becomes "dead pressed" and no longer explodes from fire but merely burns. Dead-pressed fulminate however is exploded by loosely pressed fulminate or other initial detonating agent, and then shows a higher velocity of detonation than when compressed at a lower density.

The temperature at which mercury fulminate explodes depends upon the rate at which it is heated and, to some extent, upon the state of subdivision of the sample. Wöhler and Matter [11] experi-

---

[11] *Z. ges. Schiess- u. Sprengstoffw.*, **2**, 181, 203, 244, 265 (1907).

mented with small particles of various primary explosives, heated in copper capsules in a bath of Wood's metal. If a sample did not explode within 20 seconds, the temperature of the bath was raised 10° and a new sample was tried. The temperatures at which explosions occurred were as follows.

Mercury fulminate	190°
Sodium fulminate	150°
Nitrogen sulfide	190°
Benzenediazonium nitrate	90°
Chloratotrimercuraldehyde	130°
Silver azide	290°
Basic mercury nitromethane	160°

In a later series of experiments Wöhler and Martin [12] studied a large number of fulminates and azides. The materials were in the form of microcrystalline powders, and all were compressed under the same pressure into pellets weighing 0.02 gram. The temperatures at which explosions occurred within 5 seconds were as follows.

Mercury fulminate	215°
Silver fulminate	170°
Copper fulminate	205°
Cadmium fulminate	215°
Sodium fulminate	215°
Potassium fulminate	225°
Thallium fulminate	120°
Cobalt azide	148°
Barium azide	152°
Calcium azide	158°
Strontium azide	169°
Cuprous azide	174°
Nickel azide	200°
Manganese azide	203°
Lithium azide	245°
Mercurous azide	281°
Zinc azide	289°
Cadmium azide	291°
Silver azide	297°
Lead azide	327°

Wöhler and Martin [13] in the same year also reported determinations of the smallest amounts of certain fulminates and

[12] Z. angew. Chem., 30, 33 (1917).
[13] Z. ges. Schiess- u. Sprengstoffw., 12, 1, 18 (1917).

azides necessary to cause the detonation of various high explosives.

SMALLEST AMOUNT (GRAMS) WHICH WILL CAUSE DETONATION OF:	Tetryl	Picric Acid	Trinitro-toluene	Trinitro-anisol	Trinitro-xylene
Cadmium azide........	0.01	0.02	0.04	0.1	..
Silver azide...........	0.02	0.035	0.07	0.26	0.25
Lead azide...........	0.025	0.025	0.09	0.28	..
Cuprous azide........	0.025	0.045	0.095	0.375	0.40
Mercurous azide.......	0.045	0.075	0.145	0.55	0.50
Thallium azide........	0.07	0.115	0.335	...	..
Silver fulminate.......	0.02	0.05	0.095	0.23	0.30
Cadmium fulminate....	0.008	0.05	0.11	0.26	0.35
Copper fulminate......	0.025	0.08	0.15	0.32	0.43
Mercury fulminate.....	0.29	0.30	0.36	0.37	0.40
Thallium fulminate.....	0.30	0.43	...	...	..

From these data it is apparent that mercury fulminate is by no means the most efficient initiating agent among the fulminates and azides. Silver fulminate is about 15 times as efficient as mercury fulminate for exploding tetryl, but only about ⅓ as efficient for exploding trinitroxylene. Mercury fulminate however will tolerate a higher temperature, and is much less sensitive to shock and friction, than silver fulminate. Lead azide, which has about the same initiating power as silver fulminate, has an explosion temperature more than 100° higher than that of mercury fulminate. Many other interesting inferences are possible from the data. Among them we ought especially to note that the order of the several fulminates and azides with respect to their efficiency in detonating one explosive is not always the same as their order with respect to their efficiency in detonating another.

Silver Fulminate

Silver fulminate is so sensitive and so dangerous to handle that it has not been used for practical purposes in blasting or in the military art. It early found use in toys, in tricks, and in such devices for entertainment as those which Christopher Grotz described in 1818 in his book on "The Art of Making Fireworks, Detonating Balls, &c."

Amusements with Fulminating Silver. . . .

Segars.

Are prepared by opening the smoking end, and inserting a little of the silver; close it carefully up, and it is done.

Spiders.

A piece of cork cut into the shape of the body of a spider, and a bit of thin wire for legs, will represent with tolerable exactness this insect. Put a small quantity of the silver underneath it; and on any female espying it, she will naturally tread on it, to crush it, when it will make a loud report.

Silver fulminate is still used for similar purposes in practical jokes, in toy torpedoes (see Vol. I, p. 106), and in the snaps or pull-crackers which supply the noise for bon-boms, joy-boms, and similar favors.

Silver fulminate is insoluble in nitric acid, and is decomposed by hydrochloric acid. It darkens on exposure to light. One liter of water at 13° dissolves 0.075 gram of the salt, and at 30° 0.18 gram. The double fulminate of silver and potassium, $AgONC \cdot KONC$, is soluble in 8 parts of boiling water.

Detonators

The discovery of the phenomenon of initiation by Alfred Nobel and the invention of the blasting cap [14] stand at the beginning of the development of modern explosives, perhaps the most important discovery and invention in the history of the art. The phenomenon has supplied a basis for the definition of high explosives, that is to say, of those explosives, whether sensitive or insensitive, which are incapable, without the invention, of being used safely and controllably or perhaps even of being used at all.

Nobel's experiments quickly led him to the form of the blasting cap which is now in use, a cylindrical capsule, generally of copper but sometimes of aluminum or zinc, filled for about half of its length with a compressed charge of primary explosive. The charge is fired either by an electric igniter or by a fuse, crimped into place, its end held firmly against the charge in order that the chances of a misfire may be reduced. Its action depends upon the development of an intense pressure or shock. Fulminate of mercury was the only substance known at the time of Nobel's invention which could be prepared and loaded for the purpose with reasonable safety, and caps loaded with straight fulminate were the first to

[14] Nobel, Brit. Pat. 1345 (1867).

be manufactured. The original fulminate detonators were numbered according to the amount of fulminate which they contained,

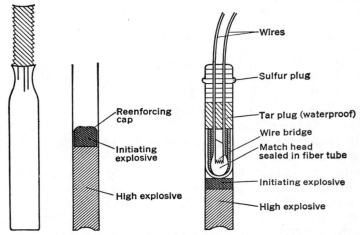

FIGURE 95. Blasting Caps. Detonator crimped to miner's fuse. Compound detonator. Compound electric detonator.

the same numbers being used throughout the world. The charges of fulminate for the various sizes are shown in the following table,

Detonator	Weight of Mercury Fulminate		External Dimensions of Capsule	
	Grams	Grains	Diameter, mm.	Length, mm.
No. 1	0.30	4.6	5.5	16
No. 2	0.40	6.2	5.5	22
No. 3	0.54	8.3	5.5	26
No. 4	0.65	10.0	6	28
No. 5	0.80	12.3	6	30–32
No. 6	1.00	15.4	6	35
No. 7	1.50	23.1	6	40–45
No. 8	2.00	30.9	6–7	50–55

along with the usual (but not universal) dimensions of the cylindrical copper capsules. The same numbers are now applied to commercial blasting caps of the same sizes, whatever the weights and

characters of the charges. A No. 6 cap, for example, is a cap of
the same size as one which contains 1 gram of straight fulminate.
No. 6 caps of different manufacturers may differ in their power

FIGURE 96. Manufacture of Detonators. (Courtesy Hercules Powder
Company.) The safe mixing of the primary explosive charge for blasting
caps is accomplished mechanically behind a concrete barricade by lifting
slowly and then lowering first one corner of the triangular rubber tray,
then the next corner, then the next, and so on. In the background, the
rubber bowl or box in which the mixed explosive is carried to the building
where it is loaded into caps.

as they differ in their composition. No. 6, 7, and 8 caps are the
only ones which are manufactured regularly in the United States,
and the No. 6 cap is the one which is most commonly used.

The fulminate in detonators was first modified by mixing it
with black powder, then with potassium nitrate, and later with

FIGURE 97. Manufacture of Detonators. (Courtesy Hercules Powder
Company.) Charging the capsules. Each of the holes in the upper steel
plate (*charging plate*) is of the right size to contain exactly enough
explosive for the charging of one detonator. The mixed explosive is
emptied onto the plate, the rubber-faced arm sweeps the material over
the charging plate filling all the holes and throwing the excess into the
box at the right. Under the charging plate is the thin *indexing plate*
which supplies a bottom to all the holes in the charging plate. The
detonator capsules, seen at the left, are placed under the indexing plate
and in line with the holes in the charging plate; the indexing plate is then
removed, the explosive falls down into the capsules, exactly the right
amount into each, and is later pressed into place.

potassium chlorate.[15] The chlorate mixtures soon attained commercial importance in the United States, and by 1910 had largely displaced straight fulminate. Detonators containing them dominated the market until recently, and are now largely, but not yet wholly, displaced by compound detonators in which use is made of the principle of the booster. Mixtures of fulminate and potassium chlorate are distinctly more hygroscopic than straight fulminate, but are cheaper and slightly safer to handle and to load. Weight for weight they make better detonators. Storm and Cope [16] in a series of experiments in the sand test bomb found that 80/20 fulminate-chlorate pulverizes more sand than the same weight of the 90/10 mixture and that this pulverizes more than straight fulminate. The results show that the sand test is an instrument of considerable precision. A difference of $\frac{1}{40}$ gram in the size of the charge of fulminate generally caused a difference of more than 1 gram in the weight of sand which was pulverized.

	WEIGHT OF SAND (GRAMS) PULVERIZED FINER THAN 30-MESH BY		
WEIGHT OF CHARGE, GRAMS	Mercury Fulminate	90/10 Fulminate-Chlorate	80/20 Fulminate-Chlorate
2.0000	56.94	58.57	59.68
1.5000	47.71	51.11	52.54
1.0000	38.33	40.13	41.42
0.7500	29.65	32.30	34.28
0.5000	22.45	23.07	23.22
0.4000	17.91	17.90	18.13
0.3500	14.16	15.13	15.94
0.3250	12.20	12.90	13.13
0.3000	10.01	12.71	12.61
0.2500	8.84	9.57	11.94
0.2250	6.93	8.71	10.29
0.2000	5.48	8.33	9.44

Storm and Cope [17] also used the sand test to determine the

[15] Detonators were manufactured abroad and sold for a time under Nobel's patent, A. V. Newton (from A. Nobel, Paris), Brit. Pat. 16,919 (1887), covering the use, instead of fulminate, of a granulated mixture of lead picrate, potassium picrate, and potassium chlorate, but the invention apparently contributed little to the advance of the explosives art.

[16] C. G. Storm and W. C. Cope, "The Sand Test for Determining the Strength of Detonators," *U. S. Bur. Mines Tech. Paper* 125, Washington, 1916, p. 43.

[17] *Ibid.*, p. 59.

minimum amounts of fulminate and of the fulminate-chlorate mixtures which were necessary to detonate several high explosives in reenforced detonators. It is necessary to specify that the tests were made with reenforced detonators, for the results would have been quite different if reenforcing caps had not been used. In an ordinary detonator TNA required 0.3125 gram of 80/20 fulminate-chlorate instead of the 0.1700 gram which was sufficient when a reenforced detonator was used.

	MINIMUM INITIATING CHARGE (GRAMS) NECESSARY FOR EXPLOSION OF 0.4 GRAM OF		
PRIMARY EXPLOSIVE	TNT	TNA	Picric Acid
Mercury fulminate	0.26	0.20	0.25
90/10 Fulminate-chlorate	0.25	0.17	0.23
80/20 Fulminate-chlorate	0.24	0.17	0.22

The reenforced detonators which were used in this work were made by introducing the weighed charge of high explosive into the detonator shell and the weighed charge of primary explosive into the small reenforcing cap while the latter was held in a cavity in a brass block which served to prevent the explosive from falling through the hole in the end of the cap. The primary explosive was then pressed down gently by means of a wooden rod, the cap was filled by adding a sufficient quantity of the high explosive from the detonator shell, this was similarly pressed down, and the reenforcing cap was then removed from the brass block and inserted carefully in the detonator shell with its perforated end upward. The detonator was then placed in a press block, a plunger inserted, and the contents subjected to a pressure of 200 atmospheres per square inch maintained for 1 minute. The pressure expanded the reenforcing cap against the detonator shell and fixed it firmly in place.

The minimum initiating charge was determined as follows. The amount of sand pulverized by a detonator loaded, say, with TNT and with fulminate insufficient to explode the TNT was determined. Another experiment with a slightly larger amount of fulminate was tried. If this showed substantially the same amount of sand pulverized, then the charge of fulminate was increased still further, and so on, until a sudden large increase in the amount of sand pulverized showed that the TNT had detonated. After this point had been reached, further increases in the amount of fulminate caused only slight increases in the amount of sand pulverized. The magnitude of the effects, and the definiteness of the results, are shown by the following data of Storm and Cope.[18]

WEIGHT OF SAND (GRAMS) PULVERIZED FINER THAN 30-MESH BY REENFORCED DETONATOR CONTAINING 0.40 GRAM TNT AND A PRIMING CHARGE (GRAMS) OF

PRIMARY EXPLOSIVE	0.3000	0.2800	0.2600	0.2500	0.2400	0.2300
Mercury fulminate	34.20	34.70	33.00	13.55	12.60	...
	...	...	31.50	...	...	...
	...	...	30.00	...	...	...
	...	...	32.70	...	...	...
	...	...	32.00	...	...	...
90/10 Fulminate-chlorate	...	...	33.55	34.45	32.95	13.90
	...	...	34.05	34.67	13.20	...
	...	...	34.35	34.07	...	...
	...	...	34.42	35.07	...	...
	...	...	34.70	33.80	...	...
80/20 Fulminate-chlorate	...	...	...	...	34.40	16.80
	...	...	...	...	34.60	...
	...	...	...	...	34.60	...
	...	...	...	...	33.80	...
	...	...	...	...	34.85	...

Fulminate owes its success as an initiating agent primarily to the fact that it explodes easily from fire—and it catches the fire more readily than do lead azide and many another primary explosive—to the fact that it quickly attains its full velocity of detonation within a very short length of material, and probably also to the fact that the heavy mercury atom which it contains enables it to deliver an especially powerful blow. Its maximum velocity of detonation is much lower than that of TNT and similar substances, and its power to initiate the detonation of high explosives is correspondingly less. Wöhler [19] in 1900 patented detonators in which a main charge of TNT or other nitro compound is initiated by a relatively small charge of fulminate.

[18] *Ibid.*, p. 55.

[19] Brit. Pat. 21,065 (1900). For an account of Wöhler's theory of initiation see *Z. ges. Schiess- u. Sprengstoffw.*, **6**, 253 (1911) and *Z. angew. Chem.*, **24**, 1111, 2089 (1911).

Detonators which thus make use of the principle of the booster are known as compound detonators and are made both with and without reenforcing caps. Some manufacturers insert the reenforcing cap with the perforated end down, others with the perforated end up.[20]

Not long after Curtius [21] had discovered and described hydrazoic (hydronitric) acid and its salts, Will and Lenze [22] experimented with the azides (hydronitrides, hydrazotates) at the military testing station at Spandau, but a fatal accident put an end to their experiments and their results were kept secret by the German war office. Wöhler and Matter [23] later studied several primary explosives in an effort to find a substitute for fulminate, and in 1907, in ignorance of the earlier work of Will and Lenze, published experiments which demonstrated the great effectiveness of the azides. At about the same time, the first attempt to use lead azide practically in the explosives industry was made by F. Hyronimus in France who secured a patent [24] in February, 1907, for the use of lead azide in detonators, to replace either wholly or in part the mercury fulminate which had theretofore been used, and this whether or not the fulminate would ordinarily be used alone or in conjunction with some other explosive substance such as picric acid or trinitrotoluene. In March of the same year Wöhler in Germany patented,[25] as a substitute for fulminate, the heavy metal salts of hydrazoic acid, "such as silver and mercury azides." He pointed out, as the advantages of these substances, that a smaller weight of them is necessary to produce detonation than is necessary of mercury fulminate, as, for example, that a No. 8 blasting cap containing 2 grams of mercury fulminate can be replaced, for use in detonating explosives, by a No. 8 copper capsule containing 1 gram of picric acid on top of which 0.023 gram of silver azide has been com-

---

[20] In addition to its other functions, the reenforcing cap tends toward greater safety by preventing actual contact between the primary explosive and the squarely cut end of the miner's fuse to which the detonator is crimped.

[21] *Ber.*, **23**, 3023 (1890); *ibid.*, **24**, 3341 (1891).

[22] Cf. Will, *Z. ges. Schiess- u. Sprengstoffw.*, **9**, 52 (1914).

[23] *Loc. cit.*

[24] French Pat. 384,792 (February 14, 1907), Supplement No. 8872 (process of manufacture), January 13, 1908.

[25] Ger. Pat. 196,824 (March 2, 1907).

pressed. In February of the next year Wöhler was granted a French patent [26] in which lead azide was specifically mentioned, but the use of this substance had already been anticipated by the patent of Hyronimus. Lead azide was soon afterwards manufactured commercially in Germany and in France, and compound detonators containing this material were used fairly generally in Europe at the time of the first World War. A few years later the manufacture of lead azide detonators was commenced in the United States. In this country compound detonators having a base charge of tetryl and primed with 80/20 fulminate-chlorate or with lead azide have been superseded in part by detonators loaded with a more powerful high-explosive charge of nitromannite, PETN, or diazodinitrophenol and primed with lead azide, alone or sensitized to flame by the addition of lead styphnate or tetracene, or with diazodinitrophenol as the primary explosive.

Testing of Detonators

Among the tests which are used for determining the relative efficiency of detonators,[27] the lead block or small Trauzl test, in which the detonators are fired in holes drilled in lead blocks and the resulting expansions of the holes are measured, and the lead or aluminum plate test in which the detonators are stood upright upon the plates and fired, and the character and extent of the effects upon the plates are observed, have already been mentioned.[28] The first of these gives results which are expressible by numbers, and in that sense quantitative, and it is evident that both methods may be applied, for example, to the determination of the minimum amount of primary explosive necessary for the initiation of a high explosive, for both show notably different effects according as the high explosive explodes or not. Another useful test is the determination of the maximum distance through which the detonator is capable of initiating the explosion of some standard material, say, a piece of cordeau loaded with TNT. In the *nail test*,[29] a wire nail is fastened to the side of the detonator, the detonator is fired, and the angle of the bend which the ex-

---

[26] French Pat. 387,640 (February 28, 1908).

[27] Clarence Hall and Spencer P. Howell, "Investigations of Detonators and Electric Detonators," *U. S. Bur. Mines Bull.* 59, Washington, 1913.

[28] Vol. I, p. 26.

[29] Hall and Howell, *op. cit.*, p. 25.

plosion imparts to the nail is measured. The sand test, in which the detonator is fired in the center of a mass of carefully screened sand contained in a suitable bomb and the sand which has been pulverized is screened off and weighed, is the most precise and significant of the tests on detonators. It is a real test of brisance, and its usefulness is not limited to the study of detonators but may be extended to the study of high explosives as well. Thus,

FIGURE 98. U. S. Bureau of Mines Sand Test Bomb No. 1. (Courtesy U. S. Bureau of Mines.) At left, assembled for making the test. At right, disassembled showing the parts. Two covers, one with a single hole for miner's fuse, the other with two holes for the two wires of an electric detonator.

two explosives may be compared by loading equal amounts in detonator shells, priming with equal amounts of the same initiator, firing in the sand test bomb, and comparing the amounts of sand pulverized.

The sand test was devised in 1910 by Walter O. Snelling, explosives chemist of the U. S. Bureau of Mines, who worked out the technique of its operation and designed the standard Bureau of Mines sand test bomb No. 1 which was used in his own investigations and in those of Storm and Cope.[30] Munroe and Taylor [31]

[30] Snelling, *Proc. Eng. Soc. Western Pennsylvania,* **28,** 673 (1912); Storm and Cope, *loc. cit.*

[31] C. E. Munroe and C. A. Taylor, "Methods of Testing Detonators," *U. S. Bur. Mines Repts. of Investigations* 2558, December, 1923.

later recommended a bomb of larger diameter, Bureau of Mines sand test bomb No. 2, as being able to differentiate more exactly between the different grades of detonators in commercial use. The test grew out of an earlier test which Snelling had developed in 1908 for measuring the strength of detonating agents. Starting

FIGURE 99. Walter O. Snelling. (Metzger & Son.) Devised the sand test. Has worked extensively with nitrostarch explosives and has patented many improvements in military and in mining explosives. Chemist at the U. S. Bureau of Mines, 1908-1916; Director of Research, Trojan Powder Company, 1917—.

with the thought that true explosives, when subjected to a sufficiently strong initiating influence, detonate in such manner as to set free more energy than that which had been applied to them by the initiating charge, he tested several materials which failed to be true explosives and, although decomposed by the detonating agent, did not give off energy enough to continue their own decomposition and to propagate a detonation wave. Copper oxalate was the best of the "near explosives" which he tried. He found it possible to measure the initiating effect of mercury fulminate and of other initial detonators by firing them in compositions consisting partly or wholly of copper oxalate, and then by chemical means determining the amount of the oxalate which had been decomposed. The experiments were carried out in a small steel bomb, the detonator was placed in the middle of a mass of oxalate or of oxalate composition, and sand was put in on top to fill the bomb completely. The fact that part of the sand was pulverized by the force of the explosion suggested that the mechanical effect of the initiator might perhaps serve as an approximate measure of the detonating efficiency; the oxalate was omitted, the bomb was filled entirely with sand, and the sand test was devised. Before Snelling left the Bureau of Mines in 1912 he had made about 40 tests on ordinary and electric detonators. Storm and Cope extended the usefulness of the test and applied it not only to the study of detonators but also to the study of the materials out of which detonators are constructed, both initial detonating agents and high explosives.

Lead Azide

Lead azide is a more efficient detonating agent than mercury fulminate. It requires a higher temperature for its spontaneous explosion, and it does not decompose on long continued storage at moderately elevated temperatures. It cannot be dead-pressed by any pressure which occurs in ordinary manufacturing operations. Lead azide pressed into place in a detonator capsule takes the fire less readily, or explodes from spark less readily, than mercury fulminate. For this reason the main initiating charge of lead azide in a blasting cap is generally covered with a layer of lead styphnate, or of styphnate-azide mixture or other *sensitizer,* which explodes more easily, though less violently, from fire, and serves to initiate the explosion of the azide.

Lead azide is not used in primers where it is desired to produce fire or flame from impact. Fulminate mixtures and certain mixtures which contain no fulminate are preferred for this purpose. Lead azide is used where it is desired to produce, either from flame or from impact, an initiatory shock for the detonation of a high explosive—in compound detonators as already described, and in the detonators of artillery fuzes. For the latter purpose, caps containing azide and tetryl (or other booster explosive) are used; the azide is exploded by impact, and the tetryl communicates the explosion to the booster or perhaps to the main charge of the shell.

Lead azide is produced as a white precipitate by mixing a solution of sodium azide with a solution of lead acetate or lead nitrate. It is absolutely essential that the process should be carried out in such manner that the precipitate consists of very small particles. The sensitivity of lead azide to shock and to friction increases rapidly as the size of the particles increases. Crystals 1 mm. in length are liable to explode spontaneously because of the internal stresses within them. The U. S. Ordnance Department specifications require that the lead azide shall contain no needle-shaped crystals more than 0.1 mm. in length. Lead azide is about as sensitive to impact when it is wet as when it is dry. Dextrinated lead azide can apparently be stored safely under water for long periods of time. The belief exists, however, that crystalline "service azide" becomes more sensitive when stored under water because of an increase in the size of the crystals.

The commercial preparation of lead azide is carried out on what is practically a laboratory scale, 300 grams of product constituting an ordinary single batch. There appear to be diverse opinions as to the best method of precipitating lead azide in a finely divided condition. According to one, fairly strong solutions are mixed while a gentle agitation is maintained, and the precipitate is removed promptly, and washed, and dried. According to another, dilute solutions ought to be used, with extremely violent agitation, and a longer time ought to be devoted to the process. The preparation is sometimes carried out by adding one solution to the other in a nickel vessel, which has corrugated sides, and is rotated around an axis which makes a considerable angle with the vertical, thereby causing turbulence in the liquid. The precipitation is sometimes carried out in the presence of dissolved colloidal material, such as gelatin or dextrin, which tends to prevent the formation of large crystals. Sometimes the lead azide is precipitated on starch or wood pulp, either of which will take up about 5 times its own weight of the material, and the impregnated starch is worked up, say, by tumbling in a sweetie barrel with a little dextrine, to form a free-flowing granular mass which can conveniently be loaded into detonators, or the impregnated wood pulp is converted into pasteboard which is cut into discs for loading. A small amount of basic salt in the lead azide makes it somewhat less sensitive to impact and slightly safer to handle, but has no appreciable effect upon its efficacy as an initiator.

The commercial preparation of the azides is carried out either by the interaction of hydrazine with a nitrite or by the interaction of sodamide with nitrous oxide. The first of these methods

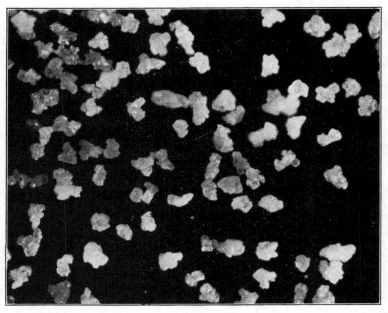

FIGURE 100. Technical Lead Azide, 90-95% pure (75×). For use in detonators. Precipitated in the presence of dextrin, it shows no crystal faces under the microscope.

follows from the original work of Curtius,[32] the second from a reaction discovered by Wislicenus [33] in 1892 and later developed for plant scale operation by Dennis and Browne.[34] Curtius first prepared hydrazoic acid by the action of aqueous or alcoholic alkali or ammonia on acyl azides prepared by the action of nitrous acid on acyl hydrazides. The hydrazides are formed by

[32] Loc. cit.; also J. prak. Chem., [2] 50, 275 (1894); Ber., 29, 759 (1896); cf. survey by Darapsky in Z. ges. Schiess- u. Sprengstoffw., 2, 41, 64 (1907).

[33] Ber., 25, 2084 (1892).

[34] Z. anorg. allgem. Chem., 40, 68 (1904).

the interaction of hydrazine with esters just as the amides are formed by the corresponding interaction of ammonia.

$$R—COOC_2H_5 + NH_2—NH_2 \longrightarrow C_2H_5OH + R—CO—NH—NH_2$$
Acyl hydrazide

$$R—CO—NH—NH_2 + HONO \longrightarrow 2H_2O + R—CO—N_3$$
Acyl azide

$$R—CO—N_3 + H_2O \longrightarrow R—COOH + HN_3$$

$$R—CO—N_3 + NH_3 \longrightarrow R—CO—NH_2 + HN_3$$
Hydrazoic acid

By acidifying the hydrolysis mixture with sulfuric acid and by fractionating the product, Curtius procured anhydrous hydrazoic as a colorless liquid which boils at 37°. Hydrazoic acid is intensely poisonous and bad smelling. It is easily exploded by flame, by a brisant explosive, or by contact with metallic mercury. The anhydrous substance is extremely dangerous to handle, but dilute solutions have been distilled without accident.

Angeli [35] obtained a white precipitate of insoluble silver azide by mixing saturated solutions of silver nitrite and hydrazine sulfate and allowing to stand in the cold for a short time. Dennstedt and Göhlich [36] later procured free hydrazoic acid by the interaction of hydrazine sulfate and potassium nitrite in aqueous solution.

$$NH_2—NH_2 \cdot H_2SO_4 + KONO \longrightarrow KHSO_4 + 2H_2O + HN_3$$
Hydrazine sulfate

The yield from this reaction is greatest if the medium is alkaline, for nitrous acid attacks hydrazoic acid oxidizing it with the liberation of nitrogen. If hydrazine sulfate [37] is used in the mixture, the resulting hydrazoic acid is not available for the preparation of lead azide until it has been distilled out of the solution. (Lead ions added to the solution would cause the precipitation of lead sulfate.) The reaction mixture may be acidified with sulfuric acid, a little ammonium sulfate may be added in order that the

[35] *Rend. acc. Lincei*, [5] 2, I, 599 (1893).

[36] *Chem.-Ztg.*, 21, 876 (1897).

[37] Hydrazine is produced commercially by treating ammonia in aqueous solution with sodium hypochlorite to form chloramine, $NH_2—Cl$, and by coupling this with another molecule of ammonia to form hydrazine and hydrochloric acid. Sulfuric acid is added to the liquid, sparingly soluble hydrazine sulfate crystallizes out, and it is in the form of this salt that hydrazine generally occurs in commerce.

ammonia may react with any unchanged nitrous acid which may be present, and the hydrazoic acid may be distilled directly into a solution of a soluble lead salt; but methods involving the distillation of hydrazoic acid present many dangers and have not found favor for commercial production. The alternative is to work with materials which contain no sulfate, and to isolate the azide by precipitation from the solution, and it is by this method that sodium azide (for the preparation of lead azide) is generally manufactured in this country and in England.

Hydrazine [38] reacts in alcohol solution with ethyl nitrite [39] and caustic soda to form sodium azide which is sparingly soluble in alcohol (0.315 gram in 100 grams of alcohol at 16°) and precipitates out.

$$NH_2—NH_2 + C_2H_5ONO + NaOH \longrightarrow NaN_3 + C_2H_5OH + 2H_2O$$

The sodium azide is filtered off, washed with alcohol, and dried. It is soluble in water to the extent of 42 grams in 100 grams of water at 18°. It is not explosive, and requires no particular precaution in its handling.

Azide has been manufactured in France and in Germany by the sodamide process. Metallic sodium is heated at about 300° while dry ammonia gas is bubbled through the molten material.

$$2Na + 2NH_3 \longrightarrow 2NaNH_2 + H_2$$

The sodamide which is formed remains liquid (m.p. 210°) and does not prevent contact between the remaining sodium and the ammonia gas. The progress of the reaction is followed by passing the effluent gas through water which absorbs the ammonia and allows the hydrogen to pass; if there is unabsorbed gas which forms an explosive mixture with air, the reaction is not yet complete. For the second step, the sodamide is introduced into a nickel or nickel-lined, trough-shaped autoclave along the bottom

---

[38] Hydrazine hydrate is actually used. It is an expensive reagent procured by distilling hydrazine sulfate with caustic soda in a silver retort. It is poisonous, corrosive, strongly basic, and attacks glass, cork, and rubber. Pure hydrazine hydrate is a white crystalline solid which melts at 40° and boils at 118°, but the usual commercial material is an 85% solution of the hydrate in water.

[39] It is necessary to use ethyl nitrite or other alcohol-soluble nitrous ester, instead of sodium nitrite, in order that advantage may be taken of a solvent from which the sodium azide will precipitate out.

of which there extends a horizontal shaft equipped with teeth. The air in the apparatus is displaced with ammonia gas, the autoclave is heated to about 230°, and nitrous oxide is passed in while the horizontal stirrer is rotated. The nitrous oxide reacts with one equivalent of sodamide to form sodium azide and water. The water reacts with a second equivalent of sodamide to form sodium hydroxide and ammonia.

$$NaNH_2 + N_2O \longrightarrow NaN_3 + H_2O$$
$$NaNH_2 + H_2O \longrightarrow NaOH + NH_3$$

The reaction is complete when no more ammonia is evolved. The product, which consists of an equimolecular mixture of sodium hydroxide and sodium azide, may be taken up in water and neutralized carefully with nitric acid, and the resulting solution may be used directly for the preparation of lead azide, or the product may be fractionally crystallized from water for the production of sodium azide. The same material may be procured by washing the product with warm alcohol which dissolves away the sodium hydroxide.

The different methods by which hydrazoic acid and the azides may be prepared indicate that the acid may properly be represented by any one or by all of the following structural formulas.

$$H-N{\overset{N}{\underset{N}{<}}}\;|\qquad H-N{\overset{N}{\underset{N}{<}}}\;\|\qquad H-N{=}N{\equiv}N$$

Hydrazoic acid is a weak acid; its ionization constant at 25°, 1.9×10^{-5}, is about the same as that of acetic acid at 25°, 1.86×10^{-5}. It dissolves zinc, iron, magesium, and aluminum, forming azides with the evolution of hydrogen and the production of a certain amount of ammonia. It attacks copper, silver, and mercury, forming azides without evolving hydrogen, and is reduced in part to ammonia and sometimes to hydrazine and free nitrogen. Its reaction with copper, for example, is closely analogous to the reaction of nitric acid with that metal.

$$Cu + 3HN_3 \longrightarrow Cu(N_3)_2 + N_2 + NH_3$$
$$3Cu + 8HNO_3 \longrightarrow 3Cu(NO_3)_2 + 2NO + 4H_2O$$

So also, like nitric acid, it oxidizes hydrogen sulfide with the liberation of sulfur.

$$H_2S + HN_3 \longrightarrow S + N_2 + NH_3$$
$$3H_2S + 2HNO_3 \longrightarrow 3S + 2NO + 4H_2O$$

Mixed with hydrochloric acid it forms a liquid, comparable to *aqua regia*, which is capable of dissolving platinum.

$$Pt + 2HN_3 + 4HCl \longrightarrow PtCl_4 + 2N_2 + 2NH_3$$
$$3Pt + 4HNO_3 + 12HCl \longrightarrow 3PtCl_4 + 4NO + 8H_2O$$

Hydrazoic acid and permanganate mutually reduce each other with the evolution of a mixture of nitrogen and oxygen. The acid and its salts give with ferric chloride solution a deep red coloration, similar to that produced by thiocyanates, but the color is discharged by hydrochloric acid.

The solubilities of the azides in general are similar to those of the chlorides. Thus, silver azide is soluble in ammonia water and insoluble in nitric acid. Lead azide, like lead chloride, is sparingly soluble in cold water, but hot water dissolves enough of it so that it crystallizes out when the solution is cooled. One hundred grams of water at 18° dissolve 0.03 gram, at 80° 0.09 gram.

The true density of lead azide is 4.8, but the loose powder has an apparent density of about 1.2.

Lead azide is dissolved by an aqueous solution of ammonium acetate, but it is not destroyed by it. The solution contains azide ions and lead ions, the latter quantitatively precipitable as lead chromate, $PbCrO_4$, by the addition of potassium dichromate solution. Lead azide in aqueous suspension is oxidized by ceric sulfate with the quantitative production of nitrogen gas which may be collected in an azotometer and used for the determination of the azide radical.

$$Pb(N_3)_2 + 2Ce(SO_4)_2 \longrightarrow PbSO_4 + 3N_2 + Ce_2(SO_4)_3$$

Nitrous acid oxidizes hydrazoic acid with the evolution of nitrogen. A dilute solution of nitric or acetic acid, in which a little sodium nitrite has been dissolved, dissolves and destroys lead azide. Such a solution may conveniently be used for washing floors, benches, etc., on which lead azide may have been spilled.

Silver Azide

Silver azide is a more efficient initiator than mercury fulminate, and about as efficient as lead azide. It melts at 251° and decomposes rapidly above its melting point into silver and nitrogen. Its temperature of spontaneous explosion varies somewhat according to the method of heating, but is considerably higher than that of mercury fulminate and slightly lower than that of lead azide. Taylor and Rinkenbach [40] reported 273°. Its sensitivity to shock, like that of lead azide, depends upon its state of subdivision.

FIGURE 101. William H. Rinkenbach. Has published many studies on the physical, chemical, and explosive properties of pure high-explosive substances and primary explosives. Research Chemist, U. S. Bureau of Mines, 1919-1927; Assistant Chief Chemist, Picatinny Arsenal, 1927-1929; Chief Chemist, 1929—.

Taylor and Rinkenbach prepared a "colloidal" silver azide which required a 777-mm. drop of a 500-gram weight to cause detonation. Mercury fulminate required a drop of 127 mm. According to the same investigators 0.05 gram of silver azide was necessary to cause the detonation of 0.4 gram of trinitrotoluene in a No. 6 detonator capsule, whether the charge was confined by a reenforcing cap or not, as compared with 0.24 gram of mercury fulminate when the charge was confined by a reenforcing cap and 0.37 gram when it was not confined. They also measured the sand-crushing power of silver azide when loaded into No. 6 detonator capsules and compressed under a pressure of 1000 pounds per square inch, and compared it with that of mercury fulminate, with the results which are tabulated below. It thus appears that

	WEIGHT OF SAND CRUSHED (GRAMS) BY	
WEIGHT OF CHARGE, GRAMS	Silver Azide	Mercury Fulminate
0.05	1.4	0.00
0.10	3.3	0.00
0.20	6.8	4.2
0.30	10.4	8.9
0.50	18.9	16.0
0.75	30.0	26.1
1.00	41.1	37.2

the sand-crushing power of silver azide is not as much greater than the sand-crushing power of mercury fulminate as the difference in their initiatory powers would suggest. Storm and Cope [41] in their studies on the sand test found that the powers of fulminate and of fulminate-chlorate mixtures to crush sand were about proportional to the initiatory powers of these materials, but the present evidence indicates that the law is not a general one.

Cyanuric Triazide

Cyanuric triazide,[42] patented as a detonating explosive by Erwin Ott in 1921, is prepared by adding powdered cyanuric chloride, slowly with cooling and agitation, to a water solution of slightly more than the equivalent quantity of sodium azide.

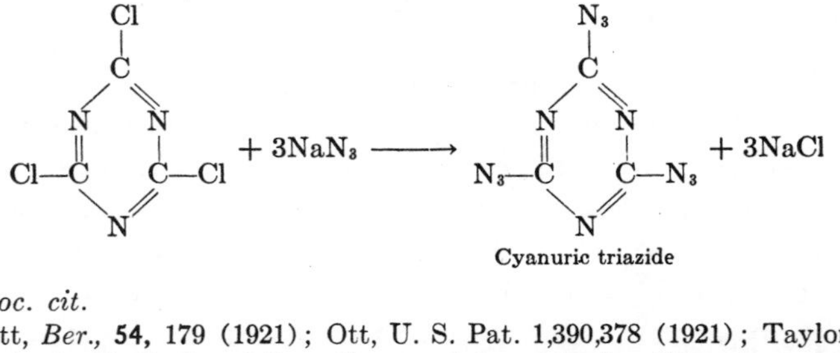

Cyanuric triazide

[41] *Loc. cit.*

[42] Ott, *Ber.*, **54**, 179 (1921); Ott, U. S. Pat. 1,390,378 (1921); Taylor and Rinkenbach, *U. S. Bur. Mines Repts. of Investigation* 2513, August, 1923; Kast and Haid, *Z. angew. Chem.*, **38**, 43 (1925).

The best results are secured if pure and finely powdered cyanuric chloride is used, yielding small crystals of pure cyanuric triazide in the first instance, in such manner that no recrystallization, which might convert them into large and more sensitive crystals, is necessary. Cyanuric chloride, m.p. 146°, b.p. 190°, is prepared by passing a stream of chlorine into a solution of hydrocyanic acid in ether or chloroform or into liquid anhydrous hydrocyanic acid exposed to sunlight. It is also formed by distilling cyanuric acid with phosphorus pentachloride and by the polymerization of cyanogen chloride, Cl—CN, after keeping in a sealed tube.

Cyanuric triazide is insoluble in water, slightly soluble in cold alcohol, and readily soluble in acetone, benzene, chloroform, ether, and hot alcohol. It melts at 94°, and decomposes when heated above 100°. It may decompose completely without detonation if it is heated slowly, but it detonates immediately from flame or from sudden heating. The melted material dissolves TNT and other aromatic nitro compounds. Small crystals of cyanuric triazide are more sensitive than small crystals of mercury fulminate, and have exploded while being pressed into a detonator capsule. Large crystals from fusion or from recrystallization have detonated when broken by the pressure of a rubber policeman.[43]

Cyanuric triazide is not irritating to the skin, and has no poisonous effects on rats and guinea pigs in fairly large doses.[43]

Taylor and Rinkenbach have reported sand test data which show that cyanuric triazide is much more brisant than mercury fulminate.[43]

WEIGHT OF EXPLOSIVE, GRAMS	WEIGHT OF SAND CRUSHED (GRAMS) BY	
	Cyanuric Triazide	Mercury Fulminate
0.050	2.6	. . .
0.100	4.8	. . .
0.200	12.2	3.8
0.400	33.2	12.2
0.600	54.4	20.1
0.800	68.9	28.2
1.000	78.6	36.8

In conformity with these results are the findings of Kast and Haid who reported that cyanuric triazide has a higher velocity of detonation than mercury fulminate. They made their measure-

43 Taylor and Rinkenbach, loc. cit., footnote 42.

ments on several primary explosives loaded into detonator capsules 7.7 mm. in internal diameter and compressed to the densities which they usually have in commercial detonators.[44]

EXPLOSIVE	DENSITY	VELOCITY OF DETONATION, METERS PER SECOND
Cyanuric triazide	1.15	5545
Lead azide	3.8	4500
Mercury fulminate	3.3	4490
Mixture: $Hg(ONC)_2$ 85%, $KClO_3$ 15%	3.1	4550
Lead styphnate	2.6	4900

Taylor and Rinkenbach found that cyanuric triazide is a more efficient initiator of detonation than mercury fulminate. This result cannot properly be inferred from its higher velocity of detonation, for there is no direct correlation between that quality and initiating efficiency. Lead azide is also a much more efficient initiator than mercury fulminate but has about the same velocity of detonation as that substance. The following results [43] were secured by loading 0.4 gram of the high explosive into detonator capsules, pressing down, adding an accurately weighed amount of the initiator, covering with a short reenforcing cap, and pressing with a pressure of 200 atmospheres per square inch. The size of the initiating charge was reduced until it was found that a further reduction resulted in a failure of the high explosive to detonate.

HIGH EXPLOSIVE	MINIMUM INITIATING CHARGE (GRAMS) OF	
	Cyanuric Triazide	Mercury Fulminate
Trinitrotoluene	0.10	0.26
Picric acid	0.05	0.21
Tetryl	0.04	0.24
Tetranitroaniline	0.09	0.20
Ammonium picrate	0.15	0.85

Cyanuric triazide is slightly more hygroscopic and distinctly more sensitive in the drop test than fulminate of mercury.[44] It is slightly volatile, and must be dried at as low a temperature as possible, preferably in vacuum.[44] Detonators in which it is used

[44] Kast and Haid, *loc. cit.*, footnote 42.

TEMPERATURE OF EXPLOSION

EXPLOSIVE	When temperature is raised 20° per minute in Glass Tube	Iron Tube	In Iron Tube Temp., °C.	Elapsed time, seconds	
Cyanuric triazide	206°	205°	200	40,	2
	208°	207°	205	0,	
Lead azide	338°	337°	335	12,	9
			340	5,	7
			345	7,	6
			350	4,	5
			355	0	
			360	0	
Mercury fulminate	175°	166°	145	480,	331
			150	275,	255
			155	135,	165
			160	64,	85
			170	40,	35
			180	15,	13
			190	10,	8
			195	8,	7
			200	7,	8
			205	5,	5
			210	1,	3
			215	0	
Mixture:	168°	169°	145	370,	365
$Hg(ONC)_2$ 85%	171°	170°	150	210,	215
$KClO_3$ 15%			155	155,	145
			160	125,	74
			170	45,	50
			180	23,	22
			190	8,	8
			195	7,	7
			200	7,	8
			205	7,	6
			210	4,	3
			215	0	
Lead styphnate	276°	275°	250	90,	85
	277°	276°	265	65,	45
		275°	270	0	

must be manufactured in such a way that they are effectively sealed.

Kast and Haid have determined the temperatures at which cyanuric triazide and certain other initiators explode spontaneously, both by raising the temperature of the samples at a constant rate and by keeping the samples at constant temperatures and noting the times which elapsed before they exploded. When no measurable time elapsed, the temperature was "the temperature of instantaneous explosion." Their data are especially interesting because they show the rate of deterioration of the materials at various temperatures.[44]

Trinitrotriazidobenzene

1,3,5-Trinitro-2,4,6-triazidobenzene [45] is prepared from aniline by the reactions indicated below.

Aniline is chlorinated to form trichloroaniline. The amino group is eliminated from this substance by means of the diazo reaction, and the resulting *sym*-trichlorobenzene is nitrated. The nitration, as described by Turek, is carried out by dissolving the material in warm 32% oleum, adding strong nitric acid, and heating at 140–150° until no more trinitrotrichlorobenzene, m.p. 187°, precipitates out. The chlorine atoms of this substance are then replaced by azido groups. This is accomplished by adding an acetone solution of the trinitrotrichlorobenzene, or better, the powdered substance alone, to an actively stirred solution of sodium azide in moist alcohol. The precipitated trinitrotriazidobenzene is filtered off, washed with alcohol and with water, and, after drying, is sufficiently pure for technical purposes. It may be

[45] Turek, *Chimie et industrie*, **26**, 781 (1931); Ger. Pat. 498,050; Brit. Pat. 298,981. Muraour, *Mém. artillerie franç.*, **18**, 895 (1939).

purified further by dissolving in chloroform and allowing to cool, greenish-yellow crystals, m.p. 131° with decomposition. It is decomposed slowly by boiling in chloroform solution.

Trinitrotriazidobenzene is readily soluble in acetone, moderately soluble in chloroform, sparingly in alcohol, and insoluble in water. It is not hygroscopic, is stable toward moisture, and does not attack iron, steel, copper, or brass in the presence of moisture. It is not appreciably volatile at 35–50°. It darkens in color superficially on exposure to the light. It decomposes on melting with the evolution of nitrogen and the formation of hexanitrosobenzene.

The same reaction occurs at lower temperatures: 0.665% of a given portion of the material decomposes in 3 years at 20°, 2.43% in 1 year at 35°, 0.65% in 10 days at 50°, and 100% during 14 hours heating at 100°. The decomposition is not self-catalyzed. The product, hexanitrosobenzene, m.p. 159°, is stable, not hygroscopic, not a primary explosive, and is comparable to tetryl in its explosive properties.

Trinitrotriazidobenzene, if ignited in the open, burns freely with a greenish flame; enclosed in a tube and ignited, it detonates with great brisance. It is less sensitive to shock and to friction than mercury fulminate. It gives a drop test of 30 cm., but it may be made as sensitive as fulminate by mixing with ground glass. The specific gravity of the crystalline material is 1.8054. Under a pressure of 3000 kilograms per square centimeter it yields blocks having a density of 1.7509, under 5000 kilograms per square centimeter 1.7526. One gram of TNT compressed in a No. 8 detonator shell under a pressure of 500 kilograms per square centimeter, with trinitrotriazidobenzene compressed on top of it under 300 kilograms per square centimeter, required 0.02 gram of the latter substance for complete detonation. Tetryl under similar conditions required only 0.01 gram. Trinitrotriazidobenzene may be dead-pressed and in that condition burns or puffs when it is ignited. It is a practical primary explosive and is prepared for loading in the granular form by mixing the moist material with nitrocellulose, adding a small amount of amyl acetate, kneading, rubbing through a sieve, and allowing to dry.

In the Trauzl test, trinitrotriazidobenzene gives 90% as much net expansion as PETN; tetryl gives 70%, TNT 60%, mercury fulminate 23%, and lead azide 16%. Used as a high explosive in compound detonators and initiated with lead azide, trinitrotriazidobenzene is about as strong as PETN and is stronger than tetryl.

Nitrogen Sulfide

Nitrogen sulfide was first prepared by Soubeiran in 1837 by the action of ammonia on sulfur dichloride dissolved in benzene.

$$6SCl_2 + 16NH_3 \longrightarrow N_4S_4 + 2S + 12NH_4Cl$$

It is conveniently prepared by dissolving 1 volume of sulfur chloride in 8 or 10 volumes of carbon disulfide, cooling, and passing in dry ammonia gas until the dark brown powdery precipitate which forms at first has dissolved and an orange-yellow solution results which contains light-colored flocks of ammonium chloride. These are filtered off and rinsed with carbon disulfide, the solution is evaporated to dryness, and the residue is extracted with boiling carbon disulfide for the removal of sulfur. The undissolved material is crude nitrogen sulfide. The hot extract on cooling deposits a further quantity in the form of minute golden-yellow crystals. The combined crude product is recrystallized from carbon disulfide.

The same product is also produced by the action of ammonia on disulfur dichloride in carbon disulfide, benzene, or ether solution.

$$6S_2Cl_2 + 16NH_3 \longrightarrow N_4S_4 + 8S + 12NH_4Cl$$

Nitrogen sulfide has a density of 2.22 at 15°. It is insoluble in water, slightly soluble in alcohol and ether, somewhat more soluble in carbon disulfide and benzene. It reacts slowly with water at ordinary temperature with the formation of pentathionic acid, sulfur dioxide, free sulfur, and ammonia.[46] It melts with sublimation at 178°, and explodes at a higher temperature which, however, is variable according to the rate at which the substance is heated. Berthelot found that it deflagrates at 207° or higher, and remarked that this temperature is about the same as the temperature of combustion of sulfur in the open air. Berthelot and Vieille [47] studied the thermochemical properties of nitrogen sulfide. Their data, recalculated to conform to our present notions of atomic and molecular weight, show that the substance is strongly endothermic and has a heat of formation of −138.8 Calories per mol. It detonates with vigor under a hammer blow, but is less sensitive to shock and less violent in its effects than mercury fulminate. Although its rate of acceleration is considerably less than that of mercury fulminate, it has been recommended as a filling for fuses, primers, and detonator caps, both alone and in mixtures with oxidizing agents such as lead peroxide, lead nitrate, and potassium chlorate.[48]

Nitrogen selenide was first prepared by Espenschied [49] by the action of ammonia gas on selenium chloride. His product was an orange-red, amorphous powder which exploded violently when heated and was dangerous to handle. Verneuil [50] studied the substance further and supplied a sample of it to Berthelot and Vieille [51] for thermochemical experiments. It detonates when brought into contact with a drop of concentrated sulfuric acid or when warmed to about 230°. It also detonates from friction, from a very gentle blow of iron on iron, and from a slightly stronger blow of wood on iron. It has a heat of formation of −169.2 Calories per mol, and, with nitrogen sulfide, illustrates the principle, as Berthelot pointed out, that in analogous series (such as that of the halides and that of the oxides, sulfides, and selenides) "the explosive character of the endothermic compounds becomes more and more pronounced as the molecular weight becomes larger."

[46] Van Valkenburgh and Bailor, *J. Am. Chem. Soc.*, **47**, 2134 (1925).

[47] Berthelot, "Sur la force des matières explosives," 2 vols., third edition, Paris, 1883, Vol. 1, p. 387.

[48] Claessen, Brit. Pat. 6057 (1913); Carl, U. S. Pat. 2,127,106 (1938).

[49] *Ann.*, **113**, 101 (1860).

[50] *Bull. soc. chim.*, [2] **38**, 548 (1882).

[51] Berthelot, *op. cit.*, p. 389.

Lead Styphnate (Lead trinitroresorcinate)

Lead styphnate is commonly prepared by adding a solution of magnesium styphnate [52] at 70° to a well-stirred solution of lead acetate at 70°. A voluminous precipitate of the basic salt separates. The mixture is stirred for 10 or 15 minutes; then dilute

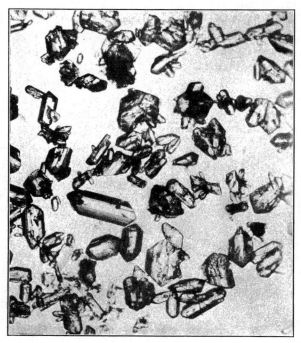

FIGURE 102. Lead Styphnate Crystals (90×).

nitric acid is added with stirring to convert the basic to the normal salt, and the stirring is continued while the temperature drops to about 30°. The product, which consists of reddish-brown, short, rhombic crystals, is filtered off, washed with water, sieved through silk, and dried.

Lead styphnate is a poor initiator, but it is easily ignited by fire or by a static discharge. It is used as an ingredient of the priming layer which causes lead azide to explode from a flash.

[52] Prepared by adding magnesium oxide to a suspension of styphnic acid in water until a clear solution results and only a very small portion of the styphnic acid remains undissolved.

A 0.05-gram sample of lead styphnate in a test tube in a bath of Wood's metal heated at a rate of 20° per minute explodes at 267–268°.

Wallbaum [53] determined the minimum charges of several primary explosives necessary for initiating the explosion of PETN. In the first series of tests, the PETN (0.4 gram) was tamped down or pressed loosely into copper capsules 6.2 mm. in inside diameter, and weighed amounts of the priming charges were pressed down loosely on top. The weights of the priming charges were decreased until one failure occurred in 10 tests with the same weight of charge. In later series, the PETN was compressed at 2000 kilograms per square centimeter. When the priming charges were pressed loosely on the compressed PETN, considerably larger amounts were generally necessary. One gram of lead styphnate, however, was not able to initiate the explosion of the compressed PETN. When the priming charges were pressed, on top of the already compressed PETN, with pressures of 500, 1000, and 1500 kilograms per square centimeter, then it was found that the tetracene and the fulminate were dead-pressed but that the amounts of lead azide and silver azide which were needed were practically the same as in the first series when both the PETN and the priming charge were merely pressed loosely. Wallbaum reports the results which are tabulated below.

Pressure on PETN, kg. per sq. cm.	0	2000	2000	2000	2000
Pressure on initiator, kg. per sq. cm.	0	0	500	1000	1500
PRIMARY EXPLOSIVE	MINIMUM INITIATING CHARGE, GRAMS				
Tetracene	0.16	0.250	dead-pressed		
Mercury fulminate (gray)	0.30	0.330	"	"	
Mercury fulminate (white)	0.30	0.340	"	"	
Lead styphnate	0.55	No detonation with 1 g.			
Lead azide (technical)	0.04	0.170	0.05	0.05	0.04
Lead azide (pure)	0.015	0.100	0.01	0.01	0.01
Silver azide	0.005	0.110	0.005	0.005	0.005

Diazonium Salts

Every student of organic chemistry has worked with diazonium salts in solution. The substances are commonly not isolated in the solid state, for the dry materials are easily exploded by shock and by friction, and numerous laboratory accidents have resulted from their unintended crystallization and drying.

[53] *Z. ges. Schiess- u. Sprengstoffw.*, **34**, 126, 161, 197 (1939).

The first volume of the *Mémorial des Poudres et Salpêtres* contains a report by Berthelot and Vieille [54] on the properties of benzenediazonium nitrate (diazobenzene nitrate). They prepared the material by passing nitrous gas into a cooled aqueous solution of aniline nitrate, diluting with an equal volume of alcohol, and precipitating in the form of white, voluminous flocks by the addition of an excess of ether.

$$2C_6H_5{-}NH_2 \cdot HNO_3 + N_2O_3 \longrightarrow 3H_2O + 2C_6H_5{-}\underset{\underset{N}{\vert\vert\vert}}{N}{-}NO_3$$

The product was washed with ether, pressed between pieces of filter paper, and dried in a vacuum desiccator. In dry air and in the dark it could be kept in good condition for many months. In the daylight it rapidly turned pink, and on longer keeping, especially in a moist atmosphere, it turned brown, took on an odor of phenol, and finally became black and swelled up with bubbles of gas.

Benzenediazonium nitrate detonates easily from the blow of a hammer or from any rubbing which is at all energetic. It explodes violently when heated to 90°. Its density at 15° is 1.37, but under strong compression gently applied it assumes an apparent density of 1.0. Its heat of formation is −47.4 Calories per mol, heat of explosion 114.8 Calories per mol.

m-Nitrobenzenediazonium perchlorate was patented by Herz [55] in 1911, and is reported to have been used in compound detonators with a high-explosive charge of nitromannite or other brisant nitric ester. It explodes spontaneously when heated to about 154°. It is sensitive to shock and to blow. Although it is very sparingly soluble in water and is stabilized to some extent by the nitro group on the nucleus, it is distinctly hygroscopic and is not exempt from the instability which appears to be characteristic of diazonium salts.

[54] *Mém. poudres*, **1**, 99 (1882–1883). Berthelot, *op. cit.*, Vol. 2, p. 35.
[55] Ger. Pat. 258,679 (1911).

Preparation of m-Nitrobenzenediazonium Perchlorate. Half a gram of *m*-nitroaniline is suspended in 5 cc. of water in a wide test tube, and 0.5 cc. of concentrated hydrochloric acid and 2.2 cc. of 20% perchloric acid solution are added. After the nitraniline has dissolved, 15 cc. of water is added and the solution is cooled by immersing the test tube in a beaker filled with a slurry of cracked ice. One-quarter of a gram of sodium nitrite dissolved in 1 or 2 cc. of water is added in 3 or 4 portions, the mixture being shaken after each addition or stirred with a stirring rod the end of which is covered with a short piece of rubber tubing. After standing in the cold for 5 minutes, the material is transferred to a filter, and the feltlike mass of pale yellow needles is washed with cold water, with alcohol, and with ether. The product is dried in several small portions on pieces of filter paper.

Diazodinitrophenol (DDNP, Dinol)

4,6-Dinitrobenzene-2-diazo-1-oxide, or diazodinitrophenol as it is more commonly called, occupies a place of some importance in the history of chemistry, for its discovery by Griess [56] led him to undertake his classic researches on the diazonium compounds and the diazo reaction. He prepared it by passing nitrous gas into an alcoholic solution of picramic acid, but it is more conveniently prepared by carrying out the diazotization in aqueous solution with sodium nitrite and hydrochloric acid.

Picramic acid, red needles, m. p. 169°, may be prepared by evaporating ammonium picrate in alcohol solution with ammonium sulfide.

Preparation of Diazodinitrophenol. Ten grams of picramic acid is suspended in 120 cc. of 5% hydrochloric acid in a beaker which stands in a basin of ice water, and the mixture is stirred rapidly with a mechanical stirrer. Sodium nitrite (3.6 grams) dissolved in 10 cc. of water is added all at once, and the stirring is continued for 20 minutes. The product is collected on a filter and washed thoroughly with ice

[56] *Ann.*, **106**, 123 (1858), **113**, 205 (1860).

water. The dark brown granular material may be used as such, or it may be dissolved in hot acetone and precipitated by the addition of a large volume of ice water to the rapidly agitated liquid, a treatment which converts it into a brilliant yellow amorphous powder.

L. V. Clark,[57] who has made an extensive study of the physical and explosive properties of diazodinitrophenol, reports that it has

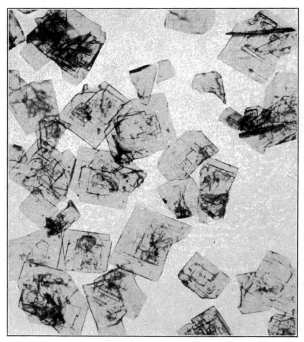

FIGURE 103. Diazodinitrophenol Crystals (90✕).

a true density at 25°/4° of 1.63. Its apparent density after being placed in a tube and tapped is only 0.27, but, when compressed in a detonator capsule at a pressure of 3400 pounds per square inch (239 kilograms per square centimeter), it has an apparent density of 0.86. It is not dead-pressed by a pressure of 130,000 pounds per square inch (9139 kilograms per square centimeter). It is soluble in nitrobenzene, acetone, aniline, pyridine, acetic acid, strong hydrochloric acid, and nitroglycerin at ordinary temperatures. Its solubility at 50° in 100 grams of solvent is: in ethyl acetate 2.45 grams, in methyl alcohol 1.25 grams, in ethyl alcohol 2.43 grams, in benzene 0.23 gram, and in chloroform 0.11 gram.

Diazodinitrophenol is less sensitive to impact than mercury fulminate and lead azide. Its sensitivity to friction is about the same as that of lead azide, much less than that of mercury fulminate. It detonates when struck a sharp blow, but, if it is ignited when it is unconfined, it burns with a quick flash, like nitrocellulose, even in quantities of several grams. This burning produces little or no local shock, and will not initiate the explosion of a high explosive. Commercial detonators containing a high-explosive charge of nitromannite and a primary explosive charge of diazodinitrophenol explode if they are crimped to a piece of miner's fuse and the fuse is lighted, but a spark falling into the open end has been reported to cause only the flashing of the diazodinitrophenol. Likewise, if an open cap of this sort falls into a fire, the diazodinitrophenol may flash, the nitromannite may later melt and run out and burn with a flash, and the detonator may be destroyed without exploding. While it is not safe to expect that this will always happen, it is an advantage of diazodinitrophenol that it sometimes occurs.

Diazodinitrophenol is darkened rapidly by exposure to sunlight. It does not react with water at ordinary temperatures, but is desensitized by it. It is not exploded under water by a No. 8 blasting cap.

Clark reports experiments with diazodinitrophenol, mercury fulminate, and lead azide in which various weights of the explosives were introduced into No. 8 detonator capsules, pressed under reenforcing caps at 3400 pounds per square inch, and fired in the No. 2 sand test bomb. His results, tabulated below, show that diazodinitrophenol is much more powerful than mercury fulminate and lead azide. Other experiments by Clark showed

WEIGHT (GRAMS) OF CHARGE	WEIGHT (GRAMS) OF SAND PULVERIZED FINER THAN 30-MESH BY		
	Diazodinitrophenol	Mercury Fulminate	Lead Azide
0.10	9.1	3.1	3.5
0.20	19.3	6.5	7.2
0.40	36.2	17.0	14.2
0.60	54.3	27.5	21.5
0.80	72.1	38.0	28.7
1.00	90.6	48.4	36.0

that diazodinitrophenol in the sand test has about the same strength as tetryl and hexanitrodiphenylamine.

Clark found that the initiatory power of diazodinitrophenol is about twice that of mercury fulminate and slightly less than that of lead azide. His experiments were made with 0.5-gram charges of the high explosives in No. 8 detonator capsules, with reenforcing caps, and with charges compressed under a pressure of 3400 pounds per square inch. He reported the results which are tabulated below.

| | MINIMUM INITIATING CHARGE (GRAMS) OF | | |
HIGH EXPLOSIVE	Mercury Fulminate	Diazo-dinitrophenol	Lead Azide
Picric acid	0.225	0.115	0.12
Trinitrotoluene	0.240	0.163	0.16
Tetryl	0.165	0.075	0.03
Trinitroresorcinol	0.225	0.110	0.075
Trinitrobenzaldehyde	0.165	0.075	0.05
Tetranitroaniline	0.175	0.085	0.05
Hexanitrodiphenylamine	0.165	0.075	0.05

One gram of diazodinitrophenol in a No. 8 detonator capsule, compressed under a reenforcing cap at a pressure of 3400 pounds per square inch, and fired in a small Trauzl block, caused an expansion of 25 cc. Mercury fulminate under the same conditions caused an expansion of 8.1 cc., and lead azide one of 7.2 cc.

Clark determined the ignition temperature of diazodinitrophenol by dropping 0.02-gram portions of the material onto a heated bath of molten metal and noting the times which elapsed between the contacts with the hot metal and the explosions: 1 second at 200°, 2.5 seconds at 190°, 5 seconds at 185°, and 10.0 seconds at 180°. At 177° the material decomposed without an explosion.

Tetracene

1-Guanyl-4-nitrosoaminoguanyltetrazene, called tetracene for short, was first prepared by Hoffmann and Roth.[58] Hoffmann and his co-workers [59] studied its chemical reactions and determined

[58] *Ber.*, **43**, 682 (1910).

[59] Hoffmann, Hock, and Roth, *ibid.*, **43**, 1087 (1910); Hoffmann and Hock, *ibid.*, **43**, 1866 (1910), **44**, 2946 (1911); Hoffmann, Hock, and Kirmreuther, *Ann.*, **380**, 131 (1911).

its structure. It is formed by the action of nitrous acid on amino-guanidine, or, more exactly, by the interaction of an aminoguani-dine salt with sodium nitrite in the absence of free mineral acid.

Aminoguanidine   Nitrous acid   Aminoguanidine   Nitrous acid

$$\text{NH}_2\!\!\diagup\!\!{}_{\text{NH}_2}^{}\!\!C\text{—NH—NH—N=N—C}\diagup\!\!{}_{\text{NH—NH—NO}}^{\text{NH}} + 3\text{H}_2\text{O}$$

1-Guanyl-4-nitrosoaminoguanyltetrazene

Tetracene is a colorless or pale yellow, fluffy material which is practically insoluble in water, alcohol, ether, benzene, and carbon tetrachloride. It has an apparent density of only 0.45, but yields a pellet of density 1.05 when it is compressed under a pressure of 3000 pounds per square inch. Tetracene forms explosive salts, among which the perchlorate is especially interesting. It is soluble in strong hydrochloric acid; ether precipitates the hydrochloride from the solution, and this on treatment with sodium acetate or with ammonia gives tetracene again. With an excess of silver nitrate it yields the double salt, $C_2H_7N_{10}OAg \cdot AgNO_3 \cdot 3H_2O$. Tetracene is only slightly hygroscopic. It is stable at ordinary temperatures both wet and dry, but is decomposed by boiling water with the evolution of $2N_2$ per molecule. On hydrolysis with caustic soda it yields ammonia, cyanamide, and triazonitroso-aminoguanidine which can be isolated in the form of a bright blue precipitate of the explosive copper salt by the addition of copper acetate to the alkaline solution. The copper salt on treatment with acid yields tetrazolyl azide (5-azidotetrazole).[60]

$$\text{NH}_2\text{—C(NH)—NH—NH—N=N—C(NH)—NH—NH—NO}$$

Triazonitrosoaminoguanidine

Tetrazolyl azide

[60] Cf. survey article by G. B. L. Smith, "The Chemistry of Aminoguani-dine and Related Substances," *Chem. Rev.*, **25**, 214 (1939).

In the presence of mineral acids, sodium nitrite reacts in a different manner with aminoguanidine, and guanyl azide is formed.

$$NH_2—C(NH)—NH—NH_2 + HONO \longrightarrow N_3—C{\overset{NH}{\underset{NH_2}{<}}} + 2H_2O$$

Guanyl azide

This substance forms salts with acids, and was first isolated in the form of its nitrate. The nitrate is not detonated by shock but undergoes a rapid decomposition with the production of light when it is heated. The picrate and the perchlorate explode violently from heat and from shock. Guanyl azide is not decomposed by boiling water. On hydrolysis with strong alkali, it yields the alkali metal salt of hydrazoic acid. It is hydrolyzed by ammoniacal silver nitrate in the cold with the formation of silver azide which remains in solution and of silver cyanamide which appears as a yellow precipitate. By treatment with acids or weak bases it is converted into 5-aminotetrazole.

$$N_3—C{\overset{NH}{\underset{NH_2}{<}}} \longrightarrow NH_2—C{\overset{N——N}{\underset{NH—N}{<}}}$$

5-Aminotetrazole

When the reaction between aminoguanidine and sodium nitrite occurs in the presence of an excess of acetic acid, still another product is formed, namely, 1,3-ditetrazolyltriazine, the genesis of which is easily understood from a consideration of the reactions already mentioned. 5-Aminotetrazole is evidently formed first; the amino group of one molecule of this substance is diazotized by the action of the nitrous acid, and the resulting diazonium salt in the acetic acid solution couples with a second molecule of the aminotetrazole.

$${\overset{N——N}{\underset{N—NH}{||}}}{>}C—NH_2 + HONO + CH_3—COOH \longrightarrow$$

$${\overset{N——N}{\underset{N—NH}{||}}}{>}C—N_2—O—CO—CH_3 + H_2O$$

$${\overset{N——N}{\underset{N—NH}{||}}}{>}C—N_2—O—CO—CH_3 + NH_2—C{\overset{N——N}{\underset{NH—N}{<}}} \longrightarrow$$

$${\overset{N——N}{\underset{N—NH}{||}}}{>}C—N=N—NH—C{\overset{N——N}{\underset{NH—N}{<}}} + CH_3—COOH$$

1,3-Ditetrazolyltriazine

Preparation of Tetracene. Thirty-four grams of aminoguanidine bicarbonate, 2500 cc. of water, and 15.7 grams of glacial acetic acid are brought together in a 3-liter flask, and the mixture is warmed on the steam bath with occasional shaking until everything has gone into solution. The solution is filtered if need be, and cooled to 30° at the tap. Twenty-seven and sixth-tenths grams of solid sodium nitrite is added. The flask is swirled to make it dissolve, and is set aside at room temperature. After 3 or 4 hours, the flask is shaken to start precipitation of the product. It is allowed to stand for about 20 hours longer (22 to 24 hours altogether). The precipitate of tetracene is washed several times by decantation, transferred to a filter, and washed thoroughly with water. The product is dried at room temperature and is stored in a bottle which is closed by means of a cork or rubber stopper.

Tetracene explodes readily from flame without appreciable noise but with the production of much black smoke. Rinkenbach and Burton,[61] who have made an extended study of the explosive properties of tetracene, report that it explodes in 5 seconds at 160° (mercury fulminate 190°). They found that it is slightly more sensitive to impact than mercury fulminate; an 8-inch drop of an 8-ounce weight was needed to explode it, a drop of 9–10 inches to explode fulminate.

The brisance of tetracene, if it is used alone and is fired by a fuse, is greatest when the explosive is not compressed at all. Thus, 0.4 gram of tetracene, if uncompressed, crushed 13.1 grams of sand in the sand test; if compressed under a pressure of 250 pounds per square inch, 9.2 grams; if under 500 pounds per square inch, 7.5 grams; and, if under 3000 pounds per square inch, 2.0 grams. The data show the behavior of tetracene as it approaches the condition of being dead-pressed.

In another series of experiments, Rinkenbach and Burton used charges of 0.4 gram of tetracene, compressed under a pressure of 3000 pounds per square inch and initiated with varying amounts of fulminate (loaded under the same pressure), and found that the tetracene developed its maximum brisance (21.1 grams of sand crushed) when initiated with 0.4 gram of fulminate. A compound primer of 0.15 gram of tetryl initiated with 0.25 gram of mercury fulminate caused 0.4 gram of tetracene to crush 22.6 grams, or substantially the same amount, of sand. It appears

61 *Army Ordnance,* **12,** 120 (1931). See also Stettbacher, *Nitrocellulose,* **8,** 141 (1936); Grottanelli, *Chimica e industria,* **18,** 232 (1936).

then that tetracene is more brisant—and presumably explodes with a greater velocity of detonation—when initiated by fulminate or tetryl than when self-initiated by fire.

Tetracene is easily dead-pressed, its self-acceleration is low, and it is not suitable for use alone as an initiating explosive.

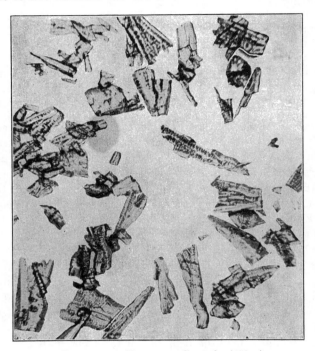

FIGURE 104. Tetracene Crystals (150×).

It is as efficient as fulminate only if it is externally initiated. It is used in detonators either initiated by another primary explosive and functioning as an intermediate booster or mixed with another primary explosive to increase the sensitivity of the latter to flame or heat. A recent patent [62] recommends the use of a mixture of tetracene and lead azide in explosive rivets. Tetracene is used in primer caps where as little as 2% in the composition results in an improved uniformity of percussion sensitivity.

[62] Brit. Pat. 528,299 (1940) to Dynamit-Aktien Gesellschaft vorm. Alfred Nobel & Co.

Hexamethylenetriperoxidediamine (HMTD)

Hexamethylenetriperoxidediamine is the only organic peroxide which has been considered seriously as an explosive. Its explosive properties commend it, but it is too reactive chemically and too unstable to be of practical use. It is most conveniently prepared by treating hexamethylenetetramine with hydrogen peroxide in the presence of citric acid which promotes the reaction by combining with the ammonia which is liberated.

$$C_6H_{12}N_4 + 3H_2O_2 \longrightarrow N\begin{array}{c} CH_2-O-O-CH_2 \\ CH_2-O-O-CH_2 \\ CH_2-O-O-CH_2 \end{array}N + 2NH_3$$

Hexamethylenetriperoxidediamine

Preparation of Hexamethylenetriperoxidediamine. Fourteen grams of hexamethylenetetramine is dissolved in 45 grams of 30% hydrogen peroxide solution which is stirred mechanically in a beaker standing in a freezing mixture of cracked ice with water and a little salt. To the solution 21 grams of powdered citric acid is added slowly in small portions at a time while the stirring is continued and the temperature of the mixture is kept at 0° or below. After all the citric acid has dissolved, the mixture is stirred for 3 hours longer while its temperature is kept at 0°. The cooling is then discontinued, the mixture is allowed to stand for 2 hours at room temperature, and the white crystalline product is filtered off, washed thoroughly with water, and rinsed with alcohol in order that it may dry out more quickly at ordinary temperatures.

Hexamethylenetriperoxidediamine is almost insoluble in water and in the common organic solvents at room temperature. It detonates when struck a sharp blow, but, when ignited, burns with a flash like nitrocellulose. Taylor and Rinkenbach [63] found its true density (20°/20°) to be 1.57, its apparent density after being placed in a tube and tapped 0.66, and its density after being compressed in a detonator capsule under a pressure of 2500 pounds per square inch only 0.91. They found that it required a 3-cm. drop of a 2-kilogram weight to make it explode, but that fulminate required a drop of only 0.25 cm. In the sand test it pulverized 2½ to 3 times as much sand as mercury fulminate, and slightly more sand than lead azide. It is not dead-pressed by a pressure of 11,000 pounds per square inch. It is considerably

[63] *Army Ordnance,* **5,** 463 (1924).

more effective than mercury fulminate as an initiator of detonation. Taylor and Rinkenbach, working with 0.4-gram portions of the high explosives and with varying weights of the primary explosives, compressed in detonator capsules under 'a pressure of 1000 pounds per square inch, found the minimum charges necessary to produce detonation to be as indicated in the following table.

| | Fulminate with Reenforcing Cap | MINIMUM INITIATING CHARGE (GRAMS) OF Hexamethylenetriperoxidediamine | |
| | | With Reenforcing Cap | Without Reenforcing Cap |
HIGH EXPLOSIVE			
Trinitrotoluene	0.26	0.08	0.10
Picric acid	0.21	0.05	0.06
Tetryl	0.24	0.05	0.06
Ammonium picrate	0.8–0.9	0.30	0.30
Tetranitroaniline	0.20	0.05	0.05
Guanidine picrate	0.30	0.13	0.15
Trinitroresorcinol	0.20	0.08	0.10
Hexanitrodiphenylamine	...	0.05	0.05
Trinitrobenzaldehyde	...	0.08	0.10

Taylor and Rinkenbach found that 0.05-gram portions of hexamethylenetriperoxidediamine, pressed in No. 8 detonator capsules under a pressure of 1000 pounds per square inch and fired by means of a black-powder fuse crimped in the usual way, caused the detonation of ordinary 40% nitroglycerin dynamite and of a gelatin dynamite which had become insensitive after storage of more than a year. The velocity of detonation of HMTD, loaded at a density of 0.88 in a column 0.22 inch in diameter, was found by the U. S. Bureau of Mines Explosives Testing Laboratory to be 4511 meters per second.

A small quantity of HMTD decomposed without exploding when dropped onto molten metal at 190°, but a small quantity detonated instantly when dropped onto molten metal at 200°. A 0.05-gram sample ignited in 3 seconds at 149°. At temperatures which are only moderately elevated the explosive shows signs of volatilizing and decomposing. Taylor and Rinkenbach report the results of experiments in which samples on watch glasses were heated in electric ovens at various temperatures, and weighed and examined from time to time, as shown below. The sample which had been heated at 60° showed no evidence of decomposition. The sample which had been heated at 75° was unchanged in color but had a faint odor of methylamine and appeared slightly moist. At 100° the substance gave off an amine odor. The residue which remained after 24 hours of heating at 100° consisted of a colorless liquid and needle crystals which were soluble in water.

% WEIGHT LOST AT	60°	75°	100°
In 2 hrs.	0.10	0.25	3.25
In 8 hrs.	0.35	0.60	29.60
In 24 hrs.	0.50	1.30	67.95
In 48 hrs.	0.50	2.25	....

When hexamethylenetriperoxidediamine is boiled with water, it disappears fairly rapidly, oxygen is given off, and the colorless solution is found to contain ammonia, formaldehyde, ethylene glycol, formic acid, and hexamethylenetetramine.

Friction Primers

Friction primers (friction tubes, friction igniters) are devices for the production of fire by the friction of the thrust, either push or pull, of a roughened rod or wire through a pellet of primer composition. They are used for firing artillery in circumstances where the propelling charge is loaded separately and is not enclosed in a brass case supplied with a percussion primer. They are sometimes crimped to an end of Bickford fuse for the purpose of lighting it. They are sometimes used for lighting flares, etc., which are thrown overboard from airplanes. For this use, the pull element of the primer is attached to the airplane by a length of twine or wire which the weight of the falling flare first pulls and then breaks off entirely.

The following table shows three compositions which have been widely used in friction primers for artillery. All the materials

Potassium chlorate	2	56.2	44.6
Antimony sulfide	1	24.6	44.6
Sulfur	..	9.0	3.6
Meal powder	..	...	3.6
Ground glass	..	10.2	3.6

are in the powdered condition except in the first mixture where half of the potassium chlorate is powdered and half of it is granular. The first mixture is probably the best. The sulfur which is contained in the second and third mixtures makes them more sensitive, but also makes them prone to turn sour after they have been wet-mixed, and these mixtures ought to be made up with a small amount of anti-acid (calcium carbonate, trimethylamine, etc., not mentioned in the table). All the mixtures are wet-mixed with 5% gum arabic solution, loaded wet, and dried out *in situ* to form pellets which do not crumble easily.

In a typical friction primer for an airplane flare, ignition is secured by pulling a loop of braided wire coated with red phosphorus and shellac through a pellet, made from potassium chlorate (14 parts) and charcoal (1.6 parts), hardened with dextrin (0.3 part).

Percussion Primers

Percussion primers produce fire or flame from the impact of the trigger or firing pin of a pistol, rifle, or cannon, or of the inertia-operated device in a fuze which functions when the projectile starts on its flight (the so-called concussion element, the primer of which is called a concussion primer) or of that which functions when the projectile strikes its target (the percussion element). A typical primer composition consists of a mixture of mercury fulminate (a primary explosive which produces the first explosion with heat and flame), antimony sulfide (a combustible material which maintains the flame for a longer time), and potassium chlorate (an oxidizing agent which supplies oxygen for the combustion). Sometimes no single primary explosive substance is present; the mixture itself is the primary explosive. Sometimes the compositions contain explosives such as TNT, tetryl, or PETN, which make them hotter, or ground glass which makes them more sensitive to percussion. Hot particles of solid (glass or heavy metal oxide) thrown out by a primer will set fire to black powder over a considerable distance, but they will fall onto smokeless powder without igniting it. The primers which produce the hottest gas are best suited for use with smokeless powder.

Primer compositions are usually mixed by hand on a glass-top table by a workman wearing rubber gloves and working alone in a small building remote from others. They are sometimes mixed dry, but in this country more commonly wet, with water or with water containing gum arabic or gum tragacanth, with alcohol alone or with an alcohol solution of shellac. The caps are loaded in much the same manner that blasting caps are loaded, the mixture is pressed down by machine and perhaps covered with a disc of tinfoil, the anvil is inserted and pressed into place (unless the primer is to be used in a cartridge or fuze of which the anvil is already an integral part), and the caps are finally dried in a dry-house and stored in small magazines until needed for loading.

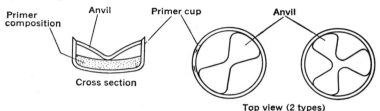

FIGURE 105. Primer Cap for Small Arms Cartridge.

For many years the standard mixture in France for all caps which were to be fired by the blow of a hammer was made from 2 parts of mercury fulminate, 1 of antimony sulfide, and 1 of saltpeter. This was mixed and loaded dry, and was considered to be safer to handle than similar mixtures containing potassium chlorate. Where a more sensitive primer was needed, the standard French composition for all concussion and percussion primers of fuzes was made from 5 parts of mercury fulminate and 9 parts each of antimony sulfide and potassium chlorate.

All the compositions listed in the following table (gum or shellac binder not included) have been used, in small arms primers or in fuze primers, by one or another of the great powers, and they illustrate the wide variations in the proportions of the ingredients which are possible or desirable according to the design of the device in which the primer is used.

Mercury fulminate.	10.0	28.0	48.8	4	5	2	11.0	32	16.5	7	19.0
Potassium chlorate.	37.0	35.5	24.4	2	9	3	52.5	45	50.0	21	33.0
Antimony sulfide...	40.0	28.0	26.2	3	3	3	36.5	23	33.5	17	43.0
Sulfur	...	...	...	..	..	..	...	..	...	..	2.5
Meal powder	...	...	...	..	..	..	...	..	...	..	2.5
Ground glass	13.0	8.5	...	5	..	..	...	..	...	5	...
Ground coke	...	...	...	..	1	..	...	..	...	..	...
Tetryl	...	...	...	..	2	..	...	..	...	..	...

A non-fulminate primer composition is probably somewhat safer to mix than one which contains fulminate. It contains no single substance which is a primary explosive, only the primary explosive mixture of the chlorate with the appropriate combustible material, or, more exactly, the explosive which exists at the point of contact between particles of the two substances. For a non-fulminate primer to perform properly, it is necessary that the composition should be mixed thoroughly and very uniformly in order that dissimilar particles may be found in contact with each other beneath the point of the anvil and may be crushed together by the blow of the trigger. It is not absolutely essential that fulminate compositions should be mixed with the same uniformity. Even if no fulminate happens to lie beneath the point of the anvil, the trigger blow sufficiently crushes the sensitive material in the neighborhood to make it explode. For mechanical reasons, the ingredients of primer composition ought not to be pulverized too finely.[64]

Several non-fulminate primer compositions are listed below.

Potassium chlorate	50	50.54	67	60	53
Antimony sulfide	20	26.31	..	30	17
Lead thiocyanate	..	...	..	..	25
Lead peroxide	25	...	..	..	..
Cuprous thiocyanate	..	...	15	3	..
TNT	5	...	..	..	5
Sulfur	..	8.76	16	7	..
Charcoal	..	...	2	..	..
Ground glass	..	12.39	..	..	..
Shellac	..	2.00	..	..	..

Sulfur ought not to be used in any primer composition, whether fulminate or non-fulminate, which contains chlorate unless an anti-acid is present. In a moist atmosphere, the sulfuric acid, which is inevitably present on the sulfur, attacks the chlorate, liberating chlorine dioxide which further attacks the sulfur, producing more sulfuric acid, and causing a self-catalyzed *souring* which results first in the primer becoming slow in its response to the trigger (hang fire) and later in its becoming inert (misfire). It is evident that the presence of fulminate in the composition will tend to nullify the effect of the souring, and that it

[64] Cf. Émile Monnin Chamot, "The Microscopy of Small Arms Primers," privately printed, Ithaca, New York, 1922.

is safest to avoid the use of sulfur with chlorate especially in non-fulminate mixtures. The second of the above-listed compositions is an undesirable one in this respect. In the third and fourth compositions, the cuprous thiocyanate serves both as a combustible and as an anti-acid, and it helps, particularly in the

Figure 106. Longitudinal Sections of Military Rifle Ammunition of the First World War. (Courtesy Émile Monnin Chamot.) The cartridge at the bottom, French 9.0-mm. Lebel rifle, the one above it, German 7.9-mm. Mauser, and the one above that, Canadian .30 caliber, all have anvils of the Berdan type integrally one with the metal of the cartridge case.

third mixture, by supplying copper oxide which is a solid vehicle for the transfer of heat. The first and the last of the above-listed mixtures are the best. They contain no sulfur, and they contain lead enough to supply plenty of solid particles of hot material.

Gunnery experts ascribe a large part of the erosion of shotgun and rifle barrels to the action of the soluble salts which are produced from the materials of the primer compositions, particularly to the chlorides which come from the chlorate, and to the sulfates which result from the combustion of the antimony sulfide. The following table lists several non-chlorate, non-erosive primer compositions. They contain no compounds of chlorine. They con-

Mercury fulminate	36	40	25	20	39
Antimony sulfide	20	25	15	20	9
Barium nitrate	..	25	25	40	41
Lead peroxide	..	..	35	10	..
Lead chromate	40	..	..	..	..
Barium carbonate	..	6	..	..	..
Picric acid	..	..	..	..	5
Powdered glass	4	4	..	..	6
Calcium silicide	..	..	..	10	..

tain either lead or barium or both, and both of these metals form sulfates which are insoluble in water. Moreover, the soluble portions of the residues from the primers which contain barium nitrate are alkaline and are even capable of neutralizing any acidity which might arise from the smokeless powder.

INDEX OF NAMES

INDEX OF SUBJECTS

K

KI starch test, 268, 285
Kekulé oil, 228
Kerosene, 66
Kieselguhr, 235, 332, 354, 356; see
    also Guhr; Fuller's earth
Koronit, 361

L

Lachrymator, 140, 144
Lactose, 65, 123, 208, 238, 240
Lactose Hexanitrate, 242
Lactose octonitrate, 241
Lady crackers, 111
Laevulosan, 243
Laminated powder, 298, 318
Lamium, oil of, 34
Lampblack, 57, 61, 64, 65, 70, 71, 72,
    84, 85, 86, 90, 108, 118, 289, 356
Lampblack effect, 98
Lampblack stars, 86, 87
Lances, 57, 58, 63, 69
Lead acetate, 425, 440
Lead azide, 3, 4, 8, 169, 183, 209, 281,
    391, 420, 421, 424 ff., 431, 434,
    440, 450
  density, 430, 434
  drop test, 209
  efficiency as initiator, 280
  minimum initiating charge, 231,
      412, 439, 441, 446
  sand test of, 445, 451
  sensitivity to fire, 419
  temperature of explosion, 411, 435
  Trauzl test of, 438, 446
  velocity of detonation, 434
Lead chloride, 84
Lead chromate, 430, 458
Lead dinitrocresolate, 147
Lead nitrate, 63, 70, 96, 425, 439
Lead oxide, 147
Lead peroxide, 439, 456, 458
Lead picrate, 3, 6, 63, 164, 417
Lead styphnate, 3, 169, 421, 424, 440,
    441
  density, 434

Lead styphnate, temperature of ex-
    plosion, 435
  velocity of detonation, 434
Lead trinitroresorcinate, 440; see
    also Lead styphnate
Lead thiocyanate, 456
Lead block compression test (lead
    block crushing test), 24 ff., 192,
    341, 342, 364, 365, 391, 421; see
    also Small lead block test
Lead block expansion test (Trauzl
    test), 24 ff., 132, 133, 159, 175,
    192, 341, 342, 344, 353, 357, 364,
    365, 366
Lead plate test, 26, 27, 233, 421
Ligroin, 216, 240, 241, 284, 316, 321,
    322
Lilac fire, 93, 118
Lilac lances, 70
Lilac stars, 84
Lime nitrogen, 377
Linen, 245, 247
Linseed oil, 34, 35, 68, 90, 118, 121
Linters, 259
Liquid explosives, 211, 214, 284, 343;
    see also Sprengel explosives
Liquid fire, 33, 78
Liquid oxygen explosives, 355 ff.
Litharge, 6, 95
Lithium azide, 411
Lithium chloride, 325
Litmus, 204, 217, 220, 406
Loading, by compression, 167
  by pouring, 130, 166
Low explosives; see Propellants
Low-freezing (l. f.) dynamite, 333,
    334, 339, 341
Lycopodium, 86
Lyddite, 159, 166; see also Picric
    acid

M

MDN, explosive, 157
MDPC, explosive, 166
MMN, explosive, 157
MTTC, explosive, 166
Macaroni press, 302

Strength of dynamite, 338, 339, 341, 345
Strepta, 33
Strontium azide, 411
Strontium carbonate, 61, 67, 69, 72, 84, 85, 86
Strontium chlorate, 86
Strontium chloride (muriate), 60
Strontium nitrate, 60, 61, 62, 64, 65, 66, 67, 68, 69, 71, 72, 85
Strontium oxalate, 85
Strontium picrate, 63
Strontium sulfate, 70, 84
Styphnic acid, 169, 440
Sublimate, corrosive, 55
Succession of color, 69
Sucrose octonitrate, 242
Sugar, cane, 61, 64, 239, 240, 247, 333, 334, 350, 358
Sugar, nitrated mixtures, 238 ff.
Sulfur (brimstone), 2, 28, 31, 34, 35, 37, 38, 39, 40, 42, 45, 48, 52, 53, 55, 58, 61, 62, 64, 65, 66, 67, 68, 69, 70, 76, 78, 79, 83, 84, 85, 87, 88, 89, 92, 93, 97, 99, 100, 104, 105, 112, 117, 118, 124, 254, 276, 333, 334, 336, 340, 341, 345, 350, 357, 358, 388, 401, 402, 453, 454, 455, 456, 457
Sulfur chloride, 220; see also Disulfur dichloride
Sulfur dichloride, 438
Sulfuryl chloride, 395
Superattenuated ballistite, 327
Sweetie barrel, 290, 291

T

TNA, 173, 417; see also Tetranitroaniline
TNB, 134; see also Trinitrobenzene
TNT, 141; see also Trinitrotoluene
TNT oil, 145
TNX, 153; see also Trinitroxylene
T4, 396; see also Cyclonite
Table fireworks, 93
Tagus, stone of, 37

Tartar, 34
Temperature of explosion, 24, 43, 50, 182, 210, 347, 348, 350, 351, 353, 364, 365, 366, 394, 405, 411, 424, 435, 436, 441, 442
Temperature of ignition, 21, 22, 165, 206, 446
Temperature of inflammation of carbon monoxide-air mixtures, 325
Tetraamylose, 244
Tetraamylose octonitrate, 244
Tetracene, 3, 238, 383, 421, 441, 446 ff.
Tetralite, 175; see also Tetryl
Tetramethylbenzidine, 179
Tetramethylolcyclohexanol pentanitrate, 286
Tetramethylolcyclohexanone tetranitrate, 286
Tetramethylolcyclopentanol pentanitrate, 285, 286
Tetramethylolcyclopentanone tetranitrate, 285, 286
Tetramethylurea, 322
Tetranitroaniline, 129, 134, 137, 173, 174, 175, 178, 186
  sensitivity to initiation, 231, 418, 434, 446, 452
Tetranitroazoxybenzene, 136
Tetranitrobiphenyl, 159
Tetranitrocarbanilide, 189
Tetranitrodiglycerin, 222, 223, 341
Tetranitrodimethylbenzidinedinitramine, 180
Tetranitrodiphenylamine, 186
Tetranitrodiphenylethanolamine, 230, 232
Tetranitrohydrazobenzene, 189
Tetranitromethane, 126, 144
Tetranitromethylaniline, 175
Tetranitronaphthalene, 156, 157, 349
Tetranitrophenylmethylnitramine, 178
Tetraphenylurea, 322
Tetrazolyl azide, 447

Section 13. PRIMARY HIGH EXPLOSIVES

MERCURY FULMINATE   13–1

Description: Mercury fulminate is an initiating explosive, commonly appearing as white or gray crystals. It is extremely sensitive to initiation by heat, friction, spark or flame, and impact. It detonates when initiated by any of these means. It is pressed into containers, usually at 3000 psi, for use in detonators and blasting caps. However, when compressed at greater and greater pressure (up to 30,000 psi), it becomes "dead pressed." In this condition, it can only be exploded by another initial detonating agent. Mercury fulminate gradually becomes inert when stored continuously above 100° F. A dark-colored product of deterioration gives evidence of this effect. Mercury fulminate is stored underwater execpt when there is danger of freezing. Then it is stored under a mixture of water and alcohol.

Comments: This material was tested. It is effective.

References: TM 9–1900, Ammunition, General, page 59.
TM 9–1910, Military Explosives, page 98.

LEAD STYPHNATE   13–2

Description: Lead styphnate is an initiating explosive, commonly appearing in the form of orange or brown crystals. It is easily ignited by heat and static discharge but cannot be used to initiate secondary high explosives reliably. Lead styphnate is used as an igniting charge for lead azide and as an ingredient in priming mixtures for small arms ammunition. In these applications, it is usually mixed with other materials first and then pressed into a metallic container (detonators and primers). Lead styphnate is stored under water except when there is danger of freezing. Then it is stored under a mixture of water and alcohol.

Comments: This item was tested. It is effective.

References: TM 9–1900, Ammunition, General, page 59.
TM 9–1910, Military Explosives, page 107.

LEAD AZIDE   13–3

Description: Lead azide is an initiating explosive and is produced as a white to buff crystalline substance. It is a more efficient detonating agent than mercury fulminate and it does not decompose on long continued storage at moderately elevated temperatures. It is sensitive to both flame and impact but requires a layer of lead styphnate priming mixture to produce reliable initiation when it is used in detonators that are initiated by a firing pin or electrical energy. It is generally loaded into aluminum detonator housings and must not be loaded into housing of copper or brass because extremely sensitive copper azide can be formed in the presence of moisture.

Comments: This material was tested. lt is effective.

References: TM 9–1900, Ammunition, General, page 60.
TM 9–1910, Military Explosives, page 103.

DDNP   13—4

Description: DDNP (diazodinitrophenol) is a primary high explosive.
It is extensively used in commercial blasting caps that are initiated by
black powder safety fuse. It is superior to mercury fulminate in stability but is not as stable as lead azide. DDNP is desensitized by
immersion in water.

Comments: This material was tested. It is effective.

References: TM 9–1900, Ammunition, General, page 60.
TM 9–1910, Military Explosives, page 103.

Section 14.   SECONDARY HIGH EXPLOSIVES

TNT   14—1

Description: TNT (Trinitrotoluene) is produced from toluene, sulfuric
acid, and nitric acid. It is a powerful high explosive. It is well suited
for steel cutting, concrete breaching, general demolition, and for underwater demolition. It is a stable explosive and is relatively insensitive
to shock. It may be detonated with a blasting cap or by primacord.
TNT is issued in 1-pound and ½-pound containers and 50-pounds to
a wooden box.

Comments: This material was tested. It is effective. TNT is toxic and
its dust should not be inhaled or allowed to contact the skin.

References: TM 9–1900, Ammunition, General, page 263.
FM 5–25, Explosives and Demolitions, page 3.

NITROSTARCH   14—2

Description: Nitrostarch is composed of starch nitrate, barium nitrate,
and sodium nitrate. It is more sensitive to flame, friction, and impact
than TNT but is less powerful. It is initiated by detonating cord.
Nitrostarch is issued in 1-pound and 1½-pound blocks. The 1-pound
packages can be broken into ¼-pound blocks. Fifty 1-pound packages
and one hundred 1½-pound packages are packed in boxes.

Comments: This material was tested. It is effective.

Reference: TM 9–1900, Ammunition, General, page 263.

TETRYL   14—3

Description: Tetryl is a fine, yellow, crystalline material and exhibits
a very high shattering power. It is commonly used as a booster in ex-

plosive trains. It is stable in storage. Tetryl is used in detonators. It is pressed into the bottom of the detonator housing and covered with a small priming charge of mercury fulminate or lead azide.

Comments: This material was tested. It is effective.

References: TM 9–1900, Ammunition, General, page 52.
TM 31–201–1, Unconventional Warfare Devices and Techniques, para 1509.

RDX   14–4

Description: RDX (cyclonite) is a white crystalline solid that exhibits very high shattering power. It is commonly used as a booster in explosive trains or as a main bursting charge. It is stable in storage, and when combined with proper additives, may be cast or press loaded. It may be initiated by lead azide or mercury fulminate.

Comments: This material was tested. It is effective.

References: TM 9–1900, Ammunition, General, page 52.
TM 31–201–1, Unconventional Warfare Devices and Techniques, para 1501.

NITROGLYCERIN   14–5

Description: Nitroglycerin is maufactured by treating glycerin with a nitrating mixture of nitric and sulfuric acid. It is a thick, clear to yellow-brownish liquid that is an extremely powerful and shock-sensitive high explosive. Nitroglycerin freezes at 56° F., in which state it is less sensitive to shock than in liquid form.

Comments: This material was tested. It is effective.

References: TM 9–1910, Military Explosives, page 123.
TM 31–201–1, Unconventional Warfare Devices and Techniques, para 1502.

COMMERCIAL DYNAMITE   14–6

Description: There are three principal types of commercial dynamite: straight dynamite, ammonia dynamite, and gelatin dynamite. Each type is further subdivided into a series of grades. All dynamites contain nitroglycerin in varying amounts and the strength or force of the explosive is related to the nitroglycerin content. Dynamites range in velocity of detonation from about 4000 to 23,000 feet per second and are sensitive to shock. The types and grades of dynamite are each used for specific purposes such as rock blasting or underground explosives. Dynamite is initiated by electric or nonelectric blasting caps. Although dynamites are furnished in a wide variety of packages, the most common unit is the ½ pound cartridge. Fifty pounds is the maximum weight per case.

Comments: This material was tested. It is effective.

References: TM 9–1900, Ammunition, General, page 265.
FM 5–25, Explosives and Demolitions, page 8.

MILITARY DYNAMITE   14–7

Description: Military (construction) dynamite, unlike commercial dynamite, does not absorb or retain moisture, contains no nitroglycerine, and is much safer to store, handle, and transport. It comes in standard sticks 1¼ inches in diameter by 8 inches long, weighing approximately ½ pound. It detonates at a velocity of about 20,000 feet per second and is very satisfactory for military construction, quarrying, and demolition work. It may be detonated with an electric or nonelectric military blasting cap or detonating cord.

Comments: This material was tested. It is effective.

References: FM 5–25, Explosives and Demolitions, page 7.
TM 9–1910, Military Explosives, page 204.

AMATOL   14–8

Description: Amatol is a high explosive, white to buff in color. It is a mixture of ammonium nitrate and TNT, with a relative effectiveness slightly higher than that of TNT alone. Common compositions vary from 80% ammonium nitrate and 20% TNT to 40 % ammonium nitrate and 60% TNT. Amatol is used as the main bursting charge in artillery shell and bombs. Amatol absorbs moisture and can form dangerous compounds with copper and brass. Therefore it should not be housed in containers of such metals.

Comments: This material was tested. It is effective.

References: FM 5–25, Explosives and Demolitions, page 7.
TM 9–1910, Military Explosives, page 182.

PETN   14–9

Description: PETN (pentaerythrite tetranitrate), the high explosive used in detonating cord, is one of the most powerful of military explosives, almost equal in force to nitroglycerine and RDX. When used in detonating cord, it has a detonation velocity of 21,000 feet per second and is relatively insensitive to friction and shock from handling and transportation.

Comments: This material was tested. It is effective.

References: FM 5–25, Explosives and Demolitions, page 7.
TM 9–1910, Military Explosives, page 135.
TM 31–201–1, Unconventional Warfare Devices and Techniques, para 1508.

BLASTING GELATIN 14–10

Description: Blasting gelatin is a translucent material of an elastic, jellylike texture and is manufactured in a number of different colors. It is considered to be the most powerful industrial explosive. Its characteristics are similar to those of gelatin dynamite except that blasting gelatin is more water resistant.

Comments: This material was tested. It is effective.

Reference: TM 9–1910, Military Explosives, page 204.

COMPOSITION B 14–11

Description: Composition B is a high-explosive mixture with a relative effectiveness higher than that of TNT. It is also more sensitive than TNT. It is composed of RDX (59%), TNT (40%), and wax (1%). Because of its shattering power and high rate of detonation, Composition B is used as the main charge in certain models of bangalore torpedoes and shaped charges.

Comments: This material was tested. It is effective.

References: FM 5–25, Explosives and Demolitions, page 7.
TM 9–1900, Ammunition, General, page 57.
TM 9–1910, Military Explosives, page 193.

COMPOSITION C4 14–12

Description: Composition C4 is a white plastic high explosive more powerful than TNT. It consists of 91% RDX and 9% plastic binder. It remains plastic over a wide range of temperatures (—70° F. to 170° F.), and is about as sensitive as TNT. It is eroded less than other plastic explosives when immersed under water for long periods. Because of its high detonation velocity and its plasticity, C4 is well suited for cutting steel and timber and for breaching concrete.

Comments: This material was tested. It is effective.

Reference: TM 9–1910, Military Explosives, page 204.

AMMONIUM NITRATE 14–13

Description: Ammonium nitrate is a white crystalline substance that is extremely water absorbent and is therefore usually packed in a sealed metal container. It has a low velocity of detonation (3600 fps) and is used primarily as an additive in other explosive compounds. When it is used alone, it must be initiated by a powerful booster or primer. It is only 55% as powerful as TNT, hence larger quantities are required to produce similar results.

Comments: This material was tested. It is effective.

Caution: Never use copper or brass containers because ammonium nitrate reacts with these metals.

References: TM 9–1900, Ammunition, General, page 264.
TM 9–1910, Military Explosives, page 119.

Description: HMX is a solid high explosive commonly used as a booster, and sometimes used as a main charge where its shattering effect is needed. It is a white substance and has a rather high melting point; hence it is usually pressed into its container. It may be initiated by lead azide or mercury fulminate.

Comments: This material was tested. It is effective.

Reference: TM 9–1910, Military Explosives, page 181.

PENTOLITE **14—15**

Description: Pentolite is a high explosive mixture of equal proportions of PETN and TNT. It is light yellow and is used as the main bursting charge in grenades, small shells, and shaped charges. Pentolite may be melted and cast in the container. Pentolite should not be drilled to produce cavities, forming tools should be used.

Comments: This material was tested. It is effective.

Reference: TM 9–1900, Ammunition, General, page 55.

PICRIC ACID **14—16**

Description: Picric acid is a yellow crystalline, high explosive bursting charge. It is initiated by lead azide or mercury fulminate. Picric acid has the same effectiveness as TNT. In loading operations, pressing is recommended; however, picric acid can be melted and poured if there is assurance of low lead content or no contact with lead. Picric acid in contact with lead produces lead picrate, a sensitive and violent explosive.

Comments: This material was tested. It is effective.

References: TM 9–1910, Military Explosives, page 159.
TM 31–201–1, Unconventional Warfare Devices and Techniques, para 1510.

GUN COTTON **14—17**

Description: Gun cotton is a nitrocellulose explosive made from cotton fibers, containing 13% or more of nitrogen. Although primarily considered a propellant, it is sometimes used as a base charge in electric primers and electrically initiated destructors because it will detonate with proper confinement.

Comments: This material was tested. It is effective.

References: TM 9–1910, Military Explosives, page 127.
TM 9–1900, Ammunition, General, page 40.

AMMONAL

AMMONAL 14–18

Description: Ammonal is a high explosive mixture composed of 22% ammonium nitrate, 67% TNT, and 11% flaked or powdered aluminum. It is sometimes used as a filler for artillery shell. The composition is 83% as effective as TNT and it explodes with a bright flash on detonation.

Comments: This material was tested. It is effective.

Reference: TM 9–1910, Military Explosives, pages 184 and 214.

IMPROVISED PLASTIC EXPLOSIVE FILLER 14–19

Description: Plastic explosive filler can be made from potassium chlorate and petroleum jelly. The potassium chlorate crystals are ground into a very fine powder and then mixed with the petroleum jelly. This explosive can be detonated with a No. 8 commercial blasting cap or with any military blasting cap. The explosive must be stored in a waterproof container until ready to use.

Comments: This material was tested. It is effective.

Reference: TM 31–210, Improvised Munitions, sec I, No. 1.

TETRYTOL 14–20

Description: Tetryol is a high explosive bursting charge containing 75% tetryl and 25% TNT. It is used as a demolition explosive, a bursting charge for mines, and in artillery shell. The explosive force of tetrytol is approximately the same as that of TNT. It may be initiated by a blasting cap. Tetrytol is usually loaded by casting.

Comments: This material was tested. It is effective.

References: TM 9–1900, Ammunition, General, page 55.
TM 9–1910, Military Explosives, page 188.

NAPALM INCENDIARY   02–1

Description: This incendiary consists of a liquid fuel which is gelled by the addition of soap powder or chips. It can be initiated by means of ignition delay systems or directly by a match flame. This incendiary is adhesive, long burning, and is suitable for setting fire to wooden and other combustible targets.

Comments: This material was tested. It is effective.

Reference: TM 31–201–1, Unconventional Warfare Devices and Techniques, para 0301.

PARAFFIN—SAWDUST INCENDIARY   02–2

Description: This incendiary consists of a mixture of paraffin wax and sawdust. It can be initiated by means of ignition delay systems or directly by a match flame. It is used for setting fire to wooden and other combustible materials. Beeswax may be substituted for paraffin wax if desired. This incendiary is slow starting but a few minutes after initiation, vigorous burning occurs.

Comments: This materials was tested. It is effective.

Reference: TM 31–201–1, Unconventional Warfare Devices and Techniques, para 0304.

GELLED GASOLINE INCENDIARY (EXOTIC THICKENERS)   02–3

Description: This item consists of gasoline that is gelled by the addition of organic chemicals. It can be initiated by means of ignition delay systems or directly by a match flame. This incendiary is adhesive and long burning and is suitable for setting fire to wooden and other combustible targets. The following gasoline gelling systems were used. Numbers in parentheses are grams added per gallon of gasoline. The first ingredient is stirred into the gasoline at room temperature. When the second is added, the gasoline will gell within a few minutes.

System	Ingredients
1	Lauryl amine (55), toluene diisocyanate (27)
2	Coco amine (55), toluene diisocyanate (27)
3	Lauryl amine (57), hexamethylene diisocyanate (25)
4	Oleyl amine (59), hexamethylene diisocyanate (23)
5	t-Octyl amine (51), toluene diisocyanate (31)
6	Coco amine (51), naphthyl isocyanate (31)
7	Delta-aminobutylmethyldiethoxysilane (51), Hexamethylene diisocyanate (31)

Section 12. LOW EXPLOSIVES

BLACK POWDER 12–1

Description: Black powder is a mixture of approximately 10% sulfur, 15% charcoal, and 75% sodium or potassium nitrate. It is manufactured in granular and pellet form. In the granular form it is a loose, free flowing, grained material and its burning rate is controlled by grain size; the finer the granulation the faster the burning. Pellets are produced when granular powder is compressed into pellet form. Each pellet has a hole through its center. Pellets are less dangerous to handle, are more efficient, and more economical to use than granular powder. Black powder is extremely sensitive to flame or spark and may be initiated with a time fuse or electric squib. The granular form of black powder is issued in 25-pound drums. The pellet form is issued in 50-pound wooden cases.

Comments: This material was tested. It is effective.

References: TM 9–1900, Ammunition, General, page 45.
TM 9–1910, Military Explosives, page 36.

FRENCH AMMONAL 12–2

Description: French ammonal is an easily improvised low explosive mixture of 86% ammonium nitrate, 6% stearic acid, and 8% aluminum powder. It is generally less effective than an equal weight of TNT. The material is loaded by pressing it into a suitable container. Initiation by an Engineer's special blasting cap is recommended.

Comments: This material was tested. It is effective.

Reference: TM 31–201–1, Unconventional Warfare Devices and Techniques, para 1401.

THERMITE INCENDIARY

Description: Thermite is composed of iron oxide and aluminum powder. It may be obtained as a commercial material or be improvised by mixing these two ingredients (three parts iron oxide and two parts aluminum powder, by volume). Thermite requires high heat for initiation and specific igniters must be used. This incendiary is used to attack metal targets by applying localized heat. It causes holes to be burned through metal and to drip molten metal on interior components. It is also useful for welding together machinery parts or steel plates (see illustrations for set-up). Thermite is safe to handle and transport because of its high ignition temperature. It burns well in cold and windy weather.

Comments: This material was tested. It is effective.

Reference: TM 31–201–1, Unconventional Warfare Devices and Techniques, para 0307.

NUMBER ONE

THE WORLD WIDE CHEMICAL TRANSLATION SERIES
Edited by E. EMMET REID
Professor of Organic Chemistry, The Johns Hopkins University

A New Reprint on the Same Subject History of Nitro - 1998

————

NITROGLYCERINE AND NITROGLYCERINE EXPLOSIVES

BY

PHOKION NAOÚM, Ph.D.
Director of Scientific Laboratories, Alfred Nobel Dynamite Company, Hamburg, Germany

1928

AUTHORIZED ENGLISH TRANSLATION WITH NOTES AND ADDITIONS

BY

E. M. SYMMES
Hercules Powder Company Wilmington, Delaware

ANGRIFF PRESS
PO Box 208
Las Vegas, Nevada   89125
New Printing 1998
I.S.B.N. 0-913022-46-2

———————————————

**A New Reprint on the Same Subject
History of Nitro - 1998 -
500 pages**